Maximum Likelihood Estimation and Inference

Statistics in Practice

Series Advisors

Human and Biological Sciences
Stephen Senn
University of Glasgow, UK

Earth and Environmental Sciences
Marian Scott
University of Glasgow, UK

Industry, Commerce and Finance
Wolfgang Jank
University of Maryland, USA

Statistics in Practice is an important international series of texts which provide detailed coverage of statistical concepts, methods and worked case studies in specific fields of investigation and study.

With sound motivation and many worked practical examples, the books show in down-to-earth terms how to select and use an appropriate range of statistical techniques in a particular practical field within each title's special topic area.

The books provide statistical support for professionals and research workers across a range of employment fields and research environments. Subject areas covered include medicine and pharmaceutics; industry, finance and commerce; public services; the earth and environmental sciences, and so on.

The books also provide support to students studying statistical courses applied to the above areas. The demand for graduates to be equipped for the work environment has led to such courses becoming increasingly prevalent at universities and colleges.

It is our aim to present judiciously chosen and well-written workbooks to meet everyday practical needs. Feedback of views from readers will be most valuable to monitor the success of this aim.

A complete list of titles in this series appears at the end of the volume.

Maximum Likelihood Estimation and Inference

With Examples in R, SAS and ADMB

Russell B. Millar

Department of Statistics, University of Auckland, New Zealand

WILEY

A John Wiley & Sons, Ltd., Publication

This edition first published 2011

Registered office
John Wiley & Sons Ltd, The Atrium, Southern Gate, Chichester, West Sussex, PO19 8SQ, United Kingdom

For details of our global editorial offices, for customer services and for information about how to apply for permission to reuse the copyright material in this book please see our website at www.wiley.com.

Library of Congress Cataloging-in-Publication Data

Millar, R. B. (Russell B.)
Maximum likelihood estimation and inference : with examples in R, SAS, and ADMB / Russell B. Millar.
p. cm.
Includes bibliographical references and index.
ISBN 978-0-470-09482-2 (hardback)
1. Estimation theory. 2. Chance–Mathematical models. I. Title.
QA276.8.M55 2011
519.5'44–dc22

2011013225

A catalogue record for this book is available from the British Library.

Print ISBN: 978-0-470-09482-2
ePDF ISBN: 978-0-470-09483-9
oBook ISBN: 978-0-470-09484-6
ePub ISBN: 978-1-119-97771-1
Mobi ISBN: 978-1-119-97772-8

Set in 10.25/12pt Times by Thomson Digital, Noida, India

Contents

Preface

Likelihood has a fundamental role in the field of statistical inference, and this text presents a fresh look at the pragmatic concepts, properties, and implementation of statistical estimation and inference based on maximization of the likelihood. The supporting theory is also provided, but for readability is kept separate from the pragmatic content.

The properties of maximum likelihood inference that are presented herein are from the point of view of the classical frequentist approach to statistical inference. The Bayesian approach provides another paradigm of likelihood-based inference, but is not covered here, though connections to Bayesian methodology are made where relevant. Leaving philosophical arguments aside (but see Chapter 14), one of the basic choices to be made before any analysis is to determine the most appropriate paradigm to use in order to best answer the research question and to meet the needs of scientific colleagues or clients. This text will aid this choice, by showing the best of what can be done using maximum likelihood under the frequentist paradigm.

The level of presentation is aimed at the reader who has already been exposed to an undergraduate course on the standard tools of statistical inference such as linear regression, ANOVA and contingency table analysis, but who has discovered, through curiosity or necessity, that the world of real data is far more diverse than that assumed by these models. For this reason, these standard techniques are not given any special attention, and appear only as examples of maximum likelihood inference where applicable. It will be assumed that the reader is familiar with basic concepts of statistical inference, such as hypothesis tests and confidence intervals.

Much of this text is focused on the presentation of tools, tricks, and bits of R, SAS and ADMB code that will be useful in analyzing real data, and these are demonstrated through numerous examples. Pragmatism is the key motivator throughout. So, for example, software utilities have been provided to ease the computational burden of the calculation of likelihood ratio confidence intervals.

Explanation of SAS and R code is made at a level that assumes the reader is already familiar with basic programming in these languages, and hence is comfortable with their general syntax, and with tasks such as data manipulation. ADMB is a somewhat different beast, and (at the present time) will be totally unfamiliar to the majority of readers. It is used sparingly. However, when the desired model is sufficiently complex or non-standard, ADMB provides a powerful choice for its implementation.

This text is divided into three parts:

Part I: Preliminaries: Chapters 1–2

The preliminaries in this part can be skimmed by the reader who is already familiar with the basic notions and properties of maximum likelihood. However, it should be noted that the simple binomial example in Chapter 1 is used to introduce several key tools, including the Wald and likelihood ratio methods for tests and confidence intervals. Their implementation in R, SAS and ADMB is via general purpose code that is easily extended to more challenging models in later chapters. Chapter 2 looks at examples of maximum likelihood modelling of independent and identically distributed data. Despite being iid data, some of these examples are nonstandard and demonstrate curious phenomena, including likelihoods that have no maximum or have multiple maxima. This chapter also sets up the basic notation employed throughout subsequent chapters.

Part II: Pragmatics: Chapters 3–10

This part covers the relevant practical application of maximum likelihood, including cutting-edge developments in methodology for coping with nuisance parameters (e.g., GREML – generalized restricted maximum likelihood) and latent variable models. The well-established methodology for construction of hypothesis tests and confidence intervals is presented in Chapter 3. But, knowing how to do the calculations isn't the same as actually working with real data, and it is Chapter 4 that really explains how it should be done. This chapter includes model selection, bootstrapping, prediction, and coverage of techniques to handle nonstandard situations. Chapter 5 looks at methods for maximizing the likelihood (especially stubborn ones), and Chapter 6 gives a flavour of some common applications, including survival analysis, and mark–recapture models. Generalized linear models are covered in Chapter 7, with some attention to variants such as the simple over-dispersion form of quasi-likelihood, and the use of nonstandard link functions. Chapter 8 covers some of the general variants of likelihood that are in common use, including quasi-likelihood and generalized estimating equations. Chapter 9 looks at modified forms of likelihood in the presence of nuisance parameters, including conditional, restricted and integrated likelihood. Chapter 10 looks at the use of latent-variable models (e.g., mixed-effects and state-space models). For arbitrary forms of such models, this is one place where ADMB comes to the fore.

Part III: Theoretical foundations: Chapters 11–14

The theory and associated tools that are required to formally establish the properties of maximum likelihood methodology are provided here. This part provides completeness for those readers who wish to understand the true meaning of statistical concepts such as efficiency and large-sample asymptotics. In addition, Chapter 14 looks at some of the fundamental issues underlying a statistical paradigm based on likelihood.

Chapter 15 contains a collection of notation, descriptions of common statistical distributions, and details of software utilities. This text concludes with partial solutions to a selection of the exercises from the end of each chapter.

This book includes an accompanying website. Please visit `www.wiley.com/go/Maximum_likelihood`

Acknowledgements

I am extremely thankful to the many cohorts of statistics students at the University of Auckland who have perused and critiqued the parts of this text that have been used in my statistical inference course. This work was greatly assisted by a University of Auckland Research Fellowship. My greatest thanks are for the unwavering support of Professor Marti Anderson at Massey University, Auckland, and for her dedication at reading through the entire first draft.

Russell B. Millar
Auckland, March 2011

Part I

PRELIMINARIES

1

A taste of likelihood

When it is not in our power to follow what is true, we ought to follow what is most probable. – René Descartes

1.1 Introduction

The word *likelihood* has its origins in the late fourteenth century (Simpson and Weiner 1989), and examples of its usage include as an indication of probability or promise, or grounds for probable inference. In the early twentieth century, Sir Ronald Fisher (1890–1962) presented the 'absolute criterion' for parameter estimation (Fisher 1912), and some nine years later he gave this criterion the name *likelihood* (Fisher 1921, Aldrich 1997). Fisher's choice of terminology was ideal, because the centuries-old interpretation of the word *likelihood* is also applicable to the formal statistical definition of likelihood that is used throughout this book.

Here, likelihood is used within the traditional framework of frequentist statistics, and maximum likelihood (ML) is presented as a general-purpose tool for inference, including the evaluation of statistical significance, calculation of confidence intervals (CIs), model assessment, and prediction. The frequentist theory underpinning the use of maximum likelihood is covered in Part III, where it is seen that maximum likelihood estimators (MLEs) have optimal properties for sufficiently large sample sizes. It is for this reason that maximum likelihood is the most widely used form of traditional parametric inference. The pragmatic use of ML inference is the primary focus of this book and is covered in Part II. The reader who is already comfortable with the concept of likelihood and its basic properties can proceed to Part II directly.

Maximum Likelihood Estimation and Inference: With Examples in R, SAS and ADMB, First Edition. Russell B. Millar.

Likelihood is also a fundamental concept underlying other statistical paradigms, especially the Bayesian approach. Bayesian inference is not considered here, but consideration of the philosophical distinctions between frequentist and Bayesian statistics is examined in Chapter 14. In addition, it is seen that some maximum likelihood methodology can be motivated using Bayesian considerations. This includes techniques for prediction (Section 4.6), and the use of integrated likelihood (Section 9.3).

A simple binomial example (Example 1.1) is used in Section 1.2 to motivate and demonstrate many of the essential properties of likelihood that are developed in later chapters. In this example, the likelihood is simply the probability of observing $y = 10$ successes from 100 trials. The fundamental conceptual point is that likelihood expresses the probability of observing 10 successes as a function of the unknown success probability p. That is, the likelihood function does not consider other values of y. It takes the knowledge that $y = 10$ was the observed number of successes and it uses the binomial probability of the outcome $y = 10$, evaluated at different possible values of p, to judge the relative likelihood of those different values of p.

1.2 Motivating example

Throughout this book, adding a zero subscript to a parameter (e.g. p_0) is used generically to denote a specified value of the parameter. This is typically either its true unknown value, or a hypothesized value.

1.2.1 ML estimation and inference for the binomial

Example 1.1 applies ML methodology to the binomial model in order to obtain the MLE of the binomial probability, the standard error of the MLE, and confidence intervals. This example is revisited and extended in subsequent chapters. For example, Sections 4.2.2 and 4.3.1 look at issues concerning approximate normality of the MLE, and Example 4.10 considers prediction of a new observation from the binomial distribution.

Example 1.1. Binomial. A random sample of one hundred trials was performed and ten resulted in success. What can be inferred about the unknown probability of success, p_0?

For any potential value of p ($0 \leq p \leq 1$) for the probability of success, the probability of y successes from n trials is given by the binomial probability formula (Section 15.4.1). With $y = 10$ successes from $n = 100$ trials, this is

$$\begin{aligned} L(p) &= \text{Prob}(10 \text{ successes}) \\ &= \frac{100!}{90!\,10!}\, p^{10}(1-p)^{90} \\ &= 1.731 \times 10^{13} \times\ p^{10}(1-p)^{90}\,, \quad 0 \leq p \leq 1\,. \end{aligned} \tag{1.1}$$

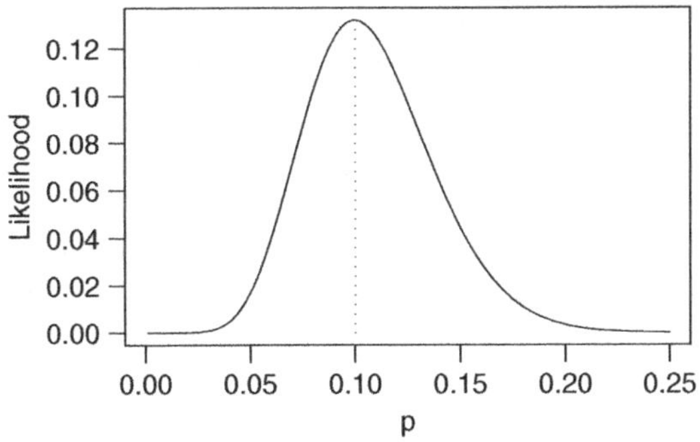

Figure 1.1 Binomial likelihood for 10 successes from 100 trials.

The above probability is the likelihood, and has been denoted $L(p)$ to make its dependence on p explicit.

A plot of $L(p)$ (Figure 1.1) shows it to be unimodal with a peak at $p = 0.1$. This is the MLE and will be denoted $\widehat{p}$. For the binomial model, the MLE of the probability of success is always the observed proportion of successes $\widehat{p} = y/n$ (Example 2.5). □

> The curve in Figure 1.1 looks somewhat like the bell-shaped curve of the normal density function. However, it is not a density (it is a likelihood function) and nor is it bell-shaped. On close inspection it can be seen that the curve is slightly right-skewed.

Box 1.1

In the above example, the MLE $\widehat{p}$ is simply a point-estimate of p_0, and is of limited use without any sense of how reliable it is. For example, it would be more meaningful to have a range of plausible values of the unknown p_0, or to know if some pre-specified value, e.g. $p_0 = 0.5$, was reasonable. Such questions can be addressed by examining the shape of the likelihood function, or more usually, the shape of the log-likelihood function.

The (natural) log of the likelihood function is used far more predominantly in likelihood inference than the likelihood function itself, for several good reasons:

1. The likelihood and log-likelihood are both maximized by the MLE.
2. Likelihood values are often extremely small (but can also be extremely large) depending on the model and amount of data. This can make numerical optimization of the likelihood highly problematic, compared to optimization of the log-likelihood.

3. The plausibility of parameter values is quantified by ratios of likelihood (Section 2.3), corresponding to a difference on the log scale.
4. It is from the log-likelihood (and its derivatives) that most of the theoretical properties of MLEs are obtained (see Part III).

The theoretical properties alluded to in Point 4 are the basis for the two most commonly used forms of likelihood inference – inference based on the likelihood ratio (LR) and inference based on asymptotic normality of the MLE. These two forms of likelihood-based inference are asymptotically equivalent (Section 12.5) in the sense that they lead to the same conclusions for sufficiently large sample sizes. However, in real situations there can be a non-negligible difference between these two approaches (Section 4.3).

Using the likelihood ratio approach in the context of Example 1.1, an interval of plausible values of the unknown parameter p_0 is obtained as all values p for which the log-likelihood is above a certain threshold. In Section 3.4 it is shown that the threshold can be chosen so that the resulting interval has desirable frequentist properties. In the continuation of Example 1.1 below, the threshold is chosen so that the resulting interval is a (approximate) 95 % confidence interval for parameter p.

The curvature of the log-likelihood is of fundamental importance in both the theory and practice of likelihood inference. The curvature is quantified by the second derivative, that is, the change in slope. When evaluated at the MLE, the second derivative is negative (because the slope changes from being positive for $p < \widehat{p}$ to negative for $p > \widehat{p}$) and the larger its absolute value the more sharply curved the log-likelihood is at its maximum. Intuitively, a sharply curved log-likelihood is desirable because this narrows the range over which the log-likelihood is close to its maximum value, that is, it narrows the range of plausible parameter values. In Section 3.2 it is seen that the variance of the MLE can be estimated by the inverse of the negative of the second derivative of the log-likelihood. This is particularly convenient in practice because some optimization algorithms evaluate the second derivative of the objective function as part of the algorithmic calculations (see Section 5.2). In the maximum likelihood context, the objective function is the log-likelihood, and the estimated variance of the MLE is an easily-calculated byproduct from such optimizers. The approximate normality of MLEs enables confidence intervals and hypothesis tests to be performed using well-established techniques.

The likelihood ratio and curvature-based methods of likelihood inference are demonstrated in the following continuation of Example 1.1.

Example 1.1 continued. The log-likelihood function for p, $0 < p < 1$, is

$$\begin{aligned} l(p) &= \log L(p) \\ &= \log\left(\frac{100!}{90!\,10!}\right) + 10\log p + 90\log(1-p) \\ &= 30.48232 + 10\log p + 90\log(1-p)\,, \end{aligned} \tag{1.2}$$

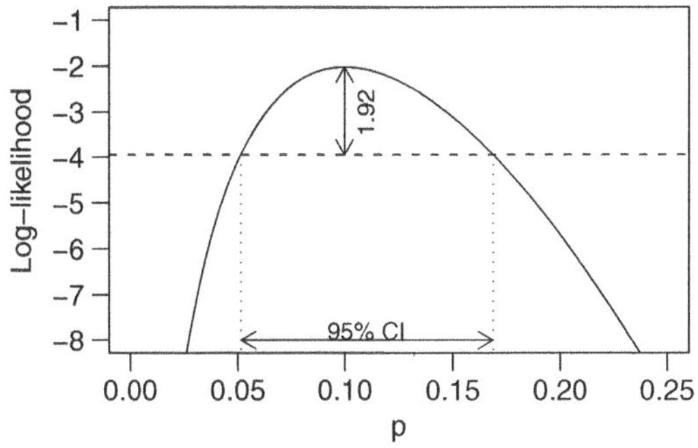

Figure 1.2 Binomial log-likelihood for 10 successes from 100 trials, and 95 % likelihood ratio confidence interval.

and the maximized value of this log-likelihood is $l(\widehat{p}) = l(0.1) \approx -2.03$.

In Section 3.4 it is seen that an approximate 95 % likelihood ratio confidence interval for parameter p is given by all values p_0 for which $l(p_0)$ is within about 1.92 of the maximized value of the log-likelihood. (The value 1.92 arises as one half of the 0.95 quantile of a chi-square distribution with one degree of freedom.) So, in this case, the interval is given by all values of p_0 for which $l(p_0)$ is -3.95 or higher. The confidence interval can be read from Figure 1.2, or obtained numerically for greater accuracy. This interval is $(0.051, 0.169)$ to the accuracy of three decimal places. From the equivalence between confidence intervals and hypothesis tests (Section 13.2) it can be concluded that the null hypothesis $H_0 : p = p_0$ will be rejected at the 5 % level for any value of p_0 outside of the interval $(0.051, 0.169)$.

To perform inference based on the curvature of the log-likelihood, the second derivative of the log-likelihood is required. This second derivative is given in Equation (11.15), and for $n = 100$ trials and $y = 10$ successes it is

$$l''(p) = \frac{\partial^2 l(p)}{\partial p^2} = -\frac{10}{p^2} - \frac{90}{(1-p)^2} . \qquad (1.3)$$

Evaluating this second derivative at the MLE $\widehat{p} = 0.1$ gives

$$l''(\widehat{p}) = -\frac{10}{0.01} - \frac{90}{0.81} \approx -1111.111 .$$

The inverse of the negative of $l''(0.1)$ is exactly 0.0009, and according to likelihood theory (Sections 3.2 and 12.2), this is the approximate variance of $\widehat{p}$. The approximate standard error is therefore $\sqrt{0.0009} = 0.03$.

Recall that for a binomial experiment, the true variance of $\widehat{p}$ is $p_0(1 - p_0)/n$, which is estimated by $\widehat{p}(1 - \widehat{p})/n$. This estimate of variance is also 0.0009, the same as that obtained from using $-1/l''(0.1)$. (In fact, for the binomial the two variance estimates are always the same, for any values of n and y.)

For sufficiently large n, the distribution of $\widehat{p}$ can be approximated by a normal distribution, thereby permitting approximate tests and confidence intervals for p to be performed using familiar techniques. These are often called Wald tests or intervals, due to the influential work of Abraham Wald in establishing the large-sample approximate normality of MLEs (e.g. Wald 1943). The $(1-\alpha)100\,\%$ Wald confidence interval for p can be obtained using the familiar formula that calculates the upper (or lower) bounds as the point estimate plus (or minus) $z_{1-\alpha/2}$ times the estimated standard error ($\widehat{\text{se}}$), where $z_{1-\alpha/2}$ is the $(1-\alpha/2)$ quantile of the standard normal distribution. Thus, the approximate 95 % Wald confidence interval is

$$\widehat{p} \pm z_{0.975}\ \widehat{\text{se}}(\widehat{p}), \tag{1.4}$$

where $z_{0.975} \approx 1.96$ and $\widehat{\text{se}}(\widehat{p}) = 0.03$. This interval is $(0.041, 0.159)$. Equivalently, this interval is the collection of the values of p_0 such the null hypothesis $H_0 : p = p_0$ is not rejected at the 5 % level by the Z-statistic. This is the values of p_0 that satisfy the inequality

$$|Z| = \left| \frac{\widehat{p} - p_0}{\widehat{\text{se}}(\widehat{p})} \right| < z_{0.975}\ . \tag{1.5}$$

□

> Although the Wald CI and test statistic in (1.4) and (1.5) may be the most commonly taught and used methods of such inference for the binomial model, it is hoped that this text will convince the reader to avoid Wald (i.e. approximate normality) methodology whenever it is practicably feasible. See the next section for more on this.

Box 1.2

1.2.2 Approximate normality versus likelihood ratio

The Wald form of confidence interval used in (1.4) is based on the approximate normal distribution of $\widehat{p}$. This is the most commonly used method for constructing approximate confidence intervals because of its intuitive appeal and computational ease. It was shown earlier that the likelihood ratio can be used as an alternative method for constructing confidence intervals – which should be used?

From a pragmatic point of view, there is considerable intuitive appeal in the Wald construction of a 95 % (say) confidence interval, with bounds given by 1.96 standard errors each side of the point estimate. This form of CI will be the most familiar to anyone with a basic grounding in frequentist statistics. However, when the LR and Wald intervals differ substantially, it is generally the case that the LR approach is superior, in the sense that the CIs obtained using likelihood ratio will have actual coverage probability closer to the a priori chosen value of (1-α) (see

Section 4.3.1). In fact, the results of Brown *et al.* (2001) question the popular usage of the Wald CI for binomial inference because of its woeful performance, even for some values of n and p for which the normal approximation to the binomial distribution is generally considered reasonable (typically, $\min(n\widehat{p}, n(1-\widehat{p})) \geq 5$). Unfortunately, the LR confidence interval is not as widely used because it requires (a little) knowledge of likelihood theory, but more importantly because it can not generally be calculated explicitly.

Application of Wald tests and construction of CIs extends to multi-parameter inference, but becomes more cumbersome and unfamiliar when simultaneous inference about two or more parameters is required. It is then that LR-based inference tends to be more commonly used. In particular, multi-parameter inference is typical of model selection problems, and in this area LR-based inference dominates. Also, it should be noted that model selection criterion such as Akaike's Information Criterion (AIC) (Section 4.4.1) make direct use of the likelihood.

In addition to the Wald and LR intervals, there are several other competing methods for constructing approximate confidence intervals for the probability parameter p in a binomial experiment. These include the Wilson score (see Box 3.1, Example 12.10, and Exercise 12.7), Agresti-Coull, and the misnamed 'exact' CIs. The comparisons performed by Agresti and Coull (1998) and Brown *et al.* (2002) suggest that the LR and Wilson score CIs are to be preferred.

Box 1.3

Summary

To conclude, Example 1.1 demonstrates likelihood inference in a nutshell. Much of the rest of this book is devoted to providing pragmatic guidance on the use (and potential abuse) of inferential methods based on likelihood ratios and approximate normality of MLEs, and their application to more complex and realistic models. These concepts extend naturally to models with two or more parameters, although the implementation can become challenging. For example, in a model where the number of parameters is $s > 2$, the second derivative of the log-likelihood is an s-dimensional square matrix (the Hessian) and the negative of its inverse provides an approximate variance matrix for the MLEs.

1.3 Using SAS, R and ADMB

This book is not just about understanding maximum likelihood inference, it is also very much about *doing* it with real data. Examples in SAS and R (Ihaka and Gentleman 1996, R Development Core Team 2010) are provided throughout Part II, along with a smattering of examples demonstrating Automatic Differentiation Model Builder (ADMB, ADMB-project (2008a, or any later version)).

Unlike the SAS and R environments, ADMB is a tool specifically designed for complex optimization problems. Due to the learning curve required to use ADMB, its use is difficult to justify if existing functionality within SAS or R can be used instead. Other than the quick demonstration of ADMB later in this chapter, it is used sparingly until Chapter 10 where it becomes the best choice for the general-purpose fitting of latent variable models. Some of its additional capabilities are noted in Sections 4.2.3 and 5.4.2.

The SAS examples presented in this text were implemented using SAS for Windows version 9.2. The SAS procedures used throughout are found in the statistics module SAS/STAT (SAS Institute 2008), with the exception that occasional use was made of the nonlinear optimizer `PROC NLP` which is in the operations research module SAS/OR. Some users of SAS/STAT may find that their licence does not extend to SAS/OR and hence will not be able to use `PROC NLP`. For this reason, `PROC NLP` is used sparingly and alternative SAS code is given where possible.

SAS procedures typically produce a lot of output by default. The output often includes a lot of superfluous information such as details about the contents of the data-set being used, computational information, and unwanted summary statistics. Throughout, the Output Delivery System (ODS) in the SAS software has been used to select only the required parts of the output produced by the SAS procedure.

Delwiche and Slaughter (2003, or any later edition) provides an excellent introduction to SAS. For ease of readability, the SAS code presented herein follows their typographical convention. This convention is to write SAS keywords in uppercase, and to use lowercase for variable names, data-set names, comments, etc. Note that SAS code is not case sensitive.

The R examples were run using R for Windows version 2.12.0. R is freely available under the terms of the Free Software Foundation's GNU General Public License (see `http://www.R-project.org`. Most of the R functions used herein are incorporated in the default installation of R. Others are available within specified R library packages, and can be easily loaded from within the R session.

ADMB is freely available via the ADMB project (`http://www.admb-project.org`), where full instructions for using ADMB can also be found. A short description of automatic differentiation is given in Section 15.6. In brief, ADMB is implemented by programming the objective function within an ADMB template file. The objective function is just the (negative) log-likelihood (and in latent variable models the density function of the latent variables also needs to be specified). An executable file is then created from the template file. Fortunately, much of the detail in creating the executable can be hidden behind convenient user interfaces. The ADMB examples in this book were run from within R using the interface provided by the `PBSadmb` package.

In many applications of ML inference it will be possible to make use of existing SAS procedures and R functions that are appropriate to the type of data being modelled, notwithstanding that this convenience often comes at the loss of flexibility. Rather than using existing functionality that is specific to the binomial model, the implementations of Example 1.1 presented below demonstrate a selection of the general-purpose tools available in SAS and R, and the use of ADMB. In particular, calculation of likelihood ratio confidence intervals is an application of profile likelihood (Section 3.6), and the examples below make use of general-purpose code for this purpose.

1.3.1 Software resources

Several small pieces of code have been written to facilitate techniques described in this text. These are listed in Section 15.5, along with a brief description of their functionality. These software resources are freely available for download from `http://www.stat.auckland.ac.nz/~millar`. This web resource also contains the complete code, and the data, for all examples used in this text.

1.4 Implementation of the motivating example

The code used below demonstrates how an explicit log-likelihood function is maximized within each of SAS, R and ADMB, and the calculation of the Wald and likelihood-ratio confidence intervals. Some efficiencies could have been gained by taking advantage of built-in functionality within the software. For example, in the SAS example, the binomial model could have been expressed using the statement `MODEL y ~ BINOMIAL(n,p)`, but the general-purpose likelihood specification has been used here for illustration. In R, various functionality (e.g. the `mle` function in package `stat4`, or `maxLik` function in the package of the same name) could have been used to shortcut some of the required code. However, the savings are minimal, and it is instructive to see the individual programming steps.

The first term of the binomial log-likelihood given in Equation (1.2) is a constant, and hence is irrelevant to maximization of the log-likelihood. However, it is good practice always to include the constant terms because it removes possible sources of confusion when fits of different model types are being compared (e.g. using Akaike's information criterion), or when verifying the fit of a model by using an alternative choice of software. Inclusion of the constant terms in the log-likelihood is becoming standard in most software applications of ML, but do not take this for granted.

The description of the code that is presented below is relatively complete, but this level of explanation is too unwieldy to be used throughout the remainder of this text. For more explanation on programming details and syntax, the reader should refer to the abundant online resources and documentation for each of these software.

1.4.1 Binomial example in SAS

The SAS code below uses `PROC NLMIXED` to implement Example 1.1, and produces the output shown in Figure 1.3.

Parameter estimates									
Parameter	Estimate	Standard Error	DF	tValue	Pr>|t|	Alpha	Lower	Upper	Gradient
p	0.1000	0.03000	1E6	3.33	0.0009	0.05	0.04120	0.1588	8.566E−7

Figure 1.3 The parameter estimates table from PROC NLMIXED*, including the 95 % Wald confidence interval (0.0412,0.1588).*

```
DATA binomial;
  y=10; n=100;
RUN;

*Select only the parameter estimates table;
ODS SELECT ParameterEstimates;

PROC NLMIXED DF=1E6 DATA=binomial;
  PARMS p=0.5;
  BOUNDS 0<p<1;
  loglhood=LOG(COMB(n,y))+y*log(p)+(n-y)*log(1-p);
  MODEL y~GENERAL(loglhood);
RUN;
```

Some features of the above code are:

- The default output includes several tables, including tables of log-likelihood values and fit statistics. The Output Delivery System statement `ODS SELECT ParameterEstimates;` is used to select only the required table.
- By default, `NLMIXED` calculates Wald intervals using a t-distribution with degrees of freedom equal to the number of observations (rows in the dataset). To get the normal-based Wald interval in (1.4), the value for the degrees of freedom needs to be set to a large number. In this case, it was set to one million using the procedure option `DF=1E6`.
- The `PARMS` statement is an optional statement used to explicitly list the parameters and their initial values.
- The `BOUNDS` statement is an optional statement used to specify the range of the parameter values (i.e. the parameter space).
- The model is specified using the `MODEL` statement. Here, the model is given as `GENERAL(loglhood)` to specify that `PROC NLMIXED` should maximize the value of the log-likelihood, `loglhood`, as specified by the preceding programming statement.
- In the SAS output in Figure 1.3, `Gradient` gives the slope of the log-likelihood upon termination of the optimization. It should be near zero. If not, then convergence of the optimizer to a maximum of the log-likelihood may not have been achieved.

- The `t-Value` and `Pr>|t|` columns in Figure 1.3 should be ignored. They are the Wald test statistic and p-value for the null hypothesis $H_0 : p = 0$. This is not a relevant hypothesis here.

One current limitation (in SAS 9.2) is that `PROC NLMIXED` does not produce likelihood ratio confidence intervals. A general-purpose macro called `Plkhci` has been written for this purpose.

```
%INCLUDE "PlkhciMacro.sas";

%MACRO BinomialProfile(p);
  PROC NLMIXED DF=1E6 DATA=Binomial; TECH=NONE;
    loglhood=LOG(COMB(n,y))+y*log(&p)+(n-y)*log(1-&p);
    MODEL y~GENERAL(loglhood);
  RUN;
%MEND;
%Plkhci(BinomialProfile,0.0,0.1,-2.0259739,side="L");
%Plkhci(BinomialProfile,0.1,1.0,-2.0259739,side="R");
```

- The user-defined macro `BinomialProfile` contains a modified version of the `NLMIXED` code that was used to produce the output in Figure 1.3, and this is passed as an argument to the profile likelihood macro `Plkhci`. More description of these macros is found in Sections 3.4.1 and 15.5.3. Note that macro commands are specified using the `%` symbol.
- The `Plkhci` macro finds the likelihood ratio confidence bounds. It writes the following lines to the log window of the SAS session:

```
Left-sided 95% LR CI bound is 0.051413
Right-sided 95% LR CI bound is 0.168779
```

For SAS installations that include the operations research OR module, `PROC NLP` provides an easier option for obtaining the likelihood ratio confidence interval, via its `PROFILE` statement. Figure 1.4 shows the table that is produced from running the following code.

```
*Select only the desired table;
ODS SELECT WaldPLLimits;
PROC NLP COV=2 VARDEF=N;
  MAX loglhood;
  PROFILE p / alpha=0.05;
  PARMS p=0.5;
  BOUNDS 0<p<1;
  n=100; y=10;
  loglhood=LOG(COMB(n,y))+y*LOG(p)+(n-y)*LOG(1-p);
RUN;
```

Wald and PL confidence limits							
N	Parameter	Estimate	Alpha	Profile likelihood confidence limits		Wald confidence limits	
1	p	0.100000	0.050000	0.051414	0.168773	0.041201	0.158799

Figure 1.4 Likelihood ratio and Wald confidence limits from `PROC NLP`.

- `PROC NLP` provides a choice of several different estimates of variance and the option `COV=2` specifies use of the curvature-based estimate employed in the motivating example. Also, by default, `PROC NLP` makes a degrees-of-freedom adjustment to the estimate of variance. This adjustment is not appropriate in the maximum likelihood context, and the procedure option `VARDEF=N` prevents this.
- The `MAX loglike` statement specifies that the value of `loglike` is to be maximized.
- The `PROFILE` statement requests calculation of a likelihood ratio confidence interval for parameter p, with confidence level $(1-\alpha)100\,\%$.

1.4.2 Binomial example in R

The R code presented below uses the general-purpose minimizer `optim`, and hence the objective function to be minimized is the negative of the log-likelihood. This is explicitly defined as function `nloglhood`, with argument p. The likelihood ratio confidence interval is obtained using the `plkhci` function (from the `Bhat` package) for profile likelihood confidence intervals.

```
> #Define the negative log-likelihood function
> nloglhood=function(p)
+   return( -(log(choose(100,10))+10*log(p)+90*log(1-p)) )
> #Minimize the negative log-likelihood
> binom.fit=optim(0.5,nloglhood,lower=0.0001,upper=0.9999,
+                   hessian=T)
> phat=binom.fit$par #The MLE
> phat.var=1/binom.fit$hessian #Variance is inverse hessian
> #Calculate approximate 95% Wald CI
> phat+c(-1,1)*qnorm(0.975)*sqrt(phat.var)
[1] 0.04120779 0.15879813

> library(Bhat) #Loading package Bhat
> #Set up list for input into plkchi function
> control.list=list(label="p",est=phat,low=0,upp=1)
> #Calculate approximate 95% likelihood ratio CI
> plkhci(control.list,nloglhood,"p")
[1] 0.05141279 0.16877909
```

- In the call of `optim`, the first argument specifies that the initial parameter value to be used by the optimizer is 0.5. The `lower` and `upper` arguments specify the parameter space – in this case they were set to 0.0001 and 0.9999 because computational error occurred if bounds of 0 and 1 were used due to `nloglhood` being undefined at these values. The `hessian=T` argument requests that the value of the second derivative of the negative log-likelihood (calculated at the MLE) be included in `binom.fit`.
- The list object `binom.fit` has several components, including the value of the calculated MLE, `binom.fit$par`, and the second derivative of the negative log-likelihood, `binom.fit$hessian`.
- The first argument to the profile likelihood function `plkhci` is a list with elements giving the parameters of `nloglhood`, the MLE, and lower and upper bounds of the parameter space.

1.4.3 Binomial example in ADMB

The following ADMB template file, `BinomialMLE.tpl`, is used to find the MLE and its approximate standard error.

```
DATA_SECTION
  init_number y
  init_number n

PARAMETER_SECTION
  init_bounded_number p(0,1)
  objective_function_value nloglhood

PROCEDURE_SECTION
  nloglhood=-(lgamma(n+1)-lgamma(y+1)-lgamma(n-y+1)); // Constant term
  nloglhood=nloglhood-(y*log(p)+(n-y)*log(1-p));
```

- ADMB requires a data section, and the data are contained in a file with name `BinomialMLE.dat`. This text file contains a single row, the contents of which is `10 100`.
- The parameter section specifies the parameter and that it is bounded between 0 and 1. It also specifies the variable that will contain the value of the negative log-likelihood, and this is calculated in the procedure section.
- The `lgamma` function is the log-gamma function, and for non-negative integers x, `lgamma`$(x+1) = \log(x!)$.

An executable program is generated from the template file, and executed, using the following R code. This uses functions from the `PBSadmb` package.

```
library(PBSadmb)
readADopts()
makeAD("BinomialMLE")
runAD("BinomialMLE")
```

- `readADopts` reads a text file containing information about the ADMB installation.
- `makeAD` creates a C++ file from the template file, compiles it, and links it to produce an executable with name `BinomialMLE.exe`.
- `runAD` runs the executable. A number of files are produced, including text file `BinomialMLE.std` containing the MLE and its standard error, and text file `BinomialMLE.par` containing the value of the negative log-likelihood at the MLE.

A few additional lines of code are required to obtain the likelihood ratio confidence interval. The parameter section requires addition of a `likeprof_number` specification to name the quantity of interest. In this case it is p, but this name is already in use, so the variable `pcopy` is used to copy the value of p. The preliminary calculations section is used to set options for the grid of `pcopy` values over which the objective function is evaluated.

```
DATA_SECTION
  init_number y
  init_number n

PARAMETER_SECTION
  init_bounded_number p(0,1)
  objective_function_value nloglhood
  likeprof_number pcopy

PRELIMINARY_CALCS_SECTION
  pcopy.set_stepnumber(500);
  pcopy.set_stepsize(0.01);

PROCEDURE_SECTION
  nloglhood=-(lgamma(n+1)-lgamma(y+1)-lgamma(n-y+1)); // Constant term
  nloglhood=nloglhood-(y*log(p)+(n-y)*log(1-p));
  pcopy=p;
```

Within R, the executable is created as before. The `runAD` function now requires the optional `lprof` argument to pass to the executable, to force calculation of the likelihood ratio confidence interval. If the template file is named `BinomialLRCI.tpl`, then the `runAD` call looks like

```
runAD("BinomialLRCI",argvec="-lprof > RunWindow.txt")
```

which also redirects a copious log file into the text file `RunWindow.txt`. This produces a text file `pcopy.plt`, in which the bounds of the 95 % confidence interval, calculated to be `0.0514` and `0.1687` can be found.

1.5 Exercises

1.1 The Poisson distribution is the default distribution for the modelling of count data. If the value $y = 3$ is observed from a Poisson distribution with unknown parameter λ then the log-likelihood is $l(\lambda) = -\lambda + 3\log\lambda - \log(6)$. This log-likelihood is maximized by $\widehat{\lambda} = 3$.

1. Plot $l(\lambda)$ for values of λ from 0.1 to 15.

2. By suitable modification to the program code in Section 1.3, use R, SAS or ADMB to verify that $\widehat{\lambda} = 3$, and to calculate the 95 % Wald and likelihood ratio confidence intervals for λ.

2

Essential concepts and iid examples

All models are wrong, but some are useful. – George E. P. Box[1]

2.1 Introduction

The binomial distribution used in the motivating example (Example 1.1) is a discrete distribution and the likelihood function was simply the probability of observing ten successes from 100 trials, regarded as a function of parameter p. It was natural to estimate p using the value $\widehat{p}$ that maximized the probability of the observed outcome. However, this logic does not immediately extend to data observed from continuous distributions, because then there are infinitely many possible outcomes, each with probability zero.

For continuous data, the likelihood is defined to be the density function evaluated at the observed data, and is regarded as a function of the unknown parameter (see Definition 2.2 for the formal statement). This has intuitive appeal, and if any justification is required, it can be argued that the measured value of a continuous random variable is subject to rounding accuracy, and therefore it is more relevant to consider the probability of realizing a value close to the measured value. For a single real-valued observation, $y \in \mathbb{R}$, this might be the probability of the random variable taking a value in the interval $(y - \epsilon, y + \epsilon)$, where ϵ depended on the

[1] See Box 2.2.

Maximum Likelihood Estimation and Inference: With Examples in R, SAS and ADMB, First Edition. Russell B. Millar.

measurement accuracy. For small ϵ this probability is approximately proportional to the likelihood (see Box 2.1).

Thanks to mathematical measure theory (e.g. Billingsley 1979), it is not necessary to make any distinction between discrete and continuous data when using the terminology 'density function'. This terminology can be used to refer to both the density function of a continuous random variable, or the probability function of a discrete random variable.

In practice, continuous data will be measured to a certain accuracy. For example, if a person's weight is rounded to the nearest 100 g, then a recorded weight of $y = 74.2$ kg is actually the event that the weight is between 74.15 and 74.25 kg. To a close degree of approximation, the probability of this event is proportional to the value of the density function (of weights, for adults randomly chosen from some hypothetical population, say) evaluated at the fixed value of 74.2. This is the likelihood, and will be a function of parameters associated with the experimental circumstances under which the weight was measured.

Box 2.1

2.2 Some necessary notation

Throughout, italic bold notation is used for vectors. If the data consist of the observation of n real values $y_i, i = 1, ..., n$, then these observations will be denoted by the vector $\boldsymbol{y} = (y_1, ..., y_n)$. By algebraic convention, vectors are column vectors, and so it would be more appropriate to write $\boldsymbol{y} = (y_1, ..., y_n)^T$. For ease of notation this convention is avoided, except where it becomes necessary within relevant algebraic formulae. Matrices will be denoted in regular upper case bold, and so $\mathbf{X}$ might be used to denote the design matrix in a linear regression, say.

When it is necessary to distinguish between observations and random variables, the random variable will be denoted in upper case. Thus, $\boldsymbol{Y} = (Y_1, ..., Y_n)$ denotes a random vector, and can be regarded as the potential outcomes from a random process (such as rolling a die n times). However, this distinction between observations and random variables can result in cumbersome notation in some places, in which case it will be overlooked where there is little chance of confusion. For example, the statement that y is distributed $N(\mu, \sigma^2)$ is a shorthand way of saying that y is a realization of a $N(\mu, \sigma^2)$ distributed random variable Y.

It is assumed that the random experiment (or process) that generated the data can be specified by a statistical distribution with (joint) density function $f(\boldsymbol{y}; \boldsymbol{\theta}_0)$, where $\boldsymbol{\theta}_0$ denotes the true unknown value (or a hypothesized value) of the parameter, which in general is a vector in $\mathbb{R}^s$. This density function may depend on covariates (i.e. explanatory variables) associated with the random vector $\boldsymbol{Y}$. The set of all possible parameter values is the parameter space, denoted Θ, and is a subset of s-dimensional real space $\mathbb{R}^s$. The collection of distributions that is formed from all values of $\boldsymbol{\theta} \in \Theta$ comprises the model. Expressing this formally gives the following definition.

Definition 2.1 Parametric statistical model. A parametric statistical model is a collection of joint density functions, $f(\mathbf{y}; \boldsymbol{\theta})$, indexed by $\boldsymbol{\theta} \in \Theta \subset \mathbb{R}^s$.

Example 2.1. The normal model. It is often assumed that 'Y is normally distributed'. Formally, this is specifying a parametric statistical model comprising the collection of all possible normal distributions. This is the collection of $N(\mu, \sigma^2)$ distributions over the parameter space $\boldsymbol{\theta} = (\theta_1, \theta_2) = (\mu, \sigma)$ with $\theta_1 \in \mathbb{R}$ and $\theta_2 \in \mathbb{R}^+$, where $\mathbb{R}^+$ denotes the positive reals. Alternatively, some readers might prefer to index this collection using the mean and variance parameterization, $\boldsymbol{\theta} = (\theta_1, \theta_2) = (\mu, \sigma^2)$, where again $\theta_1 \in \mathbb{R}$ and $\theta_2 \in \mathbb{R}^+$. The true distribution of Y is a member of this collection, denoted $Y \sim N(\mu_0, \sigma_0^2)$. □

In the vast majority of applications it is explicitly assumed that the form of the chosen model is correct, that is, the joint density of the data truly is $f(\mathbf{y}; \boldsymbol{\theta}_0)$ for some $\boldsymbol{\theta}_0 \in \Theta$. Inference is then obtained from knowledge about the behaviour of certain statistics (such as the MLE or likelihood ratio) given that the model is correctly specified. However, the quote at the start of this chapter belies this naive interpretation of the statistical model. In all but the simplest of random experiments, the statistical model can only be regarded as an approximation to truth. With judicious choice of model specification and evaluation, it should be a sufficiently good choice to be useful.

See Sections 4.4 and 12.2.4 for more on this topic.

Box 2.2

Statistical theory (Chapter 12) requires the statistical model to satisfy certain regularity conditions. Some of these conditions have obvious interpretation. For example, condition R2 (Section 12.2) requires the parameters of the model to be identifiable. Identifiability simply means that the densities $f(\mathbf{y}; \boldsymbol{\theta})$, $\boldsymbol{\theta} \in \Theta$ all correspond to distinct statistical distributions. That is, if $\boldsymbol{\theta}_A$ and $\boldsymbol{\theta}_B$ are two different parameter values then the density functions $f(\mathbf{y}; \boldsymbol{\theta}_A)$ and $f(\mathbf{y}; \boldsymbol{\theta}_B)$ must correspond to different statistical distributions. Clearly, if multiple values of $\boldsymbol{\theta}$ index the same statistical distribution then there is no possible way to distinguish between those parameter values on the basis of observing data from that distribution.

Example 2.2. Non-identifiable parameterization of ANOVA models. In a one-way analysis of variance model with p groups (i.e. levels) the expected value within group i is often formulated as

$$\mu_i = \mu + \alpha_i \,, \qquad i = 1, ..., p \,. \tag{2.1}$$

This parameterization is not identifiable because, for any $c \in \mathbb{R}$, $(\mu, \alpha_1, ..., \alpha_p)$ and $(\mu + c, \alpha_1 - c, ..., \alpha_p - c)$ correspond to the same values of μ_i, $i = 1, ..., p$.

To obtain an identifiable parameterization it would be natural to use the group means $(\mu_1, ..., \mu_p)$ directly as parameters. This convenient parameterization is not useful for higher-order ANOVAs because it is necessary to use a parameterization that permits the evaluation of main effects and interactions. Thus, an identifiable parameterization for ANOVA is typically achieved by some form of 'aliasing', that is, by setting one of the levels as a baseline against which other levels are compared. In the context of the one-way ANOVA, the default aliasing implemented in R uses the first level as the baseline by setting $\alpha_1 = 0$. In contrast, SAS uses the last level as the baseline, by setting $\alpha_p = 0$. Within ADMB, a parameter vector can be specified as a deviation vector, which imposes the restriction that its elements sum to zero, and hence specifying $\boldsymbol{\alpha} = (\alpha_1, ..., \alpha_p)$ as a deviation vector would ensure an identifiable parameterization. □

When the data $\boldsymbol{y} = (y_1, ..., y_n)$ are independent, with y_i having distribution with density $f_i(y_i; \boldsymbol{\theta})$, then the joint density function is simply the product of the individual densities

$$f(\boldsymbol{y}; \boldsymbol{\theta}) = \prod_{i=1}^{n} f_i(y_i; \boldsymbol{\theta}) .$$

If the data are iid (independent and identically distributed) then f_i can be replaced by f. This results in a minor abuse of notation, since it doesn't explicitly distinguish between the density function of the data $\boldsymbol{y}$, and that of a single observation y_i. However, this distinction is clear from the argument given to the function, that is, $f(\boldsymbol{y}; \boldsymbol{\theta})$ versus $f(y_i; \boldsymbol{\theta})$. In many cases, the Y_i may share a distribution of common form (e.g. normal), but differ due to dependence of Y_i on explanatory variables $\boldsymbol{x}_i$, in which case it may be more meaningful to replace $f_i(y_i; \boldsymbol{\theta})$ with the notation $f(y_i; \boldsymbol{x}_i, \boldsymbol{\theta})$.

Definition 2.2 Likelihood function. The likelihood function is the (joint) density function evaluated at the observed data, and regarded as a function of $\boldsymbol{\theta}$ alone. That is, $L(\boldsymbol{\theta}) \equiv L(\boldsymbol{\theta}; \boldsymbol{y}) = f(\boldsymbol{y}; \boldsymbol{\theta})$, $\boldsymbol{\theta} \in \Theta$.

Definition 2.3 Maximum likelihood estimate (MLE). Given observations $\boldsymbol{y} = (y_1, ..., y_n)$, any $\widehat{\boldsymbol{\theta}} = (\widehat{\theta}_1, ..., \widehat{\theta}_s) \in \Theta$ that maximizes $L(\boldsymbol{\theta})$ over Θ is called a maximum likelihood estimate (MLE) of the unknown true parameter $\boldsymbol{\theta}$. *[Note that this definition does not assume either the existence or uniqueness of the MLE.]*

For the sake of notational convenience, this text will generally make no notational distinction between *estimate* and *estimator*. Note that Definition 2.3 defines the ML *estimate*, for a given observed $\boldsymbol{y}$. The ML *estimator* is obtained from Definition 2.3 by replacing $\boldsymbol{y}$ with the random vector $\boldsymbol{Y} = (Y_1, ..., Y_n)$. In the case of the binomial model of Example 1.1 the parameter was denoted p (rather than θ)

and for $y = 10$ successes from 100 trials the ML estimate was $\widehat{p} = y/100 = 0.1$. The ML estimator is $\widehat{p} = Y/100$, where $Y \sim \text{Bin}(100, p)$. The distinction between estimate and estimator is implied by context, for example, when talking about the statistical properties of MLEs (such as their approximate normal distribution), it is the estimator that is being considered.

The log-likelihood function, denoted $l(\boldsymbol{\theta})$, is the (natural) log of the likelihood function. Any parameter value that maximizes the likelihood also maximizes the log-likelihood, and for reasons given in Chapter 1, it is usually the case that calculations are made using $l(\boldsymbol{\theta})$ rather than $L(\boldsymbol{\theta})$. In particular, if the log-likelihood function is differentiable then its partial derivatives are zero when evaluated at any local extreme value. These points correspond to solutions of the so-called *likelihood equation(s)*. This is the s-dimensional set of equations

$$\frac{\partial l(\boldsymbol{\theta})}{\partial \theta_k} = 0\,, \quad k = 1, \ldots, s. \tag{2.2}$$

To ensure that a solution to the likelihood equations corresponds to a local maximum (rather than a local minimum or saddle point) it is necessary to check that the $s \times s$-dimensional Hessian matrix (containing the second-order partial derivatives of the log-likelihood) is negative definite (Apostol 1967).

In some special cases, such as conventional linear regression, an MLE can be found in closed form by solving (2.2). More generally, numerical optimization of the log-likelihood is required to find an MLE (Chapter 5). Additional steps may be required to establish the uniqueness (or otherwise) of the MLE (Section 4.7.1).

2.2.1 MLEs of functions of the parameters

It is often the case that the parameters $\boldsymbol{\theta}$ used to specify the statistical model are not of direct interest, but rather, it may be desired to estimate a function of the model parameters. This issue is especially relevant because in many cases the parameterization used to specify the model will be selected for ease of implementation, or for purposes of improving the approximate normality of $\widehat{\boldsymbol{\theta}}$. Estimating a function of the parameters is straightforward because, in plain words, the MLE of a function of the parameters is that function of the MLE.

The above statement is intuitive, but nonetheless, it deserves a formal verification. To that end, let $g(\boldsymbol{\theta})$ (possibly vector valued) denote a function of $\boldsymbol{\theta}$. If g is a one-to-one function on Θ then the collection of densities in the statistical model can also be indexed by $\boldsymbol{\zeta} = g(\boldsymbol{\theta}) \in g(\Theta)$. That is, if f_ζ denotes the density as a function of parameter $\boldsymbol{\zeta}$, then $f_\zeta(\mathbf{y}; \boldsymbol{\zeta}) = f(\mathbf{y}; \boldsymbol{\theta})$, where $\boldsymbol{\theta} = g^{-1}(\boldsymbol{\zeta})$. The collection of density functions comprising the statistical model is the same in both cases, and the density that maximizes the likelihood within this collection is $f_\zeta(\mathbf{y}; g(\widehat{\boldsymbol{\theta}})) = f(\mathbf{y}; \widehat{\boldsymbol{\theta}})$. That is, $\widehat{\boldsymbol{\zeta}} = g(\widehat{\boldsymbol{\theta}})$ is the MLE of $\boldsymbol{\zeta} = g(\boldsymbol{\theta})$. This argument can be extended to the situation where g is not one-to-one.

Since the statistical model is invariant to one-to-one transformations of the parameters, the modeler is free to implement the model using the parameterization of his or her choice. The most obvious parameterization is not always the best. Some model parameterizations will result in MLEs with better distributional properties (e.g. being closer to normally distributed) than others – this is examined in Section 4.3. In conventional linear-normal models (i.e. regression, ANOVA, etc) it makes sense to parameterize a normal distribution using the mean and variance because of the well-established exact theory for these models. However, this is not the case more generally, especially in complex models that include multiple sources of random variation. Then, it may be preferable to use the log-variance as a parameter of normally distributed error terms (e.g. see Section 10.4.3).

Example 2.3. An iid sample $\boldsymbol{y} = (y_1, ..., y_n)$ was observed from a $N(\mu, \sigma^2)$ distribution, and resulted in the MLE $\widehat{\boldsymbol{\theta}} = (\widehat{\mu}, \widehat{\sigma}) = (5, 2)$. However, the quantities of interest were the coefficient of variation (CV), σ/μ, and the probability that a new observation would not exceed 6.

1) Denote the CV by $\zeta_1 = g_1(\theta) = \sigma/\mu$. The MLE of the CV is immediate, $\widehat{\zeta}_1 = g(\widehat{\theta}) = \widehat{\sigma}/\widehat{\mu} = 0.4$.
2) The probability that a new observation will not exceed 6 can be denoted

$$\zeta_2 = g_2(\theta) = P(Y \leq 6)\,.$$

This quantity needs to be expressed explicitly as a function of μ and σ. This can be done by standardizing Y by subtracting its mean and dividing by its standard deviation. That is, if $Z = (Y - \mu)/\sigma$, then

$$g_2(\theta) = P\left(Z \leq \frac{6-\mu}{\sigma}\right),$$

where $Z \sim N(0, 1)$ has a standard normal distribution. Then,

$$\begin{aligned}\widehat{\zeta}_2 = g_2(\widehat{\theta}) &= P\left(Z \leq \frac{6-\widehat{\mu}}{\widehat{\sigma}},\right) \\ &= P(Z \leq 0.5) = 0.6915\,.\end{aligned}$$

□

In practice, inference about $\boldsymbol{\zeta} = g(\boldsymbol{\theta})$ may require calculating the approximate standard error of $\widehat{\boldsymbol{\zeta}}$ (though this may not always be wise – see Section 4.3). Section 4.2 shows how the large-sample distribution of $\widehat{\boldsymbol{\zeta}}$ can be deduced from the large-sample distribution of $\widehat{\boldsymbol{\theta}}$.

2.3 Interpretation of likelihood

Chapter 14 presents some of the foundational concepts of statistical inference, including statement of the likelihood principle. The arguments in Chapter 14 are extremely controversial amongst philosopher-statisticians and they quickly become esoteric and flavoured by nuances and subtleties in meaning. This present section steers clear of the controversies, and takes a quick look at a long-established interpretation of likelihood, namely, that it provides a measure of the relative support for different values of $\boldsymbol{\theta}$ (Edwards 1972). For simplicity, the example below uses a statistical model where $\theta \in \mathbb{R}$ can take only two possible values.

Example 2.4. Likelihood to the rescue. You've just bought a fashionable Louise Vashon wristwatch for a bargain price, but now you're starting to wonder whether all is what it seems because the evening news just reported the arrest of an international gang that had been making imitation Louise Vashon watches.

Genuine Louise Vashon watches are very accurate. In fact, over a month, the number of seconds deviation from true time is well described by a $N(0, 1)$ distribution. The imitation watches are quite inaccurate and it has been found that their time deviation is well described by a $N(0, 100^2)$ distribution. A jeweller friend measured the accuracy of your watch, and it was found to gain exactly 2 s per month.

The scenario posed in Example 2.4 can be considered a parametric statistical model with a parameter space containing just two values, 0 and 1, say. When $\theta = 0$ then $f(y; \theta)$ is the density of a standard normal, and when $\theta = 1$ it is the density of a normal with mean zero and standard deviation of 100.

Box 2.3

Question: Is your watch genuine?
Argument 1: A genuine watch has time deviation that is described by a standard normal distribution, and the measured time deviation of 2 s is two standard deviations above the mean. Ouch, it looks like you were ripped off and have a fake watch!
Argument 2: Calculate the likelihood of the $N(0, 1)$ distribution and compare it to that of the $N(0, 100^2)$ distribution? These likelihoods are the respective densities evaluated at $y = 2$, and are 0.0540 and 0.0040 for genuine and imitation watches, respectively (Figure 2.1). Yippee, your watch is thirteen and a half times more likely to be genuine than not!

The conflicting conclusions from the above two arguments arise due to the fundamental difference in the framework used to address the question: *is your watch genuine?* The first argument invokes the logic of hypothesis testing. Under the null hypothesis that the watch is genuine, the observed value of $y = 2$ is somewhat extreme, and this null hypothesis would be formally rejected at the 5 % level.

A hypothesis test performed at the 5 % level is designed to falsely reject the null hypothesis 5 % of the time under repetition of the experiment. (In fact, some would say that those who perform such tests insist on being wrong 5 % of the time.) In this case, although $y = 2$ has higher likelihood for a genuine watch than a fake watch, it is nonetheless a mildly extreme value to observe from a standard normal

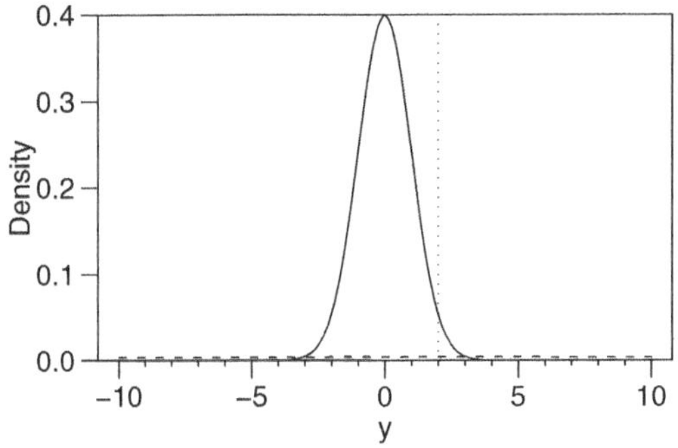

Figure 2.1 *N(0,1) (solid line) and N(0,100²) (the near-horizontal long-dashed line) densities. The likelihood principle says that it is only the values of the densities at the observed value of y = 2 (vertical dashed line) that are relevant to the question of the authenticity of your watch.*

distribution and falls within the critical region of the most powerful test of level 0.05 (Section 14.5).

The second argument is the appropriate one. The falsificationist philosophy of science (Popper 1959), as applied statistically via hypothesis testing (Neyman and Pearson 1933), is irrelevant to this example. Note that the question could have been expressed equivalently as: *is your watch a fake?* Under the null hypothesis that the watch is fake, the value $y = 2$ also results in rejection of the null hypothesis! (See the continuation of this example in Section 14.5 for full details.) □

2.4 IID examples

The examples presented here look at maximum likelihood estimation for models where the data $\boldsymbol{y}$ consist of n iid observations, and so the joint density function is $f(\boldsymbol{y};\boldsymbol{\theta}) = \prod_{i=1}^{n} f(y_i;\boldsymbol{\theta})$. Despite this simplicity, these examples do demonstrate many interesting phenomena. For example, small-sample bias of MLEs is noted in the normal and uniform examples, and it is seen that the likelihood equation for the uniform model has no solution because the likelihood is not differentiable at the MLE. In the Cauchy example, it is seen that the likelihood can have multiple maxima. In the final example, the binormal mixture model is shown to have a non-identifiable parameterization, and to have a likelihood with multiple unbounded maxima. Nonetheless, it is possible to find a sensible and meaningful local maximum.

2.4.1 IID Bernoulli (i.e. binomial)

Here, y is a a single binomial observation. Nonetheless, this can be considered an iid example because, for the purpose of making inference about p, a Bin(n, p)

experiment is equivalent to n iid Bin(1, p) experiments, that is, n iid Bernoulli(p) experiments (see Example 14.2).

Example 2.5. Binomial(n, p). Let Y be distributed Bin(n, p), $0 \leq p \leq 1$. Given that y is observed, the likelihood function is

$$L(p) = \binom{n}{y} p^y(1-p)^{n-y}.$$

The log-likelihood is[2]

$$l(p) = \log \binom{n}{y} + y\log(p) + (n-y)\log(1-p),$$

and the likelihood equation is

$$\begin{aligned}\frac{\partial l(p)}{\partial p} &= \frac{y}{p} - \frac{n-y}{1-p} \\ &= \frac{y-np}{p(1-p)} = 0, \qquad 0 < p < 1.\end{aligned}$$

The likelihood equation does not have a solution if $y = 0$ or n. For example, if $y = n$ then $\partial l(p)/\partial p = n/p$. This derivative is positive and thus the likelihood is monotone increasing, and hence $\widehat{p} = 1$. Similarly, $\widehat{p} = 0$ when $y = 0$.

If y is not equal to 0 or n then the likelihood equation has unique solution $\widehat{p} = y/n$, and it can be established that this is a maximum of $l(p)$ by showing that $\frac{\partial^2 l(p)}{\partial p^2} < 0$ when evaluated at $\widehat{p}$ (this is immediate from Equation (11.15)). Thus it follows that $\widehat{p} = y/n$ is the unique MLE for all $y = 0, 1, ..., n$. □

2.4.2 IID normal

This example demonstrates the small-sample bias that is typical of MLEs, and moreover, that this bias need not be a cause for concern.

Example 2.6. IID normal(μ, σ^2). Let Y_i, $i = 1, .., n$, be iid $N(\mu, \sigma^2)$. Here the model parameters will be taken to be the mean and standard deviation, $\mu \in \mathbb{R}$, $\sigma \in \mathbb{R}^+$, but using the mean and variance would be another obvious choice.

The normal density function for a single observation is

$$f(y; \mu, \sigma) = \frac{1}{\sqrt{2\pi}\sigma} \exp\left(-\frac{(y-\mu)^2}{2\sigma^2}\right), \quad y \in \mathbb{R}, \tag{2.3}$$

[2] To be well defined, it is assumed that $y = 0$ if $p = 0$, and $y = n$ if $p = 1$. Also, the convention that $0\log(0) = 0$ is assumed.

and so the joint density for the data vector $\boldsymbol{y} = (y_1, ..., y_n)$ is

$$f(\boldsymbol{y}; \mu, \sigma) = \prod_{i=1}^{n} f(y_i; \mu, \sigma)$$
$$= \left(\frac{1}{\sqrt{2\pi}\sigma}\right)^n \exp\left(-\frac{\sum_{i=1}^{n}(y_i - \mu)^2}{2\sigma^2}\right) .$$

The log-likelihood is therefore

$$l(\mu, \sigma) = -\frac{n}{2}\log(2\pi\sigma^2) - \frac{1}{2\sigma^2}\sum_{i=1}^{n}(y_i - \mu)^2 ,$$

and the likelihood equations are

$$\frac{\partial l(\mu, \sigma)}{\partial \mu} = \frac{1}{\sigma^2}\sum_{i=1}^{n}(y_i - \mu) = 0 , \tag{2.4}$$

and

$$\frac{\partial l(\mu, \sigma)}{\partial \sigma} = \frac{-n}{\sigma} + \frac{1}{\sigma^3}\sum_{i=1}^{n}(y_i - \mu)^2 = 0 . \tag{2.5}$$

From (2.4) it is immediate that $\widehat{\mu} = \overline{y}$.

Plugging $\widehat{\mu} = \overline{y}$ into (2.5) and solving for σ gives the ML estimate

$$\widehat{\sigma} = \sqrt{\frac{1}{n}\sum_{i=1}^{n}(y_i - \overline{y})^2} .$$

Note that the ML estimator of the variance σ^2 is

$$\widehat{\sigma}^2 = \frac{1}{n}\sum_{i=1}^{n}(Y_i - \overline{Y})^2 ,$$

and it uses a divisor that does not adjust for the loss of one degree of freedom from estimation of μ. The usual unbiased estimator is the so-called sample variance,

$$S^2 = \frac{1}{n-1}\sum_{i=1}^{n}(Y_i - \overline{Y})^2 ,$$

and since $\widehat{\sigma}^2 = (\frac{n-1}{n})S^2$, the bias of $\widehat{\sigma}^2$ is

$$E\left[\widehat{\sigma}^2\right] - \sigma^2 = \left(\frac{n-1}{n}\right)\sigma^2 - \sigma^2$$
$$= \frac{-\sigma^2}{n} .$$

The bias in $\widehat{\sigma}^2$ is typically of no great concern provided that n is reasonably large. Indeed, Press *et al.* (2007, Section 14.1 therein) have this to say on the matter,

> *We might also comment that if the difference between n and n − 1 ever matters to you, then you are probably up to no good anyway ...*

Moreover, $\widehat{\sigma}^2$ has lower mean-squared error than S^2 (Exercise 2.4) and so arguably is a better estimator of σ^2! □

The small-sample bias in the ML estimation of variance can be of greater concern for linear regression models (Box 2.4), and in the context of variance components models. For example, although the one-way random-effects model in Section 10.4 is applied to a dataset with 80 observations, these observations are taken in five groups of 16. The analysis may therefore be subject to considerable small-sample bias with respect to estimation of the between-group variability, since there are only five groups from which to estimate this variability. It is for this reason that Section 10.4 uses restricted maximum likelihood (REML) so as to incorporate a degrees-of-freedom type adjustment (Section 9.2.1).

With minor modification, the above calculations can be extended to the linear regression model where Y_i are independently distributed $N(\boldsymbol{x}_i^T\boldsymbol{\beta}, \sigma^2)$, where $\boldsymbol{x}_i$ is the covariate vector associated with observation i. It can be shown that the MLE of $\boldsymbol{\beta}$ is the usual least-squares estimator (Section 11.7.3), and that the MLE of σ^2 is

$$\widehat{\sigma}^2 = \frac{1}{n}\sum_{i=1}^{n}(Y_i - \boldsymbol{x}_i^T\widehat{\boldsymbol{\beta}})^2 .$$

In contrast, the unbiased estimator of σ^2 has $n - p$ in the denominator, where p is the dimension of $\boldsymbol{\beta}$. (It is seen in Section 9.2.1 that the unbiased variance estimator can be obtained using restricted maximum likelihood.)

Box 2.4

2.4.3 IID uniform

In this example, the log-likelihood is not differentiable at the MLE, and the likelihood equation does not have a solution. Instead, the MLE is obtained from direct inspection of the likelihood function.

Example 2.7. IID uniform$(0, m)$**.** Let $Y_i, i = 1, ..., n$, be iid from a Unif$(0, m)$ distribution, for some $m > 0$. Note that the density function is $1/m$ only for y values in the interval $[0, m]$, and is zero otherwise. Thus, the likelihood for the n

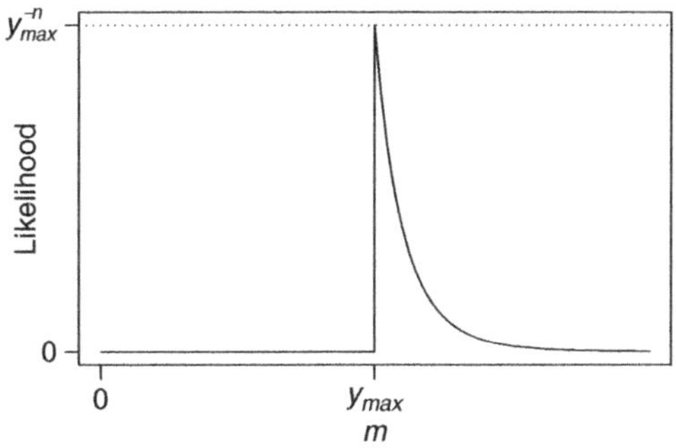

Figure 2.2 Likelihood from Unif(0,m) data.

observations is

$$L(m) = \begin{cases} \frac{1}{m^n}, & m \geq y_i \text{ for all } i \\ 0, & \text{otherwise .} \end{cases}$$

For parameter m to be greater than or equal to all y_i, it is equivalent that m be greater than or equal to the maximum y value, $y_{\max}$. That is,

$$L(m) = \begin{cases} \frac{1}{m^n}, & m \geq y_{\max} \\ 0, & \text{otherwise .} \end{cases}$$

This likelihood function is zero for $m < y_{\max}$, and is monotone decreasing for $m > y_{\max}$ (Figure 2.2), and hence it must be that $\widehat{m} = y_{\max}$.

Since $y_{\max}$ is always less than or equal to m, the estimator $Y_{\max}$ will be a negatively biased estimator of m. It can be shown that $E[Y_{\max}] = mn/(n+1)$, and hence the bias of $Y_{\max}$ is $-m/(n+1)$ (Exercise 2.3). □

2.4.4 IID Cauchy

Here, it is seen that the Cauchy log-likelihood is prone to having multiple local maxima for small sample sizes, but is typically unimodal and close to quadratic in shape for larger n.

Example 2.8. IID Cauchy(θ). Let Y_i be iid from a one-parameter Cauchy distribution with median $\theta \in \mathbb{R}$, and density function

$$f(y;\theta) = \frac{1}{\pi(1+(y-\theta)^2)}, \quad y \in \mathbb{R} .$$

This distribution has a bell-shaped density function centred at θ, but with much longer tails than the normal. (In fact, the standard Cauchy distribution is the Cauchy(0) distribution, and it is the same as a t-distribution with 1 d.o.f.)

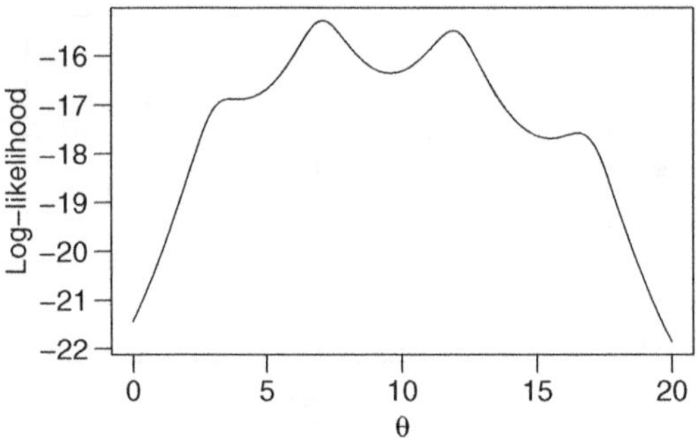

Figure 2.3 Cauchy log-likelihood function from the data $\mathbf{y} = (3,7,12,17)$.

The log-likelihood arising from the data $\mathbf{y} = (y_1, ..., y_n)$ is

$$l(\theta) = -n\log(\pi) - \sum_{i=1}^{n}\log(1 + (y_i - \theta)^2)\,, \tag{2.6}$$

and the derivative of the log-likelihood is

$$l'(\theta) = 2\sum_{i=1}^{n}\frac{y_i - \theta}{1 + (y_i - \theta)^2}\,.$$

A closed-form expression for the roots of the likelihood equation is not available. In practice, an MLE can be found by numerical optimization of $l(\theta)$, though one may have to be wary of convergence to a non-optimal local maximum.

Figure 2.3 plots the Cauchy log-likelihood function for the data $(y_1, y_2, y_3, y_4) = (3, 7, 12, 17)$ (these are the values used by Edwards (1972)), and shows a unique MLE, but multiple local maxima. With a sample size of $n = 100$, the Cauchy likelihood function appears to become better behaved, in the sense that (under repetition of the experiment) it is typically unimodal and close to quadratic in shape (Figure 2.4). □

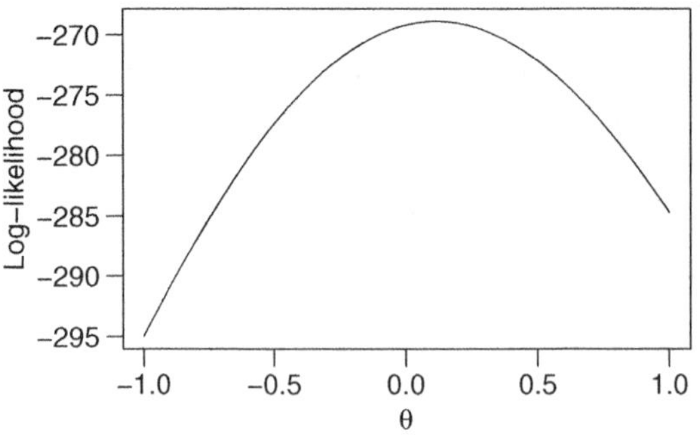

Figure 2.4 Log-likelihood of θ *for a random sample of size 100 generated from a standard (i.e.* $\theta = 0$*) Cauchy distribution.*

2.4.5 IID binormal mixture model

In this example, it is seen that the likelihood is unbounded, and hence that no global MLE can exist. In addition, it is shown that the parameterization is not identifiable. However, in practice these complications are generally pathological, and can be avoided by imposing sensible constraints on the parameter space.

Example 2.9. IID binormal($p, \mu, \sigma, \nu, \tau$) mixture model. The binormal distribution is a mixture of two normal distributions, $N(\mu, \sigma^2)$ and $N(\nu, \tau^2)$, say. Each observation, y_i, is obtained by first choosing one of the two normal distributions at random (with probabilities p and $1 - p$, respectively), and then generating a value from the chosen distribution. However, the identity of the chosen distribution is never observed. The data $\mathbf{y}$ therefore consist of some (or possibly no) observations from the $N(\mu, \sigma^2)$ distribution, and some (or possibly none) from the $N(\nu, \tau^2)$ distribution, but it is not known which observations came from which of these two distributions. The vector of parameters is $\boldsymbol{\theta} = (p, \mu, \sigma, \nu, \tau)$ where $0 \leq p \leq 1$, $\mu, \nu \in \mathbb{R}$, and $\sigma, \tau > 0$.

To express the binormal distribution formally, let B_i be (unobserved) iid Bernoulli(p) random variables. If $B_i = 1$ then Y_i is observed from the $N(\mu, \sigma^2)$ distribution, otherwise it is observed from the $N(\nu, \tau^2)$ distribution. Thus, the distribution of Y_i given B_i is

$$Y_i|(B_i = b_i) \sim \begin{cases} N(\mu, \sigma^2)\,, & b_i = 1 \\ N(\nu, \tau^2)\,, & b_i = 0\,. \end{cases} \tag{2.7}$$

The joint density of (Y_i, B_i) is therefore given by

$$\begin{aligned} f(y_i, b_i; \boldsymbol{\theta}) &= f(y_i|b_i; \mu, \sigma, \nu, \tau)P(B_i = b_i; p) \\ &= \begin{cases} \frac{p}{\sqrt{2\pi}\sigma} \exp(-(y_i - \mu)^2/2\sigma^2)\,, & b_i = 1 \\ \frac{1-p}{\sqrt{2\pi}\tau} \exp(-(y_i - \nu)^2/2\tau^2)\,, & b_i = 0\,, \end{cases} \end{aligned} \tag{2.8}$$

from which the marginal density of Y_i is obtained as

$$\begin{aligned} f(y_i; \boldsymbol{\theta}) &= \sum_{b_i \in \{0,1\}} f(y_i, b_i; \boldsymbol{\theta}) \qquad (2.9) \\ &= f(y_i, 1; \boldsymbol{\theta}) + f(y_i, 0, ; \boldsymbol{\theta}) \\ &= \frac{p}{\sqrt{2\pi}\sigma} \exp(-(y_i - \mu)^2/2\sigma^2) + \frac{1-p}{\sqrt{2\pi}\tau} \exp(-(y_i - \nu)^2/2\tau^2)\,. \qquad (2.10) \end{aligned}$$

That is, the binormal density function is a linear combination of the density functions of the $N(\mu, \sigma^2)$ and $N(\nu, \tau^2)$ distributions.

Given $\mathbf{y} = (y_1, \ldots, y_n)$, the log-likelihood function $l(\boldsymbol{\theta})$ must be maximized numerically. One complication is that $l(\boldsymbol{\theta})$ is unbounded. To see how this can happen, fix p, ν and τ at any set values you wish, with the exception that p can not be 0 or 1. Denote these fixed values by p_*, ν_* and τ_*, respectively. Now, set $\mu = y_k$ for any choice of $k \in \{1, ..., n\}$. This leaves only σ unspecified, and $\boldsymbol{\theta}_\sigma = (p_*, y_k, \sigma, \nu_*, \tau_*)$ can be used to denote the parameter vector with the values of the other parameters fixed as described. When $\boldsymbol{\theta} = \boldsymbol{\theta}_\sigma$, the binormal density function evaluated at y_k is

$$f(y_k; \boldsymbol{\theta}_\sigma) = \frac{p_*}{\sqrt{2\pi}\sigma} + \frac{1 - p_*}{\sqrt{2\pi}\tau_*} \exp(-(y_k - \nu_*)^2/2\tau_*^2) . \tag{2.11}$$

Note that (2.11) can be made arbitrarily large by making σ arbitrarily small. Since $f(\boldsymbol{y}; \boldsymbol{\theta}_\sigma) = \prod_{i=1}^n f(y_i; \boldsymbol{\theta}_\sigma)$, and each $f(y_i; \boldsymbol{\theta}_\sigma)$ is bounded away from zero (by virtue of p_*, ν_* and τ_* being fixed), it follows that $f(\boldsymbol{y}; \boldsymbol{\theta}_\sigma)$ can also be made arbitrarily large.

A further wrinkle is that the parameterization of the binormal model is not identifiable because the role of the two distributions in the mixture can be swapped. That is, the binormal distribution corresponding to parameters $(p, \mu, \sigma, \nu, \tau)$ is the same as that specified by parameters $(1 - p, \nu, \tau, \mu, \sigma)$.

The unbounded likelihood and non-identifiability issues can be eliminated by suitable restriction on the parameter space. One possibility is to constrain the ratio of the two standard deviations by requiring that $0 < c < \sigma/\tau < 1$, where c is some suitably small constant (Hathaway 1985). In practice, despite the unbounded likelihood and non-identifiability, a sensible local maximum of the likelihood function can often be found using unconstrained numerical optimization. This is especially the case if there is good separation between the two component normal distributions, and the optimizer is given a starting value of $\boldsymbol{\theta}$ that is somewhere in the general vicinity of the local maximum. Ultimately, it is the shape of the likelihood function in the neighbourhood of this local maximum that is relevant to inference. □

A well-known example of a binormal mixture distribution is the waiting time between eruptions of the Old Faithful geyser in Yellowstone National Park, Wyoming, USA. The R language contains a data-frame called `faithful`, containing 272 observed waiting times (Figure 2.5). The bimodal nature of these data suggests that two distinct geological processes are occurring within the geyser.

The 272 observations are a time series of the waiting times between successive eruptions of Old Faithful over a period of about two weeks, and it would be natural to suspect that the measurements may not be independent due to temporal autocorrelation. This is indeed the case – successive waiting times are strongly negatively correlated. However, for the sake of demonstration, this temporal autocorrelation will be overlooked, and these data will be assumed iid when they are used in later chapters.

The histogram of waiting times (Figure 2.5) shows that they look like a combination of (very roughly) 40 % from a $N(52, 25)$ distribution and 60 % from a $N(80, 25)$ distribution. The corresponding parameter values $\boldsymbol{\theta}^{(0)} = (p, \mu, \sigma, \nu, \tau) = (0.4, 52, 5, 80, 5)$ would make good starting values for finding a local MLE using numerical optimization.

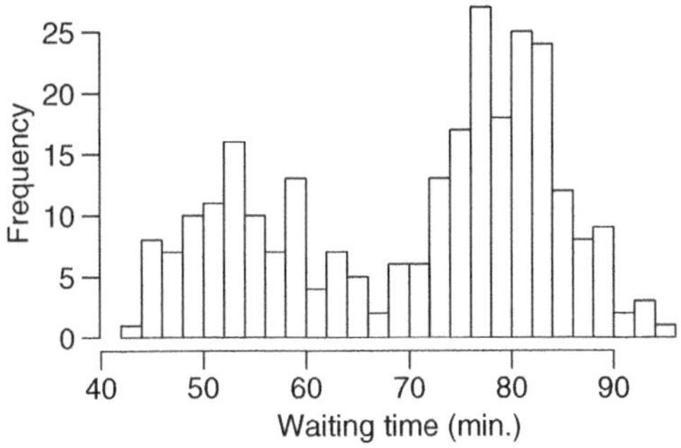

Figure 2.5 Waiting times (minutes) between eruptions of the Old Faithful geyser, from the `faithful` *data-frame.*

2.5 Exercises

2.1 If $Y_1, \ldots, Y_n$ are iid Pois(λ), show that the MLE of λ is the sample mean, $\overline{y}$.

2.2 In Example 2.3, calculate the MLE of the 0.99 quantile. That is, $\zeta = g(\theta)$, where $P(Y \leq g(\theta)) = 0.99$.

2.3 In Example 2.7 it was seen that for iid $Y_1, \ldots, Y_n$ from a Unif$(0, m)$ distribution, the ML estimator of m was the maximum observed value $Y_{\max}$. The distribution function of $Y_{\max}$ is

$$\begin{aligned} F(t) &= \text{Prob}(Y_{\max} \leq t) \\ &= \text{Prob}(Y_i \leq t,\ i = 1, \ldots, n) \\ &= \begin{cases} 0\,, & t < 0\,. \\ \left(\frac{t}{m}\right)^n, & 0 \leq t \leq m \\ 1\,, & t > m\,. \end{cases} \end{aligned}$$

Differentiate $F(t)$ to obtain the density function $f(t)$ of $Y_{\max}$, and hence show that the expected value of $Y_{\max}$ is $mn/(n+1)$.

2.4 The mean-squared error of an estimator $\widehat{\theta}$ of $\theta_0 \in \mathbb{R}$ is

$$\text{mse}(\widehat{\theta}) = E[(\widehat{\theta} - \theta_0)^2] = \text{var}(\widehat{\theta}) + \text{bias}(\widehat{\theta})^2\,,$$

and it provides an overall measure of the estimator's performance that takes into account both variability and bias.

For iid $N(\mu, \sigma^2)$ data (Example 2.6), and using the fact that var$(S^2) = 2\sigma^4/(n-1)$, calculate the mean-squared error of S^2 and $\widehat{\sigma}^2$ and hence show that mse$(\widehat{\sigma}^2) <$ mse(S^2).

2.5 Let $Y_1, \ldots, Y_n$ be iid from a geometric distribution with density function

$$f(y;p) = p^y(1-p)\,, \quad y = 0, 1, 2, \ldots\,,$$

where $0 < p < 1$. Show that the MLE of p is $\widehat{p} = \overline{y}/(1+\overline{y})$.

2.6 Let $Y_i, i = 1, \ldots, n$ be iid Gamma(α, β) (with density function parameterized as in Equation 15.5) with known α, leaving only the scale parameter β to be estimated. Show that the MLE of β is $\widehat{\beta} = \overline{y}/\alpha$.

2.7 Let $Y_1, \ldots, Y_n$ be iid from a distribution with density function

$$f(y;\theta) = \theta y^{\theta-1}, \quad 0 < y < 1,$$

where $\theta > 0$. Show that the MLE of θ is

$$\widehat{\theta} = \frac{-n}{\sum_{i=1}^{n} \log y_i}\,.$$

2.8 Let $Y_1, \ldots, Y_n$ be iid from a Pareto(α, m) distribution with density function

$$f(y;\alpha, m) = \begin{cases} \alpha\frac{m^\alpha}{y^{\alpha+1}}\,, & y \geq m \\ 0\,, & \text{otherwise}\,, \end{cases}$$

for some $\alpha > 0, m > 0$. Find the MLE of (α, m).

2.9 Let $Y_1, \ldots, Y_n$ be iid from a location-shifted exponential distribution with density function

$$f(y;\alpha,\lambda) = \frac{1}{\lambda}\exp\left(\frac{-(y-\alpha)}{\lambda}\right), \qquad y \geq \alpha, \qquad \alpha \in \mathbb{R},\ \lambda > 0\,.$$

Determine the maximum likelihood estimator of (α, λ).

2.10 Let $Y_1, \ldots, Y_n$ be iid from a Laplace distribution (also known as a double exponential distribution) with density function

$$f(y;\alpha) = \frac{1}{2}\exp(-|y-\alpha|)\,, \quad y \in \mathbb{R},\ \ \alpha \in \mathbb{R}\,.$$

Assuming that n is even (i.e. $n = 2m$ for some positive integer m), determine a maximum likelihood estimator of α. Is your MLE unique?

2.11 Let $Y_1, \ldots, Y_n$ be iid observations from a zero-inflated Poisson distribution (Section 7.6.2). This distribution arises from a mixture model where, with probability p, Y is observed from the 'distribution' which has point mass at zero (i.e. $P(Y=0) = 1$), and with probability $(1-p)$, Y is observed from a Pois(λ) distribution. The zero-inflated Poisson distribution has density function

$$f(y;p,\lambda) = \begin{cases} p + (1-p)e^{-\lambda}\,, & y = 0, \\ (1-p)\frac{e^{-\lambda}\lambda^y}{y!}\,, & y = 1, 2, 3, \ldots \end{cases}$$

Assuming that λ is known, show that the MLE of p is

$$\widehat{p} = \frac{n_0 - ne^{-\lambda}}{n(1 - e^{-\lambda})},$$

where n_0 is the number of y_i values that are equal to zero.

Part II

PRAGMATICS

3

Hypothesis tests and confidence intervals or regions

...the null hypothesis is never proved or established, but is possibly disproved ...
– Sir Ronald Fisher

3.1 Introduction

The main focus of this chapter is to demonstrate how R and SAS can be used to perform hypothesis tests and to construct confidence intervals or regions. Inference is demonstrated using the large-sample approximate normality of ML estimators (the Wald approach), and using the approximate χ^2 distribution of likelihood ratio statistics. Hypothesis tests and confidence intervals/regions are considered for a single element θ_k of the parameter vector, $\boldsymbol{\theta}$, and for a subset of (or all) elements of $\boldsymbol{\theta}$.

Chapter 4 considers more general aspects and properties of practical ML inference, and it is seen in Section 4.3.1 that inference using the likelihood ratio is generally more reliable than that based on approximate normality. However, methods based on approximate normality are popular due to their familiarity and ease of application, especially for inference about a single parameter. The most familiar is perhaps the approximate 95 % Wald CI for θ_k, calculated as $\widehat{\theta}_k$ plus or minus a couple of its standard errors. Approximate normality also has the flexibility of extending to functions of the parameters via the delta method (Section 4.2.1). However, this flexibility can be as dangerous as it is useful, because 'approximate normality'

Maximum Likelihood Estimation and Inference: With Examples in R, SAS and ADMB, First Edition. Russell B. Millar.

can often be wishful thinking rather than a reasonable approximation to the true sampling distribution of the ML estimator.

There is undeniable virtue in the simplicity of the Wald approach, and in many cases it will make little difference compared to using the likelihood ratio. However, one should be aware that approximate normality is based on assumptions of approximate linearity, and hence should be especially dubious when, for example, a parameter has a very nonlinear influence on the model.[1] To encourage use of the likelihood ratio, this chapter includes demonstration of the R function `Plkhci` and SAS macro of the same name (both described in Section 15.5) for construction of likelihood ratio confidence intervals.

The normality-based Wald approach is somewhat less appealing for joint inference about a multi-dimensional subset of the parameter vector. The formulae then require the use of general quadratic forms, and inference becomes more sensitive to the legitimacy of the assumed approximate multivariate normality. In this situation the likelihood ratio is more widely used.

3.2 Approximate normality of MLEs

Let the vector $\widehat{\boldsymbol{\theta}} = (\widehat{\theta}_1, ..., \widehat{\theta}_s)$ denote the MLE of the unknown true parameter vector $\boldsymbol{\theta}_0 = (\theta_{01}, ..., \theta_{0s}) \in \Theta \in \mathbb{R}^s$. Assuming appropriate regularity conditions and a reasonable sample size then, under repetition of observing data from the distribution with density $f(\boldsymbol{y}; \boldsymbol{\theta}_0)$, the distribution of $\widehat{\boldsymbol{\theta}}$ is approximately that of an s-dimensional (multivariate) normal random vector with expected value $\boldsymbol{\theta}_0$. (See Chapter 12 for formal statement of the regularity conditions, and Section 12.3 for the derivation of the approximate normality result.)

The approximate normality result from Section 12.3 can be written

$$\widehat{\boldsymbol{\theta}} \sim N_s(\boldsymbol{\theta}_0, \mathbf{V}(\boldsymbol{\theta}_0)) \tag{3.1}$$

where $\sim$ denotes 'approximately distributed,' and N_s denotes the s-dimensional multivariate normal distribution. The $s \times s$ matrix $\mathbf{V}(\boldsymbol{\theta}_0)$ is a function of $\boldsymbol{\theta}_0$ and the sample size, but not of $\boldsymbol{y}$. For a single element $\widehat{\theta}_k$ of $\widehat{\boldsymbol{\theta}}$, this gives

$$\widehat{\theta}_k \sim N(\theta_{0k}, v_{kk}) \tag{3.2}$$

where v_{kk} denotes the kth diagonal element of $\mathbf{V}(\boldsymbol{\theta}_0)$.

In general, $\mathbf{V}(\boldsymbol{\theta}_0)$ can not always be expressed in closed form, but it does have a convenient form in some familiar cases such as the normal (Example 3.1) and binomial models (Example 3.2).

[1] Loosely speaking, this is where a small positive change in the value of θ_k alters the model to a very different degree than a small negative change in θ_k.

Example 3.1. Let $Y_1, \ldots, Y_n$ be iid $N(\mu, \sigma^2)$. It is of course the case that $\widehat{\mu} = \overline{Y}$ has an exact normal distribution and (3.2) is the familiar result

$$\overline{Y} \sim N(\mu_0, \sigma_0^2/n)\,,$$

where (μ_0, σ_0^2) is the true parameter value. Of less familiarity will be the statement for the approximate normality of $\widehat{\sigma}^2$,

$$\widehat{\sigma}^2 \mathrel{\dot\sim} N(\sigma_0^2, 2\sigma_0^4/n)\,,$$

when n is sufficiently large. This follows from Equation (12.37). □

Example 3.2. Let Y be distributed Bin(n, p). Now, Y is equivalent to the sum of n iid Bernoulli(p) random variables and so (from Example 12.3), if n is sufficiently large,

$$\widehat{p} \mathrel{\dot\sim} N(p_0, p_0(1-p_0)/n)\,,$$

where $\widehat{p} = Y/n$, and p_0 denotes the true value of p. □

3.2.1 Estimating the variance of $\widehat{\boldsymbol{\theta}}$

The statements of approximate normality in (3.1) and (3.2) are not readily suitable for testing hypotheses or constructing confidence sets. One complication is that $\mathbf{V}(\boldsymbol{\theta})$ is not always tractable to evaluate. The second is that $\boldsymbol{\theta}_0$ is not known (but see Box 3.1).

> The parameter value is fully specified under a null hypothesis of the form $H_0 : \boldsymbol{\theta} = \boldsymbol{\theta}_0$, and then it is possible to use the value $\boldsymbol{\theta}_0$ when estimating the variance of $\widehat{\boldsymbol{\theta}}$. For example, for the Bin(n, p) model it has been shown that it is better to calculate the Wilson score test statistic of $H_0 : p = p_0$ by using $\text{var}(\widehat{p}) = p_0(1-p_0)/n$ (the variance under H_0) rather than $\widehat{\text{var}}(\widehat{p}) = \widehat{p}(1-\widehat{p})/n$ (Agresti and Coull 1998, Brown *et al.* 2001, Brown *et al.* 2002). However, the latter is more widely used because it is convenient for calculation of confidence intervals. See Example 12.10 and Exercise 12.7 for more details. This inconsistency in usage of the Wald test is another reason to use inference based on the likelihood ratio when possible.

Box 3.1

In Section 11.6.2 it is seen that $\mathbf{V}(\boldsymbol{\theta}_0)$ is the negative of the inverse of the expected (under repetition of new data from the experiment) curvature of the log-likelihood. It is natural to estimate the expected curvature of the log-likelihood using the observed curvature of the realized log-likelihood function (i.e. the log-likelihood arising from observation of the data $\boldsymbol{y}$). To be more precise, the observed curvature is the $s \times s$

Hessian matrix of second-order partial derivatives of the log-likelihood function, and will be denoted $\mathbf{H}(\boldsymbol{\theta})$. The standard estimate of $\mathbf{V}(\boldsymbol{\theta}_0)$ is the negative of the inverse of $\mathbf{H}(\boldsymbol{\theta})$, evaluated at the MLE. That is,

$$\widehat{\mathbf{V}} = -\mathbf{H}(\widehat{\boldsymbol{\theta}})^{-1} . \tag{3.3}$$

The negative of $\mathbf{H}(\widehat{\boldsymbol{\theta}})$ plays a prominent role in likelihood theory and is called the *observed* Fisher information matrix. The estimated variance matrix is just the inverse of the observed Fisher information matrix.

The estimate of variance in (3.3) is particularly convenient because many optimizing routines, including the popular Newton-Raphson algorithm (Section 5.2), calculate the Hessian matrix of the log-likelhood in the course of optimization, and hence are able to provide $\mathbf{H}(\widehat{\boldsymbol{\theta}})$. These routines typically do this calculation via numerical approximation, but the automatic differentiation capabilities of ADMB enable it to calculate $H(\boldsymbol{\theta})$ algorithmically, and hence to the accuracy of machine precision. Furthermore, it has been argued (Efron and Hinkley 1978) that the approximate normality statement in (3.1) is typically more accurate when $\widehat{\mathbf{V}}$ is used in place of $\mathbf{V}(\boldsymbol{\theta}_0)$. (This argument draws on the conditionality principle from Chapter 14, but is essentially that it is better to use the realized likelihood to determine the variance of $\widehat{\boldsymbol{\theta}}$ rather than to average over likelihoods that might have been.)

Using $\widehat{\mathbf{V}}$ in place of $\mathbf{V}(\boldsymbol{\theta}_0)$, the approximate normality of $\widehat{\boldsymbol{\theta}}$ is

$$\widehat{\boldsymbol{\theta}} \sim N_s(\boldsymbol{\theta}_0, \widehat{\mathbf{V}}) , \tag{3.4}$$

and for element k of $\widehat{\boldsymbol{\theta}}$,

$$\widehat{\theta}_k \sim N(\theta_{0k}, \widehat{v}_{kk}) , \tag{3.5}$$

where $\widehat{v}_{kk}$ denotes the kth diagonal element of $\widehat{\mathbf{V}}$. These are the statements of approximate normality that are utilized in the construction of Wald tests and confidence intervals/regions in Section 3.3,

> The approximate normality of $\widehat{\theta}_k$ in (3.5) should be interpreted as saying that its distribution function will look like that of a $N(\theta_{0k}, \widehat{v}_{kk})$ random variable. This is enough to guarantee that tests and confidence intervals that are derived under the assumption of normality will have the right properties, at least approximately. However, it does not guarantee that $\widehat{\theta}_k$ has all of the usual properties of a normal distribution. For example, it is quite possible that the mean and variance of $\widehat{\theta}_k$ do not exist, or that $\widehat{\boldsymbol{\theta}}$ does not exist for some subset of values $\boldsymbol{y}$ in the sample space that has positive probability of occurring (see Examples 4.3 and 12.2).

Box 3.2

3.3 Wald tests, confidence intervals and regions

The Wald test statistic has the same quadratic form for tests of a single parameter value, θ_k, as it does for a joint test of a subset of two or more parameter values. The general form is given by (3.11) in Section 3.3.3, but, as shown in the following section, for tests of a single parameter it is simply the square of the Z-statistic.

3.3.1 Test for a single parameter

Under $H_0 : \theta_k = \theta_{0k}$, the hypothesis can be tested using the Z-statistic obtained from standardizing (3.5). This has an approximate standard normal distribution, that is,

$$Z = \frac{\widehat{\theta}_k - \theta_{0k}}{\sqrt{\widehat{v}_{kk}}} \approx N(0, 1) . \tag{3.6}$$

The Wald test statistic, W, is the square of Z, and since the square of a standard normal distribution is the χ^2_1 distribution, it follows that

$$W = \frac{(\widehat{\theta}_k - \theta_{0k})^2}{\widehat{v}_{kk}} \approx \chi^2_1 . \tag{3.7}$$

Equation (3.7) gives the approximate distribution of W under repetition of the experiment whereby new data are generated from a distribution with density $f(\boldsymbol{y}; \boldsymbol{\theta})$, where $\boldsymbol{\theta}$ is such that its kth element is equal to θ_{0k}. Values that are too large to have plausibly come from a χ^2_1 distribution provide evidence against $H_0 : \theta_k = \theta_{0k}$. Specifically, if w is the calculated value of the Wald test statistic, the p-value is $P(X \geq w)$ where $X \sim \chi^2_1$.

An approximate level α test is given by rejecting H_0 if W exceeds the $1 - \alpha$ quantile $\chi^2_{1,1-\alpha}$, or equivalently, if $|Z| = \sqrt{W}$ exceeds the square root of $\chi^2_{1,1-\alpha}$. The square root of $\chi^2_{1,1-\alpha}$ is equal to the $1 - \alpha/2$ quantile of a standard normal, $z_{1-\alpha/2}$, by virtue of the relationship between the $N(0, 1)$ and χ^2_1 distributions.

From the equivalence of hypothesis tests and confidence intervals (Section 13.2), the approximate $(1 - \alpha)100\,\%$ confidence interval for θ_k is given by all values of θ_{0k} that are not rejected by the level α Wald test. This is the familiar interval

$$(\widehat{\theta}_k - z_{1-\alpha/2}\sqrt{\widehat{v}_{kk}},\ \widehat{\theta}_k + z_{1-\alpha/2}\sqrt{\widehat{v}_{kk}}) . \tag{3.8}$$

3.3.2 Test of a function of the parameters

If the null hypothesis is of the form $H_0 : \zeta = \zeta_0$, where $\zeta = g(\boldsymbol{\theta}) \in \mathbb{R}$ is a differentiable function of the model parameters, then the Wald test statistic is

$$W = \frac{(\widehat{\zeta} - \zeta_0)^2}{\widehat{\text{var}}(\widehat{\zeta})} , \tag{3.9}$$

and, subject to regularity conditions and sufficient sample size, is approximately χ_1^2 distributed when H_0 is true. The estimated variance of the MLE, $\widehat{\zeta} = g(\widehat{\boldsymbol{\theta}})$, can be obtained via the delta method (Section 4.2.2).

Example 3.3. Comparison of two Poisson means. Let the data consist of just the two observations $y_1 = 10$ and $y_2 = 35$, where each y_i is an independent observation from a Pois(λ_i) distribution, $i = 1, 2$. Suppose that it is of interest to test $H_0 : \lambda_2 = 2\lambda_1$. (This scenario could arise, for example, if y_1 and y_2 are hospital admissions of a certain medical condition, where hospital 2 serves twice the population size of hospital 1, and hence would be expected to have twice the admission rate.)

Here the sample size is a mere $n = 2$, which should raise some concerns about assuming approximate normality under the large sample theory of maximum likelihood. However, provided that $\lambda_i,\ i = 1, 2$ are both reasonably large, this will not be a problem since the Pois(λ_i) distributions will be well approximated by a normal distribution (Exercise 13.7).

The null hypothesis can be expressed as $H_0 : \zeta = 0$ where $\zeta = \lambda_2 - 2\lambda_1$. The Wald test statistic is calculated using (3.9) where (by Exercise 2.1) $\widehat{\lambda}_1 = 10$, $\widehat{\lambda}_2 = 35$, and so $\widehat{\zeta} = 15$. In this case the delta method simply reduces to $\widehat{\text{var}}(\widehat{\zeta}) = \widehat{\text{var}}(\widehat{\lambda}_2) + \widehat{\text{var}}(2\widehat{\lambda}_1)$, by virtue of the statistical independence of the two observations. The variance of a Poisson is its mean, and it can be verified that MLE theory gives $\widehat{\text{var}}(\widehat{\lambda}_i) = \widehat{\lambda}_i$ (from Exercise 11.2). Thus $\widehat{\text{var}}(\widehat{\zeta}) = 35 + 4 \times 10 = 75$, and the observed Wald test statistic for H_0 is

$$w = \frac{(\widehat{\zeta} - \zeta_0)^2}{\widehat{\text{var}}(\widehat{\zeta})} = \frac{(15 - 0)^2}{75} = 3\,,$$

which has a p-value of about 0.083 by reference to a χ_1^2 distribution. □

3.3.3 Joint test of two or more parameters

For notational convenience it is assumed that the null hypothesis can be stated in the form $H_0 : \boldsymbol{\psi} = \boldsymbol{\psi}_0$ where $\boldsymbol{\psi} = (\theta_1, ..., \theta_r)$ is a vector containing the first r elements of $\boldsymbol{\theta}$. That is, the value of the first $r \leq s$ parameters are fixed under H_0. There is no loss of generality in specifying H_0 in this way, because the parameter vector can always be re-ordered if necessary.

Denoting $\widehat{\boldsymbol{\psi}} = (\widehat{\theta}_1, \widehat{\theta}_2, ..., \widehat{\theta}_r)$, it follows from (3.4) that, under H_0,

$$\widehat{\boldsymbol{\psi}} \sim N_r(\boldsymbol{\psi}_0, \widehat{\mathbf{V}}_{\boldsymbol{\psi}})\,, \tag{3.10}$$

where $\widehat{\mathbf{V}}_{\boldsymbol{\psi}}$ is the approximate variance matrix of $\widehat{\boldsymbol{\psi}}$, and is given by the upper-left $r \times r$ sub-matrix of $\widehat{\mathbf{V}}$. As in the single parameter case, the Wald test statistic is obtained as a standardization of $\widehat{\boldsymbol{\psi}}$ (Seber and Lee 2003, p. 30). Noting that, by algebraic convention, $\widehat{\boldsymbol{\psi}}$ is a column vector, this standardization is written in quadratic form as

$$W = (\widehat{\psi} - \psi_0)^T \left[\widehat{\mathbf{V}}_{\psi}\right]^{-1} (\widehat{\psi} - \psi_0) \sim \chi_r^2 . \tag{3.11}$$

Hypothesis tests and p-values are obtained as in the single parameter case, except that now the degrees of freedom is r, corresponding to the number of parameters restricted by the null hypothesis. The approximate $(1-\alpha)100\,\%$ Wald confidence region for ψ is an r-dimensional ellipse, determined as all values ψ_0 such that $W \leq \chi^2_{r,1-\alpha}$.

For a Wald test of a function of the parameters, $H_0 : g(\theta) = \zeta_0$, where $g(\theta) \in \mathbb{R}^r$, $r \geq 2$, the delta method can be used to obtain the estimate of the $r \times r$ variance matrix of the MLE, $g(\widehat{\theta})$. The test statistic is then obtained analogously to (3.11).

3.3.4 In R and SAS: Old Faithful revisited

Here, the iid binormal mixture model introduced in Section 2.4.5 is fitted to waiting times between eruptions of the Old Faithful geyser (Figure 2.5) using general purpose optimizers within R and SAS. The binormal mixture model can be implemented using dedicated functions provided in several R packages. However, the direct coding and maximization of the log-likelihood is just as easy, and moreover, it provides greater flexibility for making inference and for altering the model if lack of fit is indicated.

In Section 2.4.5 the binormal model was introduced as a mixture of $N(\mu, \sigma^2)$ and $N(\nu, \tau^2)$ distributions. It has five parameters $\theta = (p, \mu, \sigma, \nu, \tau)$, where p is the probability that an observation will be generated from the $N(\mu, \sigma^2)$ distribution. From Equation (2.10), the log-likelihood arising from observations $y_1, ..., y_n$ is

$$\sum_{i=1}^{n} \log \left(pf(y_i; \mu, \sigma) + (1-p) f(y_i; \nu, \tau) \right) , \tag{3.12}$$

where $f(y; \mu, \sigma)$ and $f(y; \nu, \tau)$ are the density functions of $N(\mu, \sigma^2)$ and $N(\nu, \tau^2)$ random variables, respectively.

To begin, approximate standard errors and 95 % Wald confidence intervals are obtained for each parameter. Then, to demonstrate a joint hypothesis test, it will be posited that volcanologists have completed seismic tests which suggest that the component of the binormal with the smaller mean should have an expected waiting time of 55 s with a standard deviation of 5 s, and that this component should be responsible for the eruption one-third of the time on average. This null hypothesis is $H_0 : \psi = (1/3, 55, 5)$, where $\psi = (p, \mu, \sigma)$, and corresponds to the specification of $r = 3$ parameters.

3.3.4.1 Using R

In the R code below, the negative of the observed Fisher information matrix, $-\mathbf{H}(\widehat{\theta})$, is extracted from the list object that is returned by the `optim` optimizer. It is inverted to obtain $\widehat{\mathbf{V}}$, and the standard errors and confidence intervals are explicitly calculated. The `optim` function was briefly described in Section 1.4.2. The waiting time between eruptions is variable `waiting` in data-frame `faithful`.

```
> attach(faithful) #Make data available to the R session
> #Define the negative log-likelihood
> nllhood=function(theta,y) {
+   p=theta[1];mu=theta[2];sigma=theta[3];nu=theta[4];tau=theta[5]
+   lhood=p*dnorm(y,mu,sigma)+(1-p)*dnorm(y,nu,tau)
+   return(-sum(log(lhood))) }

> #Use start values inferred from histogram
> Faithful.fit=optim(c(0.4,52,5,80,5),nllhood,y=waiting,hessian=T)
> MLE=Faithful.fit$par
> ObsInfo=Faithful.fit$hess #Observed Fisher information matrix
> Vhat=solve(ObsInfo) #Inverse of observed Fisher information
> Std.Errors=sqrt(diag(Vhat))
> #Obtain the MLEs,estimated std errors, and approx Wald 95% CIs
> Wald.table=cbind(MLE,
+              Std.Errors,LowerBound=MLE-qnorm(0.975)*Std.Errors,
+              UpperBound=MLE+qnorm(0.975)*Std.Errors)

> parnames=c("p","mu","sigma","nu","tau")
> rownames(Wald.table)=parnames
> round(Wald.table,4) #Print to 4 decimal places.
          MLE Std.Errors LowerBound UpperBound
p      0.3609     0.0312     0.2998     0.4220
mu    54.6145     0.6995    53.2435    55.9856
sigma  5.8698     0.5370     4.8173     6.9224
nu    80.0908     0.5046    79.1017    81.0798
tau    5.8682     0.4010     5.0822     6.6542
```

Some specific points to note include:

- The additional argument `y=waiting` in the call of the `optim` function is passed to the negative log-likelihood function `nllhood`.
- Since `nllhood` is the negative log-likelihood, `Faithful.fit$hess` is the negative of $\mathbf{H}(\widehat{\boldsymbol{\theta}})$. That is, it is the observed Fisher information matrix.
- The `solve` function (used with only a single matrix argument) returns the matrix inverse, and the `diag` function returns the diagonal vector.
- The call of `optim` produced some warning messages (not shown), because it attempted to evaluate `nllhood` at parameter values outside of the parameter space (e.g. $\sigma < 0$). This can be avoided by using `lower` and `upper` bound arguments in the `optim` call.
- As previously noted in Section 1.4, a few lines of coding can be saved by using MLE convenience functions that are available in existing R packages. Alternatively, a wrapper function could easily be constructed from the above code.

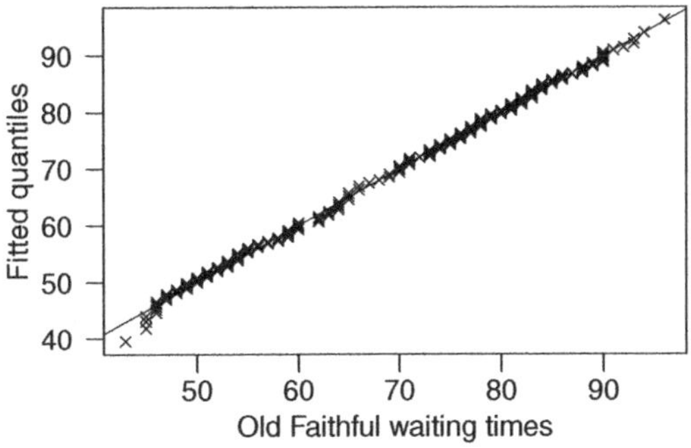

Figure 3.1 Quantile-quantile plot of the Old Faithful waiting times versus model quantiles. The identity line is also shown.

The MLE is $\widehat{\boldsymbol{\theta}} = (0.361, 54.6, 5.87, 80.1, 5.87)$, and it certainly looks very sensible with regard to the histogram of the waiting times (Figure 2.5). The waiting times were assumed to be iid, and so a quantile-quantile plot of ordered waiting times versus model quantiles will provide a strong visual assessment of model fit. However, this is not entirely straightforward because the quantiles of a binormal mixture model are a nonstandard function of the model parameters. These model quantiles were found numerically (code available at `http://www.stat.auckland.ac.nz/~millar`). The quantile-quantile plot (Figure 3.1) is very close to linear and does not indicate any obvious lack of fit of the fitted model, other than perhaps just some hint that there were fewer very short waiting times than might be predicted. (Note that this does not confirm the validity of the iid binormal model. In particular, for the sake of simplicity, the analysis herein makes no use of the temporal order of the observations, and hence can not detect the temporal autocorrelation that is known to be present.)

The additional lines of code shown below calculate the Wald test statistic and p-value for the joint null hypothesis $H_0 : (p, \mu, \sigma) = (1/3, 55, 5)$.

```
> #H0:  (p,mu,sigma)=(1/3,55,5)
> psi0=c(1/3,55,5)
> psihat=MLE[1:3]
> diff=psihat-psi0
> #Wald statistic
> W=t(diff)%*%solve(Vhat[1:3,1:3])%*%diff
> W
          [,1]
[1,] 4.428055

> cat("\n p-value is",1-pchisq(W,3))

 p-value is 0.2187981
```

So, it would appear that the observed data do not provide any real evidence against the volcanologists' hypothesis.

3.3.4.2 Using SAS

The following SAS code assumes that dataset `OldFaithful` contains the 272 observations in the variable `waiting`. The `PDF` function in this code evaluates the form of probability density function that is specified by its first argument. A brief description of other features of `PROC NLMIXED` was given in Section 1.4.1.

```
ODS SELECT ParameterEstimates;
PROC NLMIXED DATA=OldFaithful DF=1E6;
  PARMS p=0.4 mu=52 sigma=5 nu=80 tau=5;
  BOUNDS 0<p<1, 0<sigma, 0<tau;
  loglhood=log( p*PDF("NORMAL",waiting,mu,sigma)+
          (1-p)*PDF("NORMAL",waiting,nu,tau) );
  MODEL waiting ~ GENERAL(loglhood);
RUN;
```

The output table (Figure 3.2) includes $\widehat{\boldsymbol{\theta}}$, the estimated standard errors, and the lower and upper bounds of the 95 % Wald CIs.

The `CONTRAST` statement in `PROC NLMIXED` uses the Wald statistic to jointly test whether specified functions of the parameters are zero. Noting that the null hypothesis $H_0 : \boldsymbol{\psi} = \boldsymbol{\psi}_0$ can be expressed

$$H_0 : \boldsymbol{\psi} - \boldsymbol{\psi}_0 = 0 \,,$$

where $\boldsymbol{\psi} = (p, \mu, \sigma)$ and $\boldsymbol{\psi}_0 = (1/3, 55, 5)$, the above SAS code simply requires the addition of the statement

```
  CONTRAST "H0" p-1/3, mu-55, sigma-5;
```

and this produces the table of output shown in Figure 3.3.

Parameter Estimates									
Parameter	Estimate	Standard Error	DF	t Value	Pr > |t|	Alpha	Lower	Upper	Gradient
p	0.3609	0.03116	1E6	11.58	<.0001	0.05	0.2998	0.4220	-4.64E-6
mu	54.6149	0.6997	1E6	78.06	<.0001	0.05	53.2435	55.9862	2.832E-7
sigma	5.8712	0.5373	1E6	10.93	<.0001	0.05	4.8181	6.9244	-4.09E-7
nu	80.0911	0.5046	1E6	158.72	<.0001	0.05	79.1021	81.0801	-7.49E-8
tau	5.8677	0.4010	1E6	14.63	<.0001	0.05	5.0819	6.6536	-3.56E-8

Figure 3.2 Table of parameter estimates from using PROC NLMIXED to fit the binormal mixture model to the Old Faithful geyser waiting time data.

Contrasts				
Label	Num DF	Den DF	F Value	Pr > F
H0	3	1E6	1.48	0.2184

Figure 3.3 Output produced by the CONTRAST *statement. The F value is W/r.*

The F-value reported in Figure 3.3 is the value of W/r, and so it needs to be multiplied by r to obtain the Wald statistic W. In this example $r = 3$, and so the calculated value of W was 4.44. The p-value (shown as **Pr>F**) is obtained by comparison of the calculated F-value with an $F_{r,d}$ distribution. Here, d was explicitly set to one million using the `DF=1E6` option, and for such a large value of d, $F_{r,d}$ is approximately χ^2_r/r. That is, SAS is comparing the calculated value of W/r against a χ^2_r/r distribution (see Box 3.3), or equivalently, comparing W against a χ^2_r distribution, as desired.

`PROC NLMIXED` calls W/r an F-statistic because it makes an adjustment for the uncertainty in replacing $\mathbf{V}(\boldsymbol{\theta}_0)$ by $\widehat{\mathbf{V}}$ in (3.4). It is implicitly assuming that

$$\widehat{\mathbf{V}} \sim \frac{\chi^2_d}{d}\mathbf{V}(\boldsymbol{\theta}_0)\,,$$

where d denotes the degrees of freedom (by default, the number of observations in the dataset). Substituting into (3.11) gives

$$W = (\widehat{\boldsymbol{\psi}} - \boldsymbol{\psi}_0)^T \left[\widehat{\mathbf{V}}_{\boldsymbol{\psi}}\right]^{-1} (\widehat{\boldsymbol{\psi}} - \boldsymbol{\psi}_0) \stackrel{\cdot}{\sim} \frac{\chi^2_r}{\chi^2_d/d}\,,$$

and for large d,

$$F = \frac{W}{r} \stackrel{\cdot}{\sim} \frac{\chi^2_r/r}{\chi^2_d/d} \sim F_{r,d} \stackrel{\cdot}{\sim} \frac{\chi^2_r}{r}\,.$$

Box 3.3

3.4 Likelihood ratio tests, confidence intervals and regions

As in Section 3.3.3, it will be assumed for notational convenience that the null hypothesis has the form $H_0 : \boldsymbol{\psi} = \boldsymbol{\psi}_0$ where $\boldsymbol{\psi} = (\theta_1, ..., \theta_r)$ is a vector containing the first r elements of $\boldsymbol{\theta}$. This hypothesis can alternatively be expressed as

$$H_0 : \boldsymbol{\theta} \in \Theta_0 ,$$

where Θ_0 is the subset of the parameter space in which the first r elements of $\boldsymbol{\theta}$ are equal to $\boldsymbol{\psi}_0 = (\theta_{01}, ..., \theta_{0r})$. Note that this formulation can also be used to conduct likelihood ratio tests on functions $\boldsymbol{\zeta} = g(\boldsymbol{\theta}) \in \mathbb{R}^r$ of the parameters, by re-parameterizing the model such that $\boldsymbol{\zeta}$ corresponds to the first r elements of the parameter vector under the new re-parameterization.

It is seen below that the likelihood ratio test (LRT) statistic is simply twice the difference in the log-likelihoods between the unrestricted fit (obtained from maximization over the full parameter space Θ) and the fit under H_0 (obtained from maximization over the restricted parameter space Θ_0). The likelihood ratio test is particularly convenient under the so-called *simple* hypothesis under which $r = s$, since then $\boldsymbol{\theta}$ is fully specified under H_0. That is, a simple null hypothesis is of the form $H_0 : \boldsymbol{\theta} = \boldsymbol{\theta}_0$, and the restricted parameter space is zero-dimensional because it contains just the single point $\boldsymbol{\theta}_0$. Consequently, the maximized log-likelihood under H_0 is just $l(\boldsymbol{\theta}_0)$. When $r < s$ then the hypothesis is said to be a *composite* hypothesis, and Θ_0 is an $(s - r)$-dimensional subset of Θ.

It is useful to use the notation $\boldsymbol{\theta} = (\boldsymbol{\psi}, \boldsymbol{\lambda})$ where $\boldsymbol{\lambda} = (\theta_{r+1}, ..., \theta_s)$, because maximization over the restricted parameter space corresponds to maximization with respect to $\boldsymbol{\lambda}$. Then, the MLE over Θ_0 can be written $\widehat{\boldsymbol{\theta}}_0 = (\boldsymbol{\psi}_0, \widehat{\boldsymbol{\lambda}}_0)$ where $\widehat{\boldsymbol{\lambda}}_0$ is obtained by maximizing the log-likelihood with respect to $\boldsymbol{\lambda}$ when $\boldsymbol{\psi}$ is fixed at the value $\boldsymbol{\psi}_0$. That is,

$$l(\boldsymbol{\psi}_0, \widehat{\boldsymbol{\lambda}}_0; \boldsymbol{y}) = \max_{\boldsymbol{\lambda}} \; l(\boldsymbol{\psi}_0, \boldsymbol{\lambda}; \boldsymbol{y}) .$$

Regarded as a function of $\boldsymbol{\psi}$, this partially maximized (log-) likelihood is called the profile (log-) likelihood function for $\boldsymbol{\psi}$, and is re-visited in more depth in Section 3.6.

Under appropriate regularity conditions, the LRT statistic has an approximate chi-square distribution with r degrees of freedom. That is,

$$X = 2\big(l(\widehat{\boldsymbol{\theta}}; \boldsymbol{Y}) - l(\widehat{\boldsymbol{\theta}}_0; \boldsymbol{Y})\big) \stackrel{\cdot}{\sim} \chi^2_r , \tag{3.13}$$

and large values of X constitute evidence against H_0. In particular, for $r = 1$, the $(1 - \alpha)100\,\%$ likelihood ratio confidence interval (LRCI) for a single parameter, θ_k, is the collection of values θ_{0k} such that $H_0 : \theta_k = \theta_{0k}$ is not rejected by the level α LR test. That is, the LRCI is the collection of values for which

$$2\big(l(\widehat{\boldsymbol{\theta}}; \boldsymbol{Y}) - l(\theta_{0k}, \widehat{\boldsymbol{\lambda}}_0; \boldsymbol{Y})\big) < \chi^2_{1,1-\alpha} . \tag{3.14}$$

For 95 % LRCIs, the relevant quantile is $\chi^2_{1,0.95} \approx 3.84$. So, in plain words, the 95 % LRCI for θ_k is the collection of points θ_{0k} such that the 'partially maximized' log-likelihood decreases by no more than 3.84/2=1.92 compared to maximization over the full parameter space. Here, 'partially maximized' refers to the $(s - 1)$-dimensional maximization over θ_i, $i \neq k$, with θ_k fixed at the value θ_{0k}.

Likelihood ratio confidence regions for $\boldsymbol{\psi} \in \mathbb{R}^r$, where $r > 1$, are similarly obtained. These regions are the collection of $\boldsymbol{\psi}_0$ values for which maximization over Θ_0 decreases the log-likelihood by no more than half of $\chi^2_{r,1-\alpha}$ compared to $l(\widehat{\boldsymbol{\theta}}; \mathbf{y})$.

Example 3.3 continued. Recall that $y_1 = 10$ and $y_2 = 35$ were observed from Pois(λ_i) distributions, $i = 1, 2$, and it was desired to test $H_0 : \zeta = \lambda_2 - 2\lambda_1 = 0$. Unrestricted maximization results in $\widehat{\lambda}_1 = 10$ and $\widehat{\lambda}_2 = 35$ and $l(\widehat{\boldsymbol{\theta}}) = l(\widehat{\lambda}_1, \widehat{\lambda}_2; \mathbf{y}) = -4.778$. In this case it is not necessary to explicitly re-parameterize the model using (ζ, λ_1) (say), since maximization under H_0 simply equates to setting $\lambda_2 = 2\lambda_1$ and this restricted maximization can be done by hand. Under H_0, it can be shown (Exercise 3.7) that $(\widehat{\lambda}_1, \widehat{\lambda}_2) = (15, 30)$, and hence that $l(\widehat{\boldsymbol{\theta}}_0; \mathbf{y}) = -6.118$. This gives the LRT statistic $X = 2(-4.778 - (-6.118)) = 2.68$ and a p-value of approximately 0.10 by comparison with the χ^2_1 distribution. □

3.4.1 Using R and SAS: Another visit to Old Faithful

For certain classes of models, R has functions for calculating likelihood ratio confidence intervals. For example, the confidence interval function `confint` is able to calculate LR confidence intervals for some classes of models, including nonlinear least squares and generalized linear models. However, it is not applicable to the binormal mixture model, and the general-purpose function `Plkhci` is used instead.

Calculation of likelihood ratio confidence intervals is implemented within several maximum likelihood-based SAS procedures. For example, in `PROC GENMOD` (for generalized linear modelling) these are obtained by using the option `LRCI` in the `MODEL` statement. For general likelihoods, `PROC NLP` (within the OR module) contains a `PROFILE` statement for specifying the parameters for which a LRCI is required. For installations that do not include the OR module, a macro is available from `http://www.stat.auckland.ac.nz/~millar` for use with `PROC NLMIXED`, and this is demonstrated below.

To match the examples used to demonstrate the Wald approach, here it is required to find likelihood ratio confidence intervals for each of the five parameters of the binormal mixture model of the Old Faithful waiting times, and to test the joint null hypothesis $H_0 : (p, \mu, \sigma) = (1/3, 55, 5)$.

3.4.1.1 Using R

Calculation of likelihood ratio confidence intervals can be computationally intensive. For example, the 95 % LRCI for p is all values of p_0 for which the null hypothesis $H_0 : p = p_0$ is not rejected. Each such hypothesis requires a maximization over the remaining four parameters. Fortunately, the `plkhci` function (in the `Bhat` package) implements the efficient algorithm of Venzon and Moolgavkar (1988). Some explanation of the use of the `plkhci` function is given in Section 1.4.2. The R code below uses function `Plkhci` (with capital `P`), which is a very minor modification of the `plkhci` function to permit the data y to be passed as an additional argument (see Section 15.5.1).

```
> library(Bhat)
> #Control list for Plkhci, with a range (low,upp) on parameters
> plist=list(label=parnames,est=MLE,
+            low=c(0,0,0,0,0),upp=c(1,100,10,100,10))

> #Create a 5 by 2 matrix to contain the LRCIs
> LRCI.table=matrix(0,nrow=5,ncol=2,
+            dimnames=list(parnames,c("LowerBound","UpperBound")))

> source("Plkhci.R") #Defines function Plkhci
> for(i in 1:5)
+   LRCI.table[i,]=Plkhci(plist,nllhood,parnames[i],prob=0.95,
+                        eps=0.000001,y=waiting)
> round(LRCI.table,4) #Print to 4 decimal places
      LowerBound UpperBound
p         0.3013     0.4230
mu       53.2775    56.0784
sigma     4.9408     7.1116
nu       79.0478    81.0517
tau       5.1677     6.7635
```

To test $H_0 : (p, \mu, \sigma) = (1/3, 55, 5)$, function `nllhoodH0` is used to find the MLE under H_0. This function uses the fixed values of p, μ and σ, and takes $\boldsymbol{\lambda} = (\nu, \tau)$ as arguments over which `optim` performs the optimization. More generally, a wrapper function has been written to automate the maximization of the log-likelihood under H_0, and is available from `http://www.stat.auckland.ac.nz/~millar`. This function is called `Profile`, and is demonstrated in Section 3.6.1, and described in Section 15.5.2.

```
> #Negative log likelihood under H0: psi=psi0
> nllhoodH0=function(lambda,y,psi0) {
+   theta=c(psi0,lambda)
+   return(nllhood(theta,y))
+ }

> Faithful.fitH0=optim(c(80,5),nllhoodH0,y=waiting,psi0=c(1/3,55,5))
> cat("Maximized log-lhood under H0:",-Faithful.fitH0$value,"\n")
Maximized log-lhood under H0: -1036.829

> cat("Unrestricted maximized log-lhood:",-Faithful.fit$value,"\n")
Unrestricted maximized log-lhood: -1034.002

> #Calculate likelihood ratio statistic
> LRStat=2*(-Faithful.fit$value+Faithful.fitH0$value)
> cat("LRT stat is",LRStat,"& p-value is",1-pchisq(LRStat,3),"\n")
LRT stat is 5.655282 & p-value is 0.1296405
```

The LR p-value is about 0.13, and is somewhat smaller than the 0.22 p-value from the Wald test. In the Author's experience, the Wald and LR tests can give quite

Wald and PL confidence limits							
N	Parameter	Estimate	Alpha	Profile likelihood confidence limits		Wald confidence limits	
1	p	0.360886	0.050000	0.301253	0.423075	0.299804	0.421968
2	mu	54.614856	0.050000	53.277268	56.080098	53.243518	55.986194
3	sigma	5.871219	0.050000	4.941278	7.112374	4.818087	6.924352
4	nu	80.091069	0.050000	79.046875	81.051847	79.102082	81.080057
5	tau	5.867734	0.050000	5.167642	6.764319	5.081864	6.653604

Figure 3.4 Table of parameter estimates from using `PROC NLP` *to fit the binormal mixture model to the Old Faithful waiting time data. The LR confidence limits are labelled 'Profile Likelihood Confidence Limits'.*

different p-values when used to test joint null hypotheses, even for moderately large sample sizes. Here, the Wald test is particularly fallible because it requires the joint distribution of $(\widehat{p}, \widehat{\mu}, \widehat{\sigma})$ to be well approximated by a multivariate normal, and this is a much stronger assumption than requiring that the individual parameter MLEs, $\widehat{p}$, $\widehat{\mu}$, and $\widehat{\sigma}$, are approximately normally distributed. The LR p-value can require more effort to compute, but is generally preferred (see Section 4.3.1).

3.4.1.2 Using SAS

The following `PROC NLP` code can be used by users with access to the SAS operations research (OR) module, and produces the output in Figure 3.4. A brief description of some of the features of `PROC NLP` was provided in Section 1.4.1.

```
PROC NLP DATA=OldFaithful COV=2 VARDEF=N;
MAX loglhood;
PARMS p=0.5, mu=55, sigma=5, nu=80, tau=5;
BOUNDS 0<p<1, 0<sigma, 0<tau;
PROFILE p mu sigma nu tau / ALPHA=0.05;
loglhood=LOG( p*PDF("NORMAL",waiting,mu,sigma)+
          (1-p)*PDF("NORMAL",waiting,nu,tau) );
RUN;
```

If `PROC NLP` is not available then `PROC NLMIXED` can be used in combination with the `Plkhci` macro that was first seen in Section 1.4.1. The likelihood ratio CIs are computed for each parameter separately and the following code finds the CI for parameter p only. This requires the user to specify a macro that takes p as its sole argument, and containing appropriate `PROC NLMIXED` code in which p is 'hard-wired', with optimization over all remaining parameters. When this macro is invoked, the symbol `&p` is replaced by the value passed in argument p. The `Plkhci` macro is briefly described in Section 15.5.3.

```
%INCLUDE "PlkhciMacro.sas";
%MACRO OldFaithfulProfile_p(p);
PROC NLMIXED DATA=OldFaithful;
  PARMS mu=55 sigma=5 nu=80 tau=5;
  BOUNDS 0<sigma, 0<tau;
  loglhood=LOG( &p*PDF("NORMAL",waiting,mu,sigma)+
          (1-&p)*PDF("NORMAL",waiting,nu,tau) );
  MODEL waiting ~ GENERAL(loglhood);
  RUN;
%MEND;

%Plkhci(OldFaithfulProfile_p,0.0,0.3609,-1034.002,side="L");
%Plkhci(OldFaithfulProfile_p,0.3609,0.5,-1034.002,side="R");
```

The above code adds the following lines to the SAS log window.

```
Left-sided 95% LR CI bound is 0.30124929205
Right-sided 95% LR CI bound is 0.4230467286
```

For the likelihood ratio test of $H_0 : (p, \mu, \sigma) = (1/3, 55, 5)$, the maximization under H_0 can be achieved with the following code.

```
PROC NLMIXED DATA=OldFaithful DF=1E6;
  p=1/3; mu=55; sigma=5;
  PARMS nu=80 tau=5;
  BOUNDS 0<tau;
  loglhood=log( p*PDF("NORMAL",waiting,mu,sigma)+
          (1-p)*PDF("NORMAL",waiting,nu,tau) );
  MODEL waiting ~ GENERAL(loglhood);
RUN;
```

3.5 Likelihood ratio examples

3.5.1 LR inference from a two-dimensional contour plot

This example uses the log-likelihood from the logistic regression model that is fitted to the binomial data in Table 7.1 (Section 7.5). There, the experiment consisted of observing the number of fish (at given lengths) entering a trawl, and the number of those retained by it. For current purposes, it is enough to know that this is a two-parameter model, and since these parameters are regression coefficients, they will be denoted $\boldsymbol{\theta} = (\beta_1, \beta_2)$ for consistency with the notation used in Section 7.5.

The MLE is $\widehat{\boldsymbol{\theta}} = (\widehat{\beta}_1, \widehat{\beta}_2) = (-10.632, 0.304)$ and the maximized value of the log-likelihood is $l(\widehat{\beta}_1, \widehat{\beta}_2; \boldsymbol{y}) = -37.88$. A contour plot of the log-likelihood is shown in Figure 3.5. A $(1 - \alpha)100\,\%$ LR confidence region for $\boldsymbol{\theta}$ is given by all

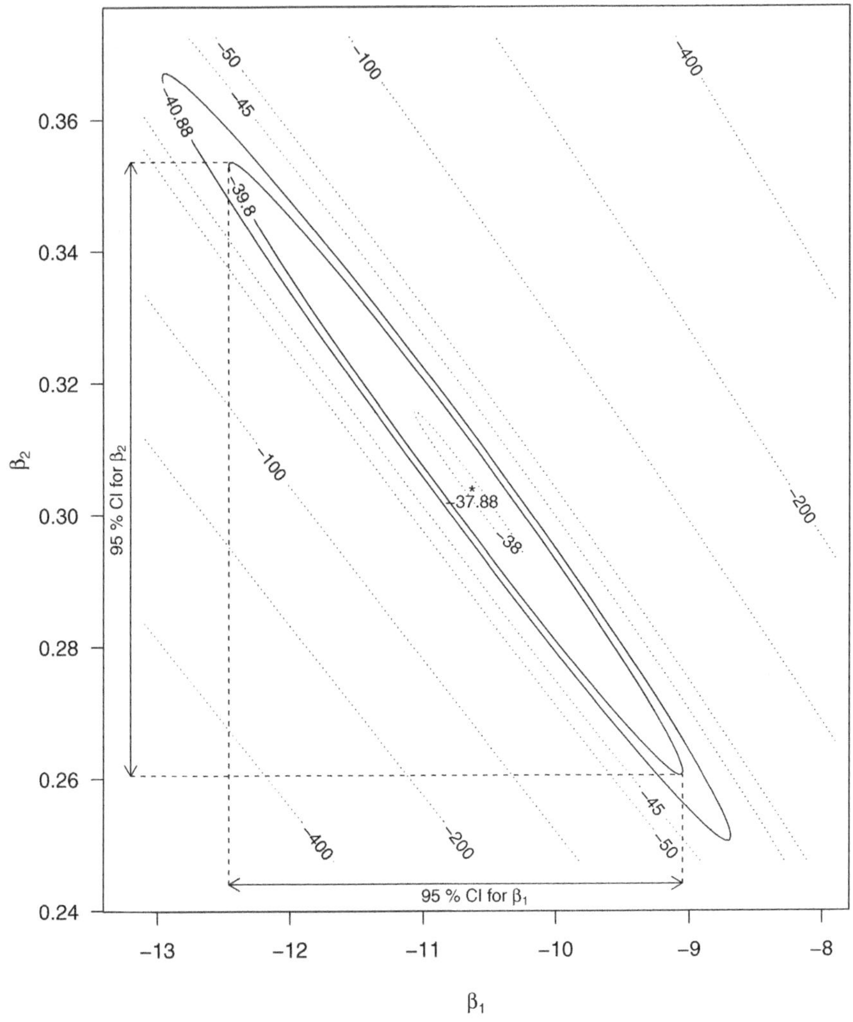

Figure 3.5 Contour plot of the two-dimensional log-likelihood $l(\beta_1, \beta_2; \mathbf{y})$. The contours corresponding to the critical values of −39.80 and −40.88 are shown using solid lines.

parameter vectors (β_1^*, β_2^*) such that the null hypothesis $H_0 : (\beta_1, \beta_2) = (\beta_1^*, \beta_2^*)$ is not rejected at significance level α. That is, such that $l(\beta_1^*, \beta_2^*; \mathbf{y}) > -37.88 - 0.5\chi^2_{2,1-\alpha}$. For a 95 % confidence region, $\chi^2_{2,0.95} \approx 5.99$, and the region is given by $l(\beta_1^*, \beta_2^*) > -40.88$.

The contour plot can also be used to obtain LR confidence intervals for β_1 and β_2. For example, an approximate 95 % LRCI for parameter β_1 is given by all values

β_1^* such that

$$\max_{\beta_2} l(\beta_1^*, \beta_2; \mathbf{y}) > l(\widehat{\beta}_1, \widehat{\beta}_2; \mathbf{y}) - 0.5\chi^2_{1,0.95} \tag{3.15}$$
$$= -37.88 - 3.84/2 = -39.80\,.$$

For any given β_1^*, the maximization over β_2 in Equation (3.15) can be determined from Figure 3.5, by scanning along the vertical line corresponding to $\beta_1 = \beta_1^*$. If this line passes through the contour $l(\beta_1, \beta_2) = -39.80$ then β_1^* is in the 95 % CI. The 95 % likelihood ratio CI for parameter β_1 is seen to be approximately $(-12.45, -9.05)$.

3.5.2 The *G*-test for contingency tables

The so-called G-test is the name given to the application of likelihood ratio tests to contingency tables. In general, the table could be one-way, two-way or multi-way, and the hypothesis could be a test of independence, or some other restriction (e.g. see Exercise 3.6). Here, it is not necessary to consider the layout of the table, and the observed counts will simply be denoted y_i, $i = 1, ..., m$, where m is the number of cells in the table.

The most convenient model for such data is the multinomial (defined in Section 15.4.1), denoted Mult$(n, p_1, ..., p_m)$ where $n = \sum_{i=1}^m y_i$. The unrestricted model has parameter space that is a subset of $\mathbb{R}^{m-1}$ because of the restriction $\sum_{i=1}^m p_i = 1$. A bit of algebra confirms that the MLEs are the observed proportions $\widehat{p}_i = y_i/n$. The maximized log-likelihood is therefore (to within a constant)

$$l(\widehat{\mathbf{p}}; y_1, ..., y_m) = \sum_{i=1}^m y_i \log(\widehat{p}_i) = \sum_{i=1}^m y_i \log(y_i/n)\,,$$

where $\widehat{\mathbf{p}} = (\widehat{p}_1, ..., \widehat{p}_m)$.

Let $\widehat{\mathbf{p}}_0 = (\widehat{p}_{01}, ..., \widehat{p}_{0m})$ denote the MLEs under H_0, and let $\widehat{y}_{0i} = n\widehat{p}_{0i}$ denote the expected cell count under H_0. Then

$$l(\widehat{\mathbf{p}}_0; y_1, ..., y_m) = \sum_{i=1}^m y_i \log(\widehat{p}_{0i}) = \sum_{i=1}^m y_i \log(\widehat{y}_{0i}/n)\,.$$

The G-test statistic is the likelihood ratio statistic

$$G = 2\big(l(\widehat{\mathbf{p}}; y_1, ..., y_m) - l(\widehat{\mathbf{p}}_0; y_1, ..., y_m)\big)$$
$$= 2\sum_{i=1}^m y_i \log(y_i/\widehat{y}_{0i})\,, \tag{3.16}$$

with degrees of freedom given by the number of parameters restricted under H_0.

Example 3.4. *G*-test of the Poisson model. In a study of micro-propagation, Marin, Jones and Hadlow (1993) measured the number of roots produced on apple cultivars. For one particular treatment, the following data were recorded from 40 cultivars.

Number of roots	0	1	2	3	4	5	6	7	8	9
Frequency	19	2	2	4	3	1	4	3	0	2

It is desired to test the null hypothesis that a Poisson model is a reasonable model for these data. One approach is to treat the frequencies as multinomial data, that is, to consider this as a multinomial experiment with 10 cells corresponding to the observed number of roots (from 0 to 9).[2] To reduce sparseness (i.e. low expected cell count), the frequencies of 8 or 9 roots can be combined so that the last cell of the multinomial will correspond to counts of 8 or more. This results in a table with 9 cells,

No. of roots	0	1	2	3	4	5	6	7	8+
Frequency, y_i	19	2	2	4	3	1	4	3	2
Expected, $\widehat{y}_{0i}$	3.45	8.46	10.36	8.46	5.18	2.54	1.04	0.36	0.15

The MLE of λ is the sample mean of the 40 observations, $\widehat{\lambda} = 2.45$, and the expected cell counts in the above table are obtained under the iid Pois($\widehat{\lambda}$) model. So, for example, the expected number of cultivars with no roots is 40 times the probability of observing $Y = 0$ from a Pois(2.45) distribution. This is $40e^{-2.45} = 3.45$.

The unrestricted multinomial model has eight parameters, whereas the restricted multinomial (obtained from the iid Poisson model) has just one, and so the G statistic is compared against a χ^2_7 distribution. Evaluating (3.16) using the values in the above table gives a G-test statistic of approximately 75.1, which is enormously extreme for a χ^2_7 distribution, and the Poisson model is well and truly rejected.

Here, the first cell of this multinomial is the frequency of cultivars that produced no roots, and it is clear that there are many more of these than expected under the Poisson model. It would therefore be natural to explore a zero-inflated model for these data. This is the objective of Exercise 3.9. □

3.6 Profile likelihood

In Section 3.4 the parameter vector was partitioned as $\boldsymbol{\theta} = (\boldsymbol{\psi}, \boldsymbol{\lambda})$ where $\boldsymbol{\psi} \in \mathbb{R}^r$, $\boldsymbol{\lambda} \in \mathbb{R}^{s-r}$, and the likelihood ratio test of $H_0 : \boldsymbol{\psi} = \boldsymbol{\psi}_0$ required a partial maximization with respect to $\boldsymbol{\lambda}$. This partial maximization can be done for any hypothesized value

[2] See the continuation of this example in Section 7.6.2 for an alternative approach.

of ψ, which leads to the notion of the profile likelihood function. Specifically, the profile log-likelihood function for ψ is

$$l^*(\psi; y) = \max_{\lambda} \; l(\psi, \lambda; y) \,. \tag{3.17}$$

When ψ is one or two-dimensional then a plot of the profile log-likelihood is an insightful tool for assessing the support given to differing values of ψ (though, it is not the only way of quantifying support for ψ – see Section 9.3 for alternative methodology). Likelihood ratio hypothesis tests of $H_0 : \psi = \psi_0$, and hence likelihood ratio confidence intervals or regions, can be evaluated directly from the plotted profile log-likelihood as was done in Figures 1.2 and 3.5. This follows immediately from noting that $l(\widehat{\theta}) = l^*(\widehat{\psi})$ and $l(\widehat{\theta}_0) = l^*(\psi_0)$, and so the LRT statistic is

$$\begin{aligned} X &= 2\big(l(\widehat{\theta}) - l(\widehat{\theta}_0)\big) \\ &= 2\big(l^*(\widehat{\psi}) - l^*(\psi_0)\big) \,. \end{aligned}$$

Box-Cox transformations are presented as a common application of profile likelihood in Section 6.2.

3.6.1 Profile likelihood for Old Faithful

Likelihood ratio confidence intervals for the five parameters of the binormal mixture model of the Old Faithful waiting times were calculated in Section 3.4.1. In practice, it can be informative to also examine the shape of the profile log-likelihood. Here, the profile log-likelihood for parameter p is obtained using utilities available in SAS and R. The profile can also be obtained in ADMB, using the `likeprof_number` declaration, as demonstrated in Section 1.4.3.

3.6.1.1 Using R

The following code uses the `Profile` function (Section 15.5.2) to calculate $l^*(p)$ over the sequence of values of p from 0.27 to 0.46 in steps of 0.005.

```
> #nllhood and parnames have been previously defined
> source("Profile.R")
> p.seq=seq(0.27,0.46,0.005)
> Profile.p=NULL
> for(i in 1:length(p.seq))
+   Profile.p[i]=Profile(parnames,nllhood,label="p",psi=p.seq[i],
+          lambda=c(50,5,80,5),y=waiting)$value

> #Display profiled log-likelihood for first 3 values in sequence
> cbind(p.seq,Profile.p)[1:3,]
     p.seq Profile.p
[1,] 0.270 -1038.584
[2,] 0.275 -1038.074
[3,] 0.280 -1037.597
```

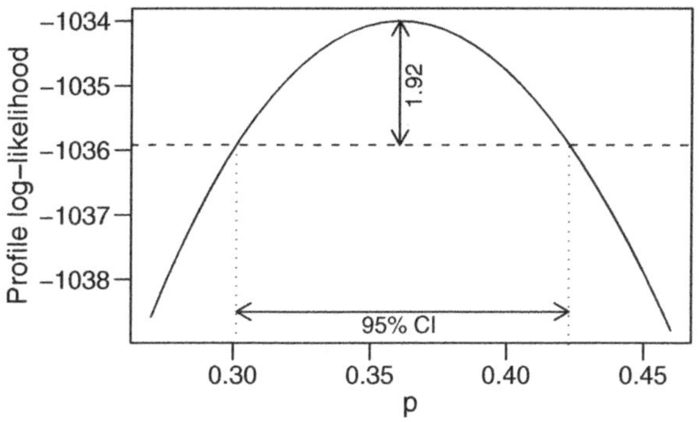

Figure 3.6 Profile log-likelihood for parameter p of the binormal mixture model for the Old Faithful waiting time data.

The profile is shown in Figure 3.6, along with the 95 % LR confidence interval of (0.3013, 0.4230) that was obtained in Section 3.4.1. The profile log-likelihood is close to quadratic in shape, which provides some assurance that the distribution of $\widehat{p}$ is close to normally distributed (see the argument in Section 4.3.1).

3.6.1.2 Using SAS

The following SAS code creates a dataset `Profile_p` containing the value of $l^*(p)$ for 39 evenly-space values of p from 0.27 to 0.46. That is 0.27 to 0.46 in steps of 0.005. This code requires the user-defined macro `OldFaithfulProfile_p` defined in Section 3.4.1.

```
%INCLUDE "ProfileMacro.sas";
%Profile(OldFaithfulProfile_p,lower=0.27,
         upper=0.46,n=39,dsname=Profile_p);
```

3.7 Exercises

3.1 For the example of Section 3.5.1, the MLE is$(\widehat{\beta}_1, \widehat{\beta}_2) = (-10.632, 0.304)$. Furthermore, the estimated variance matrix is

$$\widehat{\mathbf{V}} = \begin{pmatrix} 0.7477 & -0.02022 \\ -0.02022 & 0.0005584 \end{pmatrix}.$$

1. Calculate the p-value for $H_0 : (\beta_1, \beta_2) = (-10, 0.3)$ using the Wald test statistic.

2. Calculate an approximate 95 % Wald CI for parameter β_1.

3. Determine an approximate 99 % likelihood ratio CI for parameter β_2 using Figure 3.5.

3.2 For the Old Faithful data, use R or SAS to perform Wald and LR tests of the simple hypothesis $H_0 : (p, \mu, \sigma, \nu, \tau) = (1/3, 55, 5, 80, 5)$.

3.3 Proschan (1963) fitted an exponential distribution to the times between failures of air-conditioning systems on Boeing 720 aircraft. The twelve recorded failure times for aircraft number 8044 were

487 18 100 7 98 5 85 91 43 230 3 130

1. Assuming these data are iid Exp(μ), show that $\widehat{\mu} = \overline{y}$.

2. Calculate the 95 % Wald confidence interval for μ.

3. Draw a plot of the log-likelihood (use integer values of μ from 45 to 250, say), and add a horizontal line showing the threshold value for determining the 95 % LR confidence interval for μ.

4. Determine the 95 % LR confidence interval for μ using R or SAS.

3.4 The data below are thirty iid observations from a zero-truncated $N(\mu, \sigma^2)$ distribution.

2.59 0.93 0.70 1.83 1.46 2.04 1.31 1.55 1.19 0.16 2.09 1.32 0.18 0.12 2.17
1.52 0.84 1.85 2.29 0.94 0.01 0.89 2.76 3.15 0.52 1.91 2.11 0.81 2.46 1.30

This distribution has density function

$$f(y; \mu, \sigma^2) = \frac{1}{\sqrt{2\pi}\sigma(1 - \Phi(-\mu/\sigma))} \exp\left(-\frac{(y-\mu)^2}{2\sigma^2}\right), \quad y > 0,$$

where Φ denotes the distribution function of the standard normal distribution.

1. Produce a histogram of the data.

2. Use R or SAS to find the MLEs of μ and σ^2, and their approximate standard errors.

3. Determine the 95 % Wald intervals for μ and σ.

4. Determine the 95 % LR intervals for μ and σ.

3.5 Consider a two-by-two contingency table with cell counts denoted y_{ij} and assume that the counts are multinomial with distribution Mult$(n, p_{11}, p_{12}, p_{21}, p_{22})$ where n is the total count and p_{ij} is the probability of the cell in row i and column j. Let $r_1 = p_{11} + p_{12}$ and $c_1 = p_{11} + p_{21}$ denote the probabilities for row 1 and column 1, respectively.

1. Under the null hypothesis that rows and columns are independent, write down the multinomial log-likelihood as a function of r_1 and c_1.

2. Show that the MLEs of r_1 and c_1 are simply the proportions of the counts in row 1 and column 1, respectively.

3. It follows that $\widehat{p}_{ij} = \frac{y_{i1}+y_{i2}}{n}\frac{y_{1j}+y_{2j}}{n}$. Hence, implement a G-test of independence for the contingency table

	Outcome	
	Good	Bad
Treatment 1	54	36
Treatment 2	46	44

3.6 Under Hardy-Weinberg equilibrium, alleles A and B occur independently and thus the probabilities of genotypes AA, AB and BB are p^2, $2p(1-p)$ and $(1-p)^2$, respectively, where p is the probability of allele type A. McDonald, Verrelli and Geyer (1996) recorded genotype frequencies of 14 AA, 21 AB and 25 BB genotypes from 60 American oysters. Perform a G-test of Hardy-Weinberg equilibrium for these data.

3.7 For the continuation of Example 3.3 in Section 3.4:

1. Verify that $l(\widehat{\boldsymbol{\theta}}) = -4.778$.

2. Under H_0, write the log-likelihood as a function of λ_1 alone (using the fact that $\lambda_2 = 2\lambda_1$), and show that the maximum occurs when λ_1 is equal to 15.

3. Hence, verify that the maximized value of the log-likelihood is -6.118 under H_0.

3.8 The log-likelihood contours displayed in Figure 3.5 are from a binomial logistic model for the data in Table 7.1. For each i, n_i is the number of fish of length l_i (to the nearest cm) entering a trawl, and y_i is the number that were retained, $i = 1, \ldots, 37$. The logistic regression model that is fitted to these data assumes that $Y_i \sim \text{Bin}(n_i, p_i)$ are independent. The case study in Section 7.5 parameterizes p_i using the logistic model

$$p_i = \frac{\exp(\beta_1 + \beta_2 l_i)}{1 + \exp(\beta_1 + \beta_2 l_i)},$$

where β_1 and β_2 are the two parameters to be estimated. However, for this exercise, p_i are to be parameterized using parameters β_2 and $l_{50\%}$ in the form

$$p_i = \frac{\exp(\beta_2(l_i - l_{50\%}))}{1 + \exp(\beta_2(l_i - l_{50\%}))}.$$

Note that this is an example of an inverse prediction problem, because parameter $l_{50\,\%}$ denotes the value of the covariate l that predicts $p = 0.5$.

1. Find the MLE $(\widehat{\beta}_2, \widehat{l}_{50\,\%})$ by maximizing the likelihood (expressed as a function of β_2 and $l_{50\,\%}$) using the `optim` function in R or the `NLMIXED` or `NLP` procedures in SAS.

2. If using R, add the calculations required to give the approximate standard errors of $\widehat{\beta}_2$ and $\widehat{l}_{50\,\%}$.

3. Use R or SAS to find approximate 95 % likelihood ratio CIs for β_2 and $l_{50\,\%}$. (Note: If using R, this can be done using the `Plkhci` function. With SAS procedure `NLMIXED` it can be done in conjunction with the `Plkhci` macro in SAS, or with procedure `NLP` it simply requires inclusion of an appropriate `PROFILE` statement.) The CI for β_2 can be verified immediately from Figure 3.5.

4. Obtain the approximate 95 % likelihood ratio CI for $l_{50\,\%}$ using only Figure 3.5.

3.9 In Example 3.4 a G-test was used to show that a Poisson model was not appropriate for the apple micro-propagation data. It was seen that the data contained many more zeros than expected under the Poisson model, and hence it would be natural to consider a zero-inflated Poisson (ZIP) model for these data. The density of the ZIP model is given in Exercise 2.11 (there it was assumed that λ was known, but here both p and λ are to be estimated).

1. Verify that $(\widehat{p}, \widehat{\lambda}) = (0.4698, 4.6216)$ by maximizing the likelihood using the `optim` function in R or the `NLMIXED` or `NLP` procedures in SAS.

2. If using R, add the calculations required to give the approximate standard errors of $\widehat{p}$ and $\widehat{\lambda}$.

3. Find approximate 95 % likelihood ratio CIs for p and λ. (See the note in Question 3.8, Part III.)

4. Verify that the G test statistic for the ZIP model is 4.97, and determine the p-value for the hypothesis that these data are iid ZIP.

(Note: To check your fit of the ZIP model, this model can be fitted using SAS procedure `GENMOD` or the R package `VGAM`.)

3.10 Let $Y_1, ..., Y_n$ follow a first-order auto-regressive model (AR1) of the form

$$(Y_{i+1} - \mu) = \rho(Y_i - \mu) + \epsilon_{i+1}\,, \qquad i = 1, ..., n-1\,,$$

where $\epsilon_i \sim N(0, \sigma^2)$ are iid and $-1 < \rho < 1$. Here $\boldsymbol{\theta} = (\mu, \rho, \sigma)$. It follows that the distribution of Y_{i+1} given $y_1, ...y_i$ is

$$Y_{i+1} \mid y_1, ...y_i \sim N\big(\mu + \rho(y_i - \mu), \sigma^2\big)\,, \qquad i = 1, ..., n-1\,. \qquad (3.18)$$

Also, it can be assumed that

$$Y_1 \sim N\left(\mu, \frac{\sigma^2}{1-\rho^2}\right) . \tag{3.19}$$

Since

$$f(\mathbf{y};\boldsymbol{\theta}) = f(y_1;\boldsymbol{\theta}) \prod_{i=1}^{n-1} f(y_{i+1}|y_1, ..., y_i;\boldsymbol{\theta}) , \tag{3.20}$$

if follows that the likelihood $L(\boldsymbol{\theta}; \mathbf{y})$ is given by the product of the likelihoods arising from the $n-1$ terms in (3.18), and the term in (3.19).
One hundred observations from an AR1 model are contained in dataset `AR1.dat`, available from `http://www.stat.auckland.ac.nz/~millar`

1. Plot the data (with y_i on the vertical axis against i on the horizontal axis).
2. Maximize the log-likelihood using the `optim` function in R or the `NLMIXED` or `NLP` procedures in SAS.
3. Verify your fitted model using the `arima0` function in R, or `PROC ARIMA` in SAS.
4. Perform a likelihood ratio test of the null hypothesis that the observations are iid. That is, test $H_0 : \rho = 0$.

4

What you really need to know

Make everything as simple as possible, but no simpler. – Albert Einstein

4.1 Introduction

This chapter delves into some of the important issues that are encountered in practice, beginning in Section 4.2 where the delta method is used for determining the approximating normal distribution of functions of the MLE. Relevant examples are presented at the end of this section, including the calculation of the approximate variance of estimated shellfish biomass (obtained as the product of estimated abundance and average weight), and the approximate variance of the log odds-ratio in a two-by-two contingency table. Section 4.3 takes a closer look at ML inference based on approximate normality, and compares it to the use of likelihood ratio. It is seen that inference based on the likelihood ratio is generally more reliable, but comes at greater computational expense. Model selection criteria are briefly encountered in Section 4.4. Section 4.5 presents the bootstrap as a computationally intensive method that can be useful for inference in data-poor and nonstandard situations. The bootstrap also provides one of several practical methodologies for prediction (Section 4.6). Finally, in Section 4.7 this chapter concludes with a brief look at a variety of situations under which ML inference can go astray.

Maximum Likelihood Estimation and Inference: With Examples in R, SAS and ADMB, First Edition. Russell B. Millar.

4.2 Inference about $g(\boldsymbol{\theta})$

In practice, the research question may require inference about quantities that are a function of $\boldsymbol{\theta}$. The function $g(\boldsymbol{\theta})$ could also involve covariates, for example, $g(\boldsymbol{\theta}) \equiv g(\boldsymbol{\theta}, \boldsymbol{x}_i)$ could be the fitted value for observation i with covariate vector $\boldsymbol{x}_i$.

In some situations, inference about $\zeta = g(\boldsymbol{\theta})$ can be made by re-parameterizing the model in such a way that ζ becomes a model parameter. For example, in the logistic regression example in the case study of Section 7.5, the unknown parameter vector is $\boldsymbol{\theta} = (\beta_1, \beta_2)$. However, β_1 and β_2 are themselves of little interest to the researcher, but rather it is the value of $\zeta = -\beta_1/\beta_2$ that is relevant. This model could be re-parameterized using $\gamma = (\zeta, \beta_1)$, say, since there is a one-to-one correspondence between values of $\boldsymbol{\theta}$ and values of γ. However, one major difficulty is that re-parameterizations may prevent the use of convenient software. In the logistic regression case, the model parameterized using γ can no longer be expressed as a generalized linear model and can not be fitted using the software employed in Section 7.5.

The delta method provides a shortcut for inference about $g(\boldsymbol{\theta})$, based on the approximate normality of MLEs. It was seen in Section 2.2.1 that the maximum likelihood estimator of $g(\boldsymbol{\theta})$ is $g(\widehat{\boldsymbol{\theta}})$, and the delta method derives the approximate distribution of $g(\widehat{\boldsymbol{\theta}})$ from the approximate distribution of $\widehat{\boldsymbol{\theta}}$. The delta method can be considered the practical application of the delta theorem that is presented in more detail in Part III of this text (Section 13.6.4).

4.2.1 The delta method

The delta method is not particular to MLEs, and here it is derived for an arbitrary random variable X. In Section 4.2.2, the role of X is taken by the MLE $\widehat{\boldsymbol{\theta}}$.

4.2.1.1 One-dimensional case

Suppose that the random variable X has mean μ and variance σ^2, and let $g : \mathbb{R} \to \mathbb{R}$ be a differentiable function. The delta method provides approximations for the mean and variance of the random variable $g(X)$, and is obtained from the first-order Taylor series approximation of $g(X)$ around μ,

$$g(X) \approx g(\mu) + g'(\mu)(X - \mu) . \tag{4.1}$$

Taking the expectations on both sides of (4.1) gives

$$E[g(X)] \approx g(\mu) ,$$

and taking the variances gives

$$\text{var}(g(X)) \approx g'(\mu)^2\sigma^2 .$$

These approximations will be reasonably accurate if g is close to linear in a neighbourhood of μ in which X is sufficiently concentrated. Example 4.7 presents a contrived example where this is not the case, and consequently the delta method approximation is seen to fail. It can also be the case that $g(X)$ does not possess a mean and variance at all (e.g. see Example 4.3).

Example 4.1. Variance-stabilizing transformation for Poisson data. Let Y be Poisson distributed with mean and variance λ, and consider the random variable $g(Y) = \sqrt{Y}$. The square root function has derivative $g'(y) = 1/(2\sqrt{y})$, and so

$$\begin{aligned}\text{var}\left(\sqrt{Y}\right) &\approx g'(\lambda)^2\text{var}(Y)\\ &= \frac{1}{4\lambda}\lambda = \frac{1}{4},\end{aligned}$$

which does not depend on λ. □

Historically, count data have been modelled by applying linear regression models to the square-rooted counts. The justification is that, assuming the counts have variance proportional to their mean, the transformed data will have approximately homogeneous variance. This is just an attempt to shoe-horn data into an unnatural form so as to apply an inappropriate (albeit familiar) model, and it will be unclear how to interpret the regression parameters except in the simplest of analyses. With few exceptions, count data should be analyzed using models that appropriately describe their sampling variability, such as generalized linear models (Chapter 7) and their extensions.

Box 4.1

4.2.1.2 Multi-dimensional case

In the multi-dimensional case, suppose that the random vector $\boldsymbol{X} \in \mathbb{R}^s$ has mean vector $\boldsymbol{\mu}$ and $s \times s$ variance matrix $\boldsymbol{\Sigma}$. The function $g : \mathbb{R}^s \to \mathbb{R}^p$ can be denoted

$$g(\boldsymbol{x}) = \begin{pmatrix} g_1(\boldsymbol{x})^T \\ \cdot \\ \cdot \\ \cdot \\ g_p(\boldsymbol{x})^T \end{pmatrix},$$

where each function $g_i : \mathbb{R}^s \to \mathbb{R}$ is assumed to be differentiable with respect to every element of $\boldsymbol{x}$. The $p \times s$ Jacobian (derivative) matrix of g is

$$\mathbf{G}(\boldsymbol{x}) = \begin{pmatrix} g_1'(\boldsymbol{x})^T \\ \cdot \\ \cdot \\ \cdot \\ g_p'(\boldsymbol{x})^T \end{pmatrix} = \begin{pmatrix} \frac{\partial g_1}{\partial x_1}, \ldots, \frac{\partial g_1}{\partial x_s} \\ \cdot \\ \cdot \\ \cdot \\ \frac{\partial g_p}{\partial x_1}, \ldots, \frac{\partial g_p}{\partial x_s} \end{pmatrix},$$

and the first order Taylor series approximation of $g(\mathbf{X})$ around $\boldsymbol{\mu}$ is

$$g(\mathbf{X}) \approx g(\boldsymbol{\mu}) + \mathbf{G}(\boldsymbol{\mu})(X - \boldsymbol{\mu}) . \tag{4.2}$$

It follows that,

$$E[g(\boldsymbol{X})] \approx g(\boldsymbol{\mu}) ,$$

and

$$\text{var}(g(\boldsymbol{X})) \approx \mathbf{G}(\boldsymbol{\mu})\boldsymbol{\Sigma}\mathbf{G}(\boldsymbol{\mu})^T .$$

Example 4.2. Let Y_1 and Y_2 be independently distributed Pois(λ_i), $i = 1, 2$, respectively. Consider the random vector

$$\begin{pmatrix} T_1 \\ T_2 \end{pmatrix} = \begin{pmatrix} g_1(Y_1, Y_2) \\ g_2(Y_1, Y_2) \end{pmatrix} = \begin{pmatrix} \frac{Y_1}{Y_1+Y_2} \\ Y_1 + Y_2 \end{pmatrix} .$$

The Jacobian of this transformation is

$$\mathbf{G}(\boldsymbol{y}) = \begin{pmatrix} \frac{y_2}{(y_1+y_2)^2} & \frac{-y_1}{(y_1+y_2)^2} \\ 1 & 1 \end{pmatrix} ,$$

and so, denoting $\boldsymbol{\lambda} = (\lambda_1, \lambda_2)$, the approximate variance of (T_1, T_2) is

$$\begin{aligned} \text{var}\begin{pmatrix} T_1 \\ T_2 \end{pmatrix} &\approx \mathbf{G}(\boldsymbol{\lambda}) \begin{pmatrix} \lambda_1 & 0 \\ 0 & \lambda_2 \end{pmatrix} \mathbf{G}(\boldsymbol{\lambda})^T \\ &= \begin{pmatrix} \frac{\lambda_1\lambda_2}{(\lambda_1+\lambda_2)^3} & 0 \\ 0 & \lambda_1 + \lambda_2 \end{pmatrix} = \begin{pmatrix} \frac{p(1-p)}{n} & 0 \\ 0 & n \end{pmatrix} , \end{aligned}$$

where $n = \lambda_1 + \lambda_2$ and $p = \lambda_1/n$. □

4.2.2 The delta method applied to MLEs

In Section 3.2 the notation $\widehat{\boldsymbol{\theta}} \,\dot{\sim}\, N_s(\boldsymbol{\theta}_0, \mathbf{V}(\boldsymbol{\theta}_0))$ was used to make explicit that, subject to regularity conditions and sufficiently large sample size, the maximum likelihood estimator has a distribution that is well approximated by a normal distribution with mean $\boldsymbol{\theta}_0$ (the unknown true parameter vector) and variance depending on $\boldsymbol{\theta}_0$. This result can be used with the delta method to obtain the approximate distribution of a function of the MLE.

Analogously to Equation (4.2), the first order Taylor series approximation of $g : \mathbb{R}^s \to \mathbb{R}^p$ around $\boldsymbol{\theta}_0$ is

$$g(\widehat{\boldsymbol{\theta}}) \approx g(\boldsymbol{\theta}_0) + \mathbf{G}(\boldsymbol{\theta}_0)(\widehat{\boldsymbol{\theta}} - \boldsymbol{\theta}_0) . \tag{4.3}$$

Since normality is preserved by linear transformations, it follows that if $\widehat{\boldsymbol{\theta}} \in \mathbb{R}^s$ is approximately normally distributed then so too is $g(\widehat{\boldsymbol{\theta}}) \in \mathbb{R}^p$, provided that the higher order terms omitted from the Taylor series expansion in (4.3) can be ignored. The approximate mean and variance matrix of $g(\widehat{\boldsymbol{\theta}})$ are given by the delta method, and thus

$$g(\widehat{\boldsymbol{\theta}}) \overset{\cdot}{\sim} N_p\big(g(\boldsymbol{\theta}_0), \mathbf{G}(\boldsymbol{\theta}_0)\mathbf{V}(\boldsymbol{\theta}_0)\mathbf{G}(\boldsymbol{\theta}_0)^T\big) . \tag{4.4}$$

This can be regarded as the approximate version of the asymptotic result stated in Equation (12.21) in Part III of this text. Replacing $\mathbf{V}(\boldsymbol{\theta}_0)$ by its estimate $\widehat{\mathbf{V}}$, and approximating $\mathbf{G}(\boldsymbol{\theta}_0)$ by $\mathbf{G}(\widehat{\boldsymbol{\theta}})$, gives a more pragmatic statement for construction of Wald tests and confidence interval/regions,

$$g(\widehat{\boldsymbol{\theta}}) \overset{\cdot}{\sim} N_p\big(g(\boldsymbol{\theta}_0), \mathbf{G}(\widehat{\boldsymbol{\theta}})\widehat{\mathbf{V}}\mathbf{G}(\widehat{\boldsymbol{\theta}})^T\big) . \tag{4.5}$$

One of the caveats associated with approximate normality of MLEs (and functions of MLEs) is that the sample size must be sufficiently large. But, just how large is *sufficiently* large? The answer to that question may differ dramatically for $\widehat{\boldsymbol{\theta}}$ and $g(\widehat{\boldsymbol{\theta}})$. For a given sample size, it could be that $\widehat{\boldsymbol{\theta}}$ has distribution that is very close to normal, but that $g(\widehat{\boldsymbol{\theta}})$ does not, or vice-versa. Thus, if inference based on normal approximation is to be conducted, it is of particular relevance to choose a sensible parameterization of the model. Section 4.3.2 contains some guidance regarding this issue.

Example 4.3. Variance of log-odds. Let Y be distributed Bin(n, p) and consider estimation of $\zeta = g(p) = \text{logit}(p) = \log(\frac{p}{1-p})$. The ratio $p/(1-p)$ is the so-called odds of the 'success' event (i.e. where $P(\text{success}) = p$), and ζ is commonly called the log-odds (of 'success').

For sufficiently large n, $\widehat{p} = Y/n$ is approximately normal with mean p and variance $p(1-p)/n$. The derivative of the logit function is

$$\frac{\partial g(p)}{\partial p} = \frac{1}{p(1-p)} .$$

Hence, if $p = p_0$ is the true value of p, then $\widehat{\zeta} = \text{logit}(\widehat{p})$ has (for sufficiently large n) a distribution close to that of a $N(\eta, \tau^2)$ random variable, where $\eta = \text{logit}(p_0)$

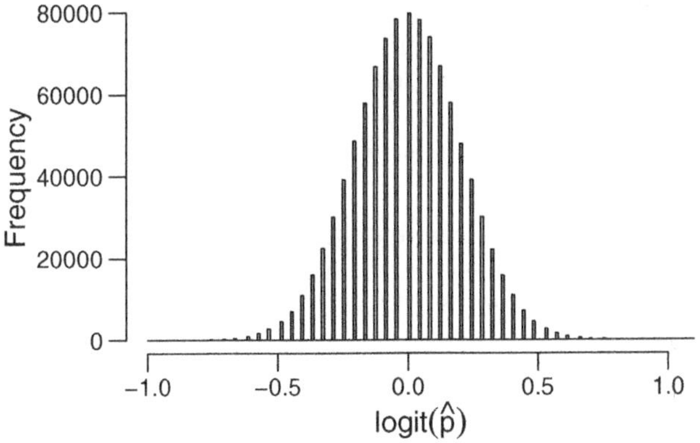

Figure 4.1 Histogram of the logit of the observed proportion from one million simulations of a binomial experiment with n = 100 and $p_0 = 0.5$.

and

$$\tau^2 = g'(p_0)^2 p_0(1-p_0)/n$$
$$= 1/(np_0(1-p_0)) .$$

Figure 4.1 shows a histogram of one million values of $\text{logit}(\widehat{p})$ where Y is distributed Bin(100, 0.5), and so $\widehat{p} = Y/100$. The approximating normal distribution has mean equal to $\text{logit}(0.5) = 0$ and variance of $1/25$, that is, standard deviation of 0.2. This approximation looks good here. However, $\text{logit}(\widehat{p})$ does not possess a mean or variance because it is undefined when $Y = 0$ or 100, which has the near-infinitesimal probability 1.6×10^{-30}. (Strictly speaking, $\text{logit}(\widehat{p})$ is not a random variable at all!)

In practice, p_0 is not known and the pragmatic version of the delta method in (4.5) is used. For the motivating example of Chapter 1 (Example 1.1), $y = 10$ was observed from a binomial experiment with $n = 100$ trials. This gives an estimate of $\widehat{\zeta} = -2.197$, and (under repetition of the binomial experiment) the approximate distribution of the estimator is taken to be

$$\widehat{\zeta} = \text{logit}(\widehat{p}) \;\dot{\sim}\; N\left(\text{logit}(p_0), \frac{1}{n\widehat{p}(1-\widehat{p})}\right)$$
$$= N\left(\text{logit}(p_0), \frac{1}{9}\right) . \qquad (4.6)$$

The approximate standard error of $\widehat{\zeta}$ is $\frac{1}{3}$, and the corresponding Wald 95 % confidence interval for ζ is therefore

$$\text{logit}(\widehat{p}) \pm z_{0.975} \times \frac{1}{3} \approx (-2.851, -1.544) .$$

This confidence interval can be back-transformed into a confidence interval for p by recognizing that the inverse of the transformation $\zeta = \text{logit}(p)$ is $p = e^\zeta/(1+e^\zeta)$, resulting in the interval $(0.055, 0.176)$. By comparison, in Chapter 1 the Wald interval calculated using the original p parameterization was $(0.041, 0.159)$, and the likelihood ratio interval was $(0.051, 0.169)$. □

In the above example, it is seen that the Wald 95 % confidence interval for p differs depending on the parameterization used. That is, the Wald approach is not parameterization invariant. Thus, Wald-based inference can be especially dangerous if an inappropriate parameterization is used. This is investigated in more depth in Section 4.3, and in particular, for the binomial model it is argued that parameter p is not a good choice for Wald-based inference, and that $\zeta = \text{logit}(p)$ is preferable (see Section 4.3.1).

4.2.3 The delta method using R, SAS and ADMB

The following snippets of code implement the delta method in the context of Example 4.3. That is, $y = 10$ is observed from a Bin(100, p) experiment, and it is desired to determine the approximate standard error of $\widehat{\zeta} = \text{logit}(\widehat{p})$. This was seen to be $\frac{1}{3} \approx 0.3333$ (from Equation (4.6)).

4.2.3.1 Using R

The msm package includes a function deltamethod for general application of the delta method to simple algebraic functions of $\boldsymbol{\theta}$. It employs the symbolic differentiation capabilities of R (specifically, the deriv function) to calculate the Jacobian $\mathbf{G}(\widehat{\boldsymbol{\theta}})$ required in (4.5), and its default action is to return the approximate standard error(s) of $g(\widehat{\boldsymbol{\theta}})$.

The first argument to deltamethod is the function $g : \mathbb{R}^s \rightarrow \mathbb{R}^p$, specified as a formula (or list of formulae if $p > 1$) using parameter names x1, . . . ,xs. The next two arguments are $\widehat{\boldsymbol{\theta}}$ and its (approximate) variance $\widehat{\mathbf{V}}$.

```
> library(msm)
> phat=0.1 #MLE
> var.phat=0.0009 #Estimated variance of phat
> deltamethod(~log(x1/(1-x1)),mean=phat,cov=var.phat)
[1] 0.3333333
```

4.2.3.2 Using SAS

The NLMIXED procedure has an ESTIMATE statement that is used to specify $g(\boldsymbol{\theta}) : \mathbb{R}^s \rightarrow \mathbb{R}$. If $g(\boldsymbol{\theta}) \in \mathbb{R}^p$, then multiple ESTIMATE statements can be used to specific each element $g_i(\boldsymbol{\theta})$, $i = 1, ..., p$. In addition to the ESTIMATE statement, the SAS code below has a second change from that seen for the binomial model in

Additional estimates								
Label	Estimate	Standard Error	DF	t Value	Pr > |t|	Alpha	Lower	Upper
Logit(p)	−2.1972	0.3333	1E6	−6.59	<.0001	0.05	−2.8505	−1.5439

Figure 4.2 The table of additional estimates from `PROC NLMIXED`*, showing the standard deviation of logit($\widehat{p}$).*

Section 1.4.1. For convenience, it uses the `BINOMIAL` model specification, thereby avoiding calculation of the log-likelihood as required using the `GENERAL` form. This code results in the output shown in Figure 4.2.

```
DATA binomial;
  y=10; n=100;

PROC NLMIXED DF=1E6 DATA=binomial;
  PARMS p=0.5;
  BOUNDS 0<p<1;
  MODEL y~BINOMIAL(n,p);
  ESTIMATE "Logit(p)" log(p/(1-p));
RUN;
```

Alternatively, the `DeltaMethod` macro (available from `http://www.stat.auckland.ac.nz/~millar` is provided for general implementation of the delta method, for $g : \mathbb{R}^s \to \mathbb{R}^p$ for $p \le s \le 2$.

```
%INCLUDE "DeltaMethodMacro.sas";
%DeltaMethod(expr1=log(x1/(1-x1)),mu1=0.1,var1=0.0009);
```

This produces a SAS output table identical to Figure 4.2.

4.2.3.3 Using ADMB

In ADMB, the `sdreport` variable type is used to specify $g(\boldsymbol{\theta}) : \mathbb{R}^s \to \mathbb{R}$. Multiple `sdreport` declarations can be used to specify each element $g_i(\boldsymbol{\theta}) \in \mathbb{R}$ if $g(\boldsymbol{\theta}) \in \mathbb{R}^p$, and the resulting standard errors (and correlation matrix) are written to text files. Example 4.3 can be implemented in ADMB via addition of the line

```
sdreport_number logitp
```

in the `PARAMETER_SECTION`, and

```
logitp=log(p/(1-p));
```

in the `PROCEDURE_SECTION` of the ADMB template file presented in Section 1.4.3 .

4.2.4 Delta method examples

Example 4.4. Variance of a product (geoduck biomass). Consider the estimation of the product of θ_i and θ_j, $g(\boldsymbol{\theta}) = \theta_i\theta_j$, $i \neq j$. This function maps from $\mathbb{R}^s$ to $\mathbb{R}$ and so its derivative, $\mathbf{G}$, is a row vector of length s, with ith element equal to θ_j, jth element equal to θ_i, and all other elements equal to zero. The approximate variance of $g(\widehat{\boldsymbol{\theta}}) = \widehat{\theta}_i\widehat{\theta}_j$ is therefore

$$\begin{aligned}\widehat{\text{var}}(\widehat{\theta}_i\widehat{\theta}_j) &= \mathbf{G}(\widehat{\boldsymbol{\theta}})\widehat{\mathbf{V}}\mathbf{G}(\widehat{\boldsymbol{\theta}})^T \\ &= \widehat{\theta}_j^2\widehat{\text{var}}(\widehat{\theta}_i) + 2\widehat{\theta}_i\widehat{\theta}_j\widehat{\text{cov}}(\widehat{\theta}_i, \widehat{\theta}_j) + \widehat{\theta}_i^2\widehat{\text{var}}(\widehat{\theta}_j) .\end{aligned} \tag{4.7}$$

Formula (4.7) is well known and widely applied. One application is provided by Gribben, Helson and Millar (2004) in an investigation of the biomass of the geoduck *Panopea zelandica* (Figure 4.3) at several bays in the North Island of New Zealand. The experiment proceeded in two independent stages, the first using SCUBA to estimate the number of geoducks in the bay, and the second weighed a random sample to estimate their mean weight. In Kennedy Bay, the number of geoducks was estimated to be 22 976 ($= \widehat{\theta}_1$) with approximate standard error of 4007, and the mean weight was estimated to be 242.2 g ($= \widehat{\theta}_2$) with approximate standard error of 8.6 g. The estimate of geoduck biomass in Kennedy Bay was $22\,976 \times 242.2 \approx 5.6 \times 10^6$ g, that is, 5.6 tonnes. The two stages of the experiment were independent and hence $\text{cov}(\widehat{\theta}_1, \widehat{\theta}_2) = 0$. The approximate variance of the estimated biomass is therefore

$$\begin{aligned}\widehat{\theta}_1^2\widehat{\text{var}}(\widehat{\theta}_2) + \widehat{\theta}_2^2\widehat{\text{var}}(\widehat{\theta}_1) &= 22\,976^2 \times 8.6^2 + 242.2^2 \times 4007^2 \\ &\approx 9.8 \times 10^{11} ,\end{aligned} \tag{4.8}$$

giving an approximate standard error of 9.9×10^5 g, that is, just under one tonne.

The above calculation is performed by the R code

```
>deltamethod(~x1*x2,mean=c(22976,242.2),cov=diag(c(4007^2,8.6^2)))
```

Figure 4.3 New Zealand geoduck Panopea zelandica. Photo reproduced by permission of Paul Gribben.

or the SAS code

```
%DeltaMethod(expr1=x1*x2,mu1=22976,var1=4007**2,
             mu2=242.2,var2=8.6**2,cov=0);
```

□

Variance of a product. It can be shown (Goodman (1960) and Exercise 4.5) that the exact variance of the product $\widehat{\theta}_1\widehat{\theta}_2$ of two independent estimators is

$$\text{var}(\widehat{\theta}_1\widehat{\theta}_2) = E[\widehat{\theta}_1]^2\text{var}(\widehat{\theta}_2) + E[\widehat{\theta}_2]^2\text{var}(\widehat{\theta}_1) + \text{var}(\widehat{\theta}_1)\text{var}(\widehat{\theta}_2) . \qquad (4.9)$$

Furthermore, if $\widehat{\text{var}}(\widehat{\theta}_1)$ and $\widehat{\text{var}}(\widehat{\theta}_2)$ are unbiased estimators of the true variances $\text{var}(\widehat{\theta}_1)$ and $\text{var}(\widehat{\theta}_2)$ then

$$\widehat{\text{var}}(\widehat{\theta}_1\widehat{\theta}_2) = \widehat{\theta}_1^2\widehat{\text{var}}(\widehat{\theta}_2) + \widehat{\theta}_2^2\widehat{\text{var}}(\widehat{\theta}_1) - \widehat{\text{var}}(\widehat{\theta}_1)\widehat{\text{var}}(\widehat{\theta}_2) , \qquad (4.10)$$

is an unbiased estimator of $\text{var}(\widehat{\theta}_1\widehat{\theta}_2)$. However, it is *not* the case that (4.10) will necessarily be a better estimator of $\text{var}(\widehat{\theta}_i\widehat{\theta}_j)$ than (4.7). Moreover, the value of (4.10) can be negative and hence this estimator is not truly unbiased because negative estimators of variance are not permissable. In Example 4.4 the difference between (4.8) and (4.10) is negligible and, to three significant figures, both result in an approximate standard error for the estimated biomass of 9.90×10^5 g.

Box 4.2

Example 4.5. Variance matrix of a vector transformation. In the context of the binormal mixture model for the Old Faithful geyser waiting time data (Section 3.3.4), suppose that it is of interest to jointly consider the ratio of the mean parameters, μ/ν, and ratio of the standard deviations, σ/τ. That is, $g(\boldsymbol{\theta}) = (\mu/\nu, \sigma/\tau)$.

The approximate variance matrix of $g(\widehat{\boldsymbol{\theta}})$ is obtained by the additional R code

```
> deltamethod(list(~x2/x4,~x3/x5),MLE,Vhat,ses=F)
             [,1]         [,2]
[1,] 7.684293e-05 0.0002495297
[2,] 2.495297e-04 0.0167564703
```

where the option `ses=F` was used to suppress the output of the standard errors, so that `deltamethod` would instead return the variance matrix. That is,

$$\begin{pmatrix} \widehat{\mu}/\widehat{\nu} \\ \widehat{\sigma}/\widehat{\tau} \end{pmatrix} \stackrel{\cdot}{\sim} N_2\left(\begin{pmatrix} \mu_0/\nu_0 \\ \sigma_0/\tau_0 \end{pmatrix}, \begin{pmatrix} 7.684 \times 10^{-5} & 2.495 \times 10^{-4} \\ 2.495 \times 10^{-4} & 0.01676 \end{pmatrix}\right)$$

The SAS code in Section 3.3.4 requires the two additional statements

```
ESTIMATE "Ratio of means" mu/nu;
ESTIMATE "Ratio of ses" sigma/tau;
```

and it is also necessary to add the procedure option `ECOV` to the `PROC NLMIXED` statement so that SAS will output the variance matrix of $g(\widehat{\boldsymbol{\theta}})$ rather than just the individual standard errors. □

Example 4.6. Variance of the log odds-ratio. Consider the two-by-two contingency table

	Col. 1	Col. 2	
Row 1	y_{11}	y_{12}	y_{1+}
Row 2	y_{21}	y_{22}	y_{2+}

Let p_i denote the probability of the column 1 event, given that the observation is in row i, $i = 1, 2$. Within row i, the odds of column 1 is defined to be $p_i/(1 - p_i)$, and the odds-ratio is the ratio of the row 1 and row 2 odds. That is,

$$OR = \frac{p_1/(1-p_1)}{p_2/(1-p_2)} .$$

The odds-ratio is commonly used to quantify the magnitude of any association between the row and column events, with an odds-ratio of unity corresponding to no association. The MLE of the odds ratio is given by replacing p_i by its MLE $\widehat{p}_i = y_{i1}/y_{i+}$, giving

$$\widehat{OR} = \frac{\widehat{p}_1/(1-\widehat{p}_1)}{\widehat{p}_2/(1-\widehat{p}_2)} = \frac{y_{11}y_{22}}{y_{12}y_{21}} .$$

Inference about the odds-ratio is usually made on the log scale, that is, by working with the log odds-ratio. The log odds-ratio can be written as the difference in log odds of the two rows

$$\log(\widehat{OR}) = \log\left(\frac{\widehat{p}_1}{1-\widehat{p}_1}\right) - \log\left(\frac{\widehat{p}_2}{1-\widehat{p}_2}\right) , \tag{4.11}$$

where the two log odds terms on the right-hand side of (4.11) are independent. This is a difference in logits, and so using the result from Example 4.3,

$$\begin{aligned}
\widehat{\text{var}}(\log(\widehat{OR})) &= \widehat{\text{var}}\left[\log\left(\frac{\widehat{p}_1}{1-\widehat{p}_1}\right)\right] + \widehat{\text{var}}\left[\log\left(\frac{\widehat{p}_2}{1-\widehat{p}_2}\right)\right] \\
&= \frac{1}{y_{1+}\widehat{p}_1(1-\widehat{p}_1)} + \frac{1}{y_{2+}\widehat{p}_2(1-\widehat{p}_2)} \\
&= \frac{y_{1+}}{y_{11}y_{12}} + \frac{y_{2+}}{y_{21}y_{22}} \\
&= \frac{1}{y_{11}} + \frac{1}{y_{12}} + \frac{1}{y_{21}} + \frac{1}{y_{22}} .
\end{aligned} \tag{4.12}$$

It is interesting to note that the above variance formula can be obtained more directly by using the equivalence between binomial (or multinomial) and Poisson models (see Exercise 14.7 and Example 9.2). In particular, for making inference about the log odds-ratio, one can model the cell counts, y_{ij} as independent realizations from a Pois(λ_{ij}) distribution. With Y_{ij}, $i = 1, 2$, $j = 1, 2$ denoting independent Pois(λ_{ij}) random variables, it is then the case that

$$\begin{aligned}\widehat{\text{var}}(\log(\widehat{OR})) &= \widehat{\text{var}}(\log Y_{11} - \log Y_{12} - \log Y_{21} + \log Y_{22}) \\ &= \widehat{\text{var}}(\log Y_{11}) + \widehat{\text{var}}(\log Y_{12}) + \widehat{\text{var}}(\log Y_{21}) + \widehat{\text{var}}(\log Y_{22}) ,\end{aligned}$$

where, from the delta method, and with the notation $g(y) = \log y$,

$$\begin{aligned}\widehat{\text{var}}(g(Y_{ij})) &= g'(\lambda_{ij})^2 \widehat{\text{var}}(Y_{ij}) \\ &\approx \frac{1}{y_{ij}^2} y_{ij} = \frac{1}{y_{ij}} .\end{aligned}$$

A confidence interval for the odds-ratios is obtained by exponentiating the Wald confidence interval for the log odds-ratio. In R, this approach is implemented in the `oddsratio` function (using the `method="wald"` option) within the `epitools` package. In SAS it is calculated by `PROC FREQ` when the `MEASURES` option is used in a `TABLES` statement. □

4.3 Wald statistics – quick and dirty?

Are they quick? – generally yes. In the scalar parameter case the use of Wald statistics is straightforward and very familiar, with an approximate 95 % confidence interval given by $\widehat{\theta}$ plus or minus a couple of standard deviations.

Are they dirty? Their lack of invariance to parameterization is one reason to suspect that they could be (Box 4.3). The following example is derived from Fears, Benichou and Gail (1996) who look at the fallibility of the Wald statistic, and it shows that the Wald statistic can be very dirty indeed.

Example 4.7. In a scalar parameter model with parameter space $\Theta = \mathbb{R}$, suppose that $\widehat{\theta} = 1$ and $\widehat{\text{var}}(\widehat{\theta}) = 0.01$. Then the Wald test statistic for $H_0 : \theta = 0$ is

$$W_\theta = \frac{(\widehat{\theta} - 0)^2}{0.01} = 100 .$$

Comparing $W_\theta = 100$ to a χ^2_1 distribution, the p-value for H_0 is a near-infinitesimal 8×10^{-24}.

Now, consider a re-parameterization of the model using $\zeta = g_a(\theta) = \theta^a$ where a is any odd positive integer, that is, $a \in \{1, 3, 5, ...\}$. (This requirement on a ensures that, as a function defined on the parameter space $\Theta = \mathbb{R}$, $g_a(\theta)$ is a one-to-one mapping.) Then, $\widehat{\zeta} = 1$ is the MLE and the null hypothesis can be expressed

$H_0 : \zeta = 0$. The derivative of the transformation is $g'(\theta) = a\theta^{a-1}$, and from the delta method

$$\widehat{\text{var}}(\widehat{\zeta}) = g'(\widehat{\theta})^2 \widehat{\text{var}}(\widehat{\theta}) = 0.01a^2 ,$$

since $g'(\widehat{\theta}) = g'(1) = a$. The Wald test statistic for the equivalent hypothesis $H_0 : \zeta = 0$ is therefore

$$W_\zeta = \frac{(\widehat{\zeta} - 0)^2}{0.01a^2} = \frac{100}{a^2} .$$

Note that the statistic W_ζ can be made arbitrarily small, and hence the p-value for H_0 can be made arbitrarily close to unity, by using increasingly large values of a. For example, when $a = 25$ then $W_\zeta = 0.16$ and the p-value for H_0 is approximately 0.69. □

The Wald test statistic is not invariant to transformation. To see this, consider the scalar parameter case and suppose that $\widehat{\theta} \dot{\sim} N(\theta_0, \widehat{v})$. Let $\zeta = g(\theta)$ be invertible, so that $H_0 : \theta = \theta_0$ is equivalent to $H_0 : \zeta = \zeta_0$ where $\zeta_0 = g(\theta_0)$. From (4.5), $g(\widehat{\theta}) \dot{\sim} N(g(\theta_0), g'(\widehat{\theta})^2\widehat{v}) = N(\zeta_0, g'(\widehat{\theta})^2\widehat{v})$. By applying a Taylor series expansion of $g(\theta_0)$ around $\widehat{\theta}$, the Wald test statistic for $H_0 : \zeta = \zeta_0$ can be written

$$\begin{aligned} \frac{(\widehat{\zeta} - \zeta_0)^2}{g'(\widehat{\theta})^2\widehat{v}} &= \left(\frac{g'(\widehat{\theta})(\widehat{\theta} - \theta_0) + \ldots}{g'(\widehat{\theta})\widehat{v}^{1/2}} \right)^2 \\ &= \frac{(\widehat{\theta} - \theta_0)^2}{\widehat{v}} + \ldots . \end{aligned} \tag{4.13}$$

The first term in (4.13) is the Wald test statistic for $H_0 : \theta = \theta_0$. The remainder terms, denoted by ..., are the cause of the non-invariance, and their effect could be substantial if g is highly nonlinear in the neighbourhood of $\widehat{\theta}$.

Box 4.3

The above example shows that the p-value can be made arbitrarily close to unity by using a suitably ridiculous re-parameterization of the model. In this case, the degeneration of the Wald statistic arises because the linear approximation of $g(\theta)$ around $g(\widehat{\theta})$ can be made arbitrarily bad (Figure 4.4), and hence the remainder terms in (4.13) are non-negligible. In contrast, inference based on the likelihood ratio would be unaffected, because the statistical model is preserved under re-parameterization.

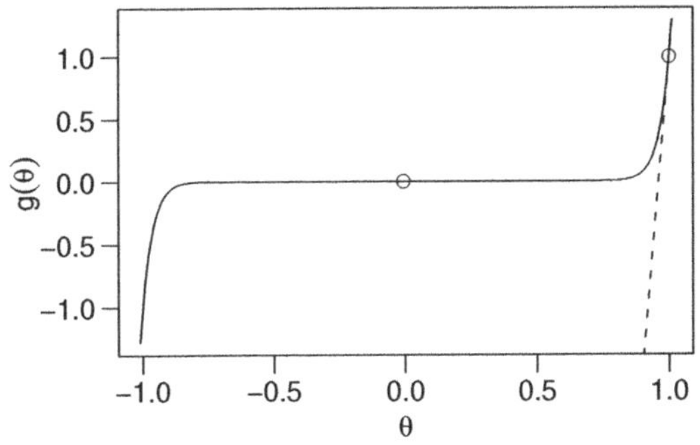

Figure 4.4 Plot of $g(\theta) = \theta^{25}$, showing the inadequacy of a linear approximation (dashed line) at $\widehat{\theta} = 1$.

4.3.1 Wald versus likelihood ratio revisited

Pawitan (2001, p. 47 therein) provides a heuristic argument to support the assertion that the likelihood ratio test statistic always performs at least as well as the Wald test statistic. The argument proceeds along the following lines – in the scalar parameter case, $\theta \in \mathbb{R}$, when the log-likelihood is quadratic then the likelihood ratio and Wald test statistics are identical (Exercise 4.8). In this situation, the likelihood function corresponds to that of an iid normal model with θ denoting the mean, and where the observations have sample mean $\widehat{\theta}$. Hence, if θ_0 is the true parameter value then it would be reasonable to suppose that

$$W = \frac{(\widehat{\theta} - \theta_0)^2}{\widehat{\mathrm{var}}(\widehat{\theta})} \sim \chi_1^2 ,$$

to a high degree of approximation, and hence also for the identical likelihood ratio statistic.

It follows from the above argument that if there exists *any* parameterization $\zeta = g(\theta)$ such that the log-likelihood is an approximately quadratic function of ζ then the likelihood ratio statistic will have good performance. Note that the particular parameterization does not need to be known, since the likelihood ratio statistic is invariant to re-parameterization. However, for inference based on the Wald test statistic to have good performance it will be necessary to use an appropriate parameterization. This parameterization may not be obvious, and may not be convenient for purposes of model implementation.

Figure 4.5 shows that the binomial log-likelihood from observing $y = 10$ from the Bin(100, p) model is very close to quadratic under the parameterization $\zeta = \arcsin\sqrt{p}$ (i.e. $\sin^{-1}\sqrt{p}$). (This happens to be the variance stabilizing transformation for binomial data, Exercise 4.1). So, using likelihood ratio should give reliable inference here. For Wald-based inference, parameterizations $\zeta = \mathrm{logit}(p)$ or $\zeta = \arcsin\sqrt{p}$ would be preferable to using the original parameter p.

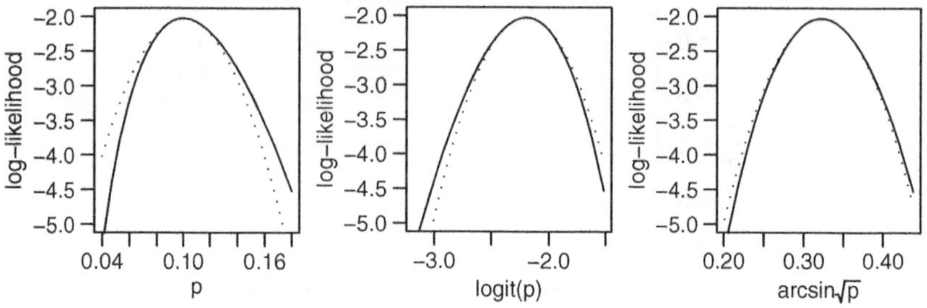

Figure 4.5 The log-likelihood function $l(\zeta)$ from observing $y = 10$ from a Bin(100, p) experiment, for $\zeta = p$, $\zeta = logit(p)$, and $\zeta = \arcsin\sqrt{p}$. The dashed line shows the quadratic approximation around the MLE.

4.3.2 Pragmatic considerations

In practice, there is a trade-off between the ease (and necessity) of providing standard errors, and the greater computational effort required to perform inference based on likelihood ratio. The latter is minimal for some standard models, thanks to the built-in capabilities within R and SAS. These include the `confint` function in R, and the `LRCI` option in the `MODEL` statement of `PROC GENMOD` in SAS. However, in general, calculation of a likelihood ratio confidence interval may require explicit coding of an appropriately parameterized log-likelihood in the software of your choice. The R function `Plkhci`, SAS macro `Plkhci`, or ADMB declaration `likeprof` can then be used to obtain the likelihood ratio confidence interval.

The geoduck example in Example 4.4 is one where using likelihood ratio to make inference about the geoduck biomass would have been challenging. Calculating a likelihood ratio confidence interval for the geoduck biomass would have required using the sum of the log-likelihood functions under the two independent experiments that were conducted to estimate geoduck abundance and average weight, respectively. This combined log-likelihood would be of nonstandard form, and would require explicit coding and maximization.

The convenience of Wald-based inference will often outweigh the preferential use of likelihood-ratio inference. Then, care must be taken to ensure that a reasonable parameterization is being employed. For example, a Wald confidence interval for the odds-ratio, OR, in a two-by-two contingency table is preferably obtained by exponentiating the Wald interval calculated for $\log OR$ (Example 4.6). In this case, use of $\log OR$ can be motivated by the need for inference to be unaffected if the order of columns 1 and 2 is reversed, because the column order is an arbitrary choice. Under column reversal, the odds ratio becomes the reciprocal of the original OR, and this is a nonlinear transformation that will not preserve Wald-based inference. Note that the sign of the log odds-ratio simply reverses under the re-ordering of columns. A reversal of sign is a linear transformation, and it follows that Wald-based inference using the log odds-ratio will be invariant to column re-ordering.

More generally, to avoid problems that can arise when parameters are near a boundary of the parameter space (Section 4.7.3), give consideration to transforming parameters that are restricted to a subset of $\mathbb{R}$, so that the transformed parameters reside on all of $\mathbb{R}$. The log transformation is often sensible for parameters that are constrained to be positive. This includes variance parameters (e.g. see Figure 10.4). Indeed, if it were not for ease of presentation, it would be better to implement the log-likelihood function of a binormal mixture model using parameters $\boldsymbol{\zeta} = (\text{logit}(p), \mu, \log\sigma, \nu, \log\tau)$, but for Wald-based inference of the Old Faithful waiting times it matters little.

In some situations there are guidelines for determining when Wald-based inference is considered to be acceptable in practice. For example, in the Bin(n, p) model it is typical to require $\min(n\widehat{p}, n(1-\widehat{p})) \geq 5$, although it should be noted that Brown *et al.* (2001) remain highly critical of the Wald statistic even when this criterion is satisfied. This is somewhat justified. The interval calculated as $\widehat{p}$ plus or minus 1.96 standard deviations is the easy way, but not a good way. Using likelihood ratio, or Wald-based inference on parameter $\zeta = \text{logit}(p)$, is preferable (see also Box 3.1).

4.4 Model selection

So far, the statistical models used in the examples and exercises have been prespecified. This specification has been based on consideration of the type of data being measured (e.g. continuous data, counts, proportions) and perhaps a preliminary inspection of the data. However, selection of an appropriate statistical model is a very important consideration because inference is conditional on the model being correctly specified.[1] The practical reality is that the statistical model will generally not be a perfectly true description of the underlying mechanisms that generated the data, but hopefully will at least be good enough to be useful. An excellent overview of the relevance of model choice is provided by Harrell (2001, Section 1.4 therein).

Model selection commonly arises in the context of regression modelling where it is necessary to choose an appropriate subset from the set of all possible explanatory covariates and their interactions. In this case the collection of all possible models can be obtained by either backward selection (sequentially removing terms from the full model that contains all possible terms) or forward selection (adding terms one at a time to the null model). Hypothesis tests could be used to decide whether a term should be removed or added, but while this may seem a reasonable approach, it has already been seen that hypothesis testing is not a tool that is suited to the task of choosing a preferred model (Example 2.4). The use of formal hypothesis testing in this context is heuristic and is not based on any principles regarding the selection of the 'best' model.

4.4.1 AIC

Akaike (1974) provided a holistic approach to model selection by using the concept of information loss. This loss is quantified by the Kullback-Leibler divergence (Kullback 1959) between the true unknown model and the fitted model, and can be

[1] But, see Chapter 8 for methodology that does not require a fully specified parametric statistical model.

regarded as the loss of predictive ability from using the fitted model rather than the true model. Under this framework, the preferred model is the one that minimizes this loss amongst the collection of all candidate models.

In the context of maximum likelihood estimation, Akaike (1974) obtained a simple formula for estimation of the predictive loss from using $f(\boldsymbol{y}; \widehat{\boldsymbol{\theta}})$ as an estimate of the true density function of $\boldsymbol{y}$. This lead to what is now known as Akaike's information criterion (AIC)

$$\text{AIC} = -2l(\widehat{\boldsymbol{\theta}}) + 2s\,, \tag{4.14}$$

where s is the number of model parameters. The fitted model with smallest AIC is preferred. AIC can be regarded as a criterion that provides a parsimonious balance between model fit, as quantified by $l(\widehat{\boldsymbol{\theta}})$, and model complexity, as quantified by the number of parameters, s.

As a general rule-of-thumb, if the difference in AIC between two models is 2 or more then it can be said that the model with smaller AIC is strongly preferred. However, if the difference is less than 2 then it could be argued that both models are worthy of consideration. See Burnham and Anderson (2002) for discussion of the notion of model averaging, whereby inference explicitly incorporates the uncertainty in choice of the preferred model.

Box 4.4

Model selection using AIC does not require the competing models to be nested. For example, the competing models could specify different error structure on the data (e.g. normal versus lognormal) or could use different functional forms to describe the effect of explanatory variables (e.g. generalized linear models with different link function).

Example 4.8. The micro-propagation count data encountered in Example 3.4 are the 40 counts of the number of roots on apple cultivars. The data are:

0	0	0	0	0	0	0	0	0	0	0	0	0	0	0	0	0	0	0	1
1	2	2	3	3	3	3	4	4	4	5	6	6	6	6	7	7	7	9	9

In Example 3.4 it was seen that the iid Poisson model was not appropriate for these data, primarily due to the preponderance of zeros. Generalizations of the Poisson model are fitted to these data in Example 7.8, where it is noted that the iid Poisson, zero-inflated Poisson (ZIP), negative binomial (NB), and zero-inflated negative binomial (ZINB) models fitted to these data have log-likelihoods of -110.555, -74.880, -81.000, and -74.598 respectively. The ZIP and NB models each have two parameters, and the ZINB has three, and the resulting AICs are therefore 223.109, 153.761, 166.000 and 155.196, respectively. The zero-inflated models are strongly preferred, and the simpler ZIP is preferred to the ZINB. □

There are numerous variants of the AIC, including a version that uses a correction for small sample size (AICC), and quasi-AIC (QAIC) for use with over-dispersed models (Section 7.6.1). Also, Bayesian application of the minimum information loss argument (Schwarz 1978) has lead to the Bayesian information criterion (BIC). The BIC is analogous to AIC, except that the $2s$ term in (4.14) is replaced by $s \log n$ where n is the sample size. See Burnham and Anderson (2002) for a comprehensive presentation of the AIC and several of its variants.

Many studies of the relative performance of AIC and its variants have appeared in the published literature, and it has been established that AIC tends to over-fit when sample size is large (e.g. Zheng and Loh 1995). Consequently, some modelers prefer BIC to AIC, because (for $n \geq 8$), BIC imposes a stronger penalty on model complexity than AIC.

Warning. Since AIC can be used to compare non-nested models, it is essential to ensure that all constant terms are used in calculation of the log-likelihoods, because these constants could differ between the models under comparison.

Box 4.5

The number of candidate models can be large, especially in the situation of determining the best set of terms to use in a regression model. For such purposes, R and SAS provide some stepwise functionality for moving through the collection of all possible models. For example, see the R function `step`, and the `SELECTION` option in the `MODEL` statements of the regression procedure `PROC REG` and logistic regression procedure `PROC LOGISTIC`.

4.5 Bootstrapping

Frequentist inferential procedures are based on the notion of repeat sampling and so, in the likelihood context, it is necessary to determine the properties and behaviour of ML-based inference under repetition of the experiment. The general tools and techniques that we have been using up to this point have been obtained from a well-established body of theory that required large doses of calculus, probability theory and mathematical statistics (see Chapters 12 and 13). The bootstrap effectively replaces this calculus and theory with pure computational effort.

The essential concept of bootstrapping is to emulate repetition of the experiment by simulating new data on the computer, followed by recalculation of the MLE using the simulated data. The appellation 'bootstrap' was chosen by Efron (1979) in the first comprehensive account of this computer-intensive methodology. The terminology was inspired by the adage 'to pull oneself up by one's bootstrap', arising from the adventures of Baron von Münchhausen (Raspe 1948). Münchausen escaped from a deep hole through his own efforts, by the brute force of pulling up on his bootstraps. The statistical bootstrap uses the brute force of the computer.

The bootstrap has many potential advantages over the large-sample tools constructed in Chapter 12 and used up to this point. First and foremost, simulation of the experiment, followed by re-fitting of the model to the simulated data, is an intuitive thing to do. This simulation-based approach allows inference to be extended to situations where the theory becomes intractable, such as including parameter uncertainty in prediction (Section 4.6.5). Moreover, bootstrapping can be used for situations where the set of regularity conditions specified in Chapter 12 do not hold. Section 4.7.3 presents such a situation, where regularity conditions do not hold because the parameter lies on the boundary of the parameter space under the null hypothesis. Bootstrapping can be used to obtain a valid p-value (see Section 4.5.4 and Exercise 4.9) in this situation.

Bootstrapping is also applicable beyond the arena of likelihood-based inference and can be used to investigate the properties of a wide spectrum of estimators. However, to be valid, the bootstrap does require that the estimator be *consistent*. This property is formally defined in Section 12.2, but an adequate working definition of consistency is that the estimator will get closer and closer to the true parameter value as sample size increases to infinity.

Application of bootstrap methodology has evolved into many variants (e.g. Efron 1987). Herein, attention will be restricted to the simplest form, because of the virtue of simplicity and because this form has the most general application. In particular, no coverage will be given here to using the bootstrap for bias adjustment. For this, and more extensive coverage of the bootstrap, see Efron and Tibshirani (1993) and Davison and Hinkley (1997). For a more applied focus, see Manly (1997) and Chernick (2008).

The R package `boot` provides extensive functionality for bootstrapping, and incorporates some of the variants not covered here. For example, the `boot.ci` function provides several alternatives to the percentile method confidence interval that is presented in Section 4.5.2.

4.5.1 Bootstrap simulation

The bootstrap emulates the sampling distribution of $\widehat{\boldsymbol{\theta}}$ by emulating the processes of data generation and model fitting. It does this by generating artificial data $\boldsymbol{y}^* = (y_1^*, ..., y_n^*)$ from a distribution that approximates the true unknown sampling distribution of the actual data, followed by recalculating the MLE using these artificial data. This is done a large number of times, B say, resulting in a large collection of bootstrap MLEs, denoted $\widehat{\boldsymbol{\theta}}^*_{(j)}$, $j = 1, ..., B$. The distribution of these artificially generated bootstrap MLEs can then be used to infer the sampling distribution of $\widehat{\boldsymbol{\theta}}$ (Section 4.5.2).

One obvious choice for a distribution to approximate the true unknown sampling distribution of the data is to use the distribution under the fitted model. That is, to generate $\boldsymbol{y}^*$ from the distribution with density $f(\cdot;\widehat{\boldsymbol{\theta}})$. This is parametric bootstrapping.

Nonparametric bootstrapping is an option when the sampling randomness in the data-generating process can be explicitly emulated. For example, if the data $y_i, i = 1, ..., n$ are iid then the *empirical distribution function*, EDF, can be used as a discrete approximation to the true unknown cumulative distribution function of Y. That is,

$$\begin{aligned}\widehat{F}(y) &= \frac{\text{Number of } y_i \leq y}{n} \\ &\approx P(Y \leq y)\,,\end{aligned}$$

where $\widehat{F}(y)$ denotes the EDF. Note that the EDF assigns probability mass $\frac{1}{n}$ to each y_i value (and if two or more data points take the same value then the probability mass for that value is summed over those data points). The nonparametric bootstrap generates new data $\boldsymbol{y}^*$ by random sampling from the EDF. Consequently, random sampling from the EDF can be accomplished by sampling *with* replacement from the observed data $(y_1, ..., y_n)$.

More generally, it is sometimes the case that the data can be partitioned into distinct subsets such that the data are iid within each subset, in which case the nonparametric bootstrap is applied separately within each subset. It would also be appropriate to re-sample the collection of subsets (with replacement) if this emulated the experimental design used to obtain the data (e.g. see Section 3.8 of Davison and Hinkley 1997).

4.5.2 Bootstrap confidence intervals

Here, discussion is restricted to the most commonly used form of bootstrap confidence interval, the percentile method. This method is the simplest, and it performs well in most situations.

Suppose that a $(1-\alpha)100\,\%$ CI is required for θ_k (component k of $\boldsymbol{\theta}$). Let $\widehat{\theta}^*_{k,\alpha/2}$ and $\widehat{\theta}^*_{k,1-\alpha/2}$ denote the $\alpha/2$ and $1-\alpha/2$ empirical quantiles of the collection of values $\widehat{\theta}^*_{k,(j)}$, $j = 1, ..., B$, where $\widehat{\theta}^*_{k,(j)}$ denotes component k of the bootstrap MLE from bootstrap simulation j. The $(1-\alpha)100\,\%$ percentile method CI is simply

$$(\widehat{\theta}^*_{k,\alpha/2}, \widehat{\theta}^*_{k,1-\alpha/2})\,. \tag{4.15}$$

4.5.2.1 Justification

The percentile method interval in (4.15) is justified with the following argument based on the parametric bootstrap. The estimator, $\widehat{\boldsymbol{\theta}}$, is the MLE obtained from observing data $\boldsymbol{y}$ distributed according to the density $f(\boldsymbol{y}, \boldsymbol{\theta}_0)$. Analogously, for each bootstrap simulation, $\widehat{\boldsymbol{\theta}}^*$ is the MLE obtained from observing simulated data

$\boldsymbol{y}^*$ distributed according to the density $f(\boldsymbol{y}, \widehat{\boldsymbol{\theta}})$. So, subject to appropriate regularity conditions, it will be reasonable to assume that the behaviour of $\widehat{\boldsymbol{\theta}}$ as an estimator of $\boldsymbol{\theta}_0$ should be well approximated by the behaviour of $\widehat{\boldsymbol{\theta}}^*$ as an estimator of $\widehat{\boldsymbol{\theta}}$. That is, for component k of the parameter vector,

$$\widehat{\theta}_k^* - \widehat{\theta}_k \approx_D \widehat{\theta}_k - \theta_{0k} , \tag{4.16}$$

where $\approx_D$ denotes approximately equal in distribution.

Suppose, for now (see Box 4.6), that the distributions of the quantities on the left and right sides of (4.16) are approximately symmetric around zero. Then,

$$\begin{aligned} \widehat{\theta}_k^* - \widehat{\theta}_k &\approx_D -(\widehat{\theta}_k - \theta_{0k}) \\ &= \theta_{0k} - \widehat{\theta}_k . \end{aligned} \tag{4.17}$$

Thus, for any interval (a, b), $a < b$,

$$P_*(a < \widehat{\theta}_k^* - \widehat{\theta}_k < b) \approx P_{\boldsymbol{\theta}_0}(a < \theta_{0k} - \widehat{\theta}_k < b) , \tag{4.18}$$

where P_* denotes probability with respect to the bootstrap distribution of $\widehat{\boldsymbol{\theta}}^*$, and $P_{\boldsymbol{\theta}_0}$ denotes probability with respect to the sampling distribution of $\widehat{\boldsymbol{\theta}}$ from repetition of observing data from the true model $f(\boldsymbol{y}, \boldsymbol{\theta}_0)$.

Equation (4.18) can be rewritten as

$$P_*(a + \widehat{\theta}_k < \widehat{\theta}_k^* < b + \widehat{\theta}_k) \approx P_{\boldsymbol{\theta}_0}(a + \widehat{\theta}_k < \theta_{0k} < b + \widehat{\theta}_k) ,$$

and so, if a and b are such that $P_*(a + \widehat{\theta}_k < \widehat{\theta}_k^* < b + \widehat{\theta}_k) = 1 - \alpha$ then it is also the case that $P_{\boldsymbol{\theta}_0}(a + \widehat{\theta}_k < \theta_{0k} < b + \widehat{\theta}_k) = 1 - \alpha$. Since a and b are arbitrary, this establishes that any interval containing $\widehat{\theta}_k^*$ with probability $(1 - \alpha)$ is also an interval containing θ_{0k} with probability $(1 - \alpha)$.

> The assumption of symmetry used to obtain (4.17) can be weakened by virtue of the invariance of confidence intervals and quantiles to monotone transformations That is, if $\zeta = g(\theta_k)$ is monotone increasing then the percentile method confidence interval $(\widehat{\theta}_{k,\alpha/2}^*, \widehat{\theta}_{k,1-\alpha/2}^*)$ for θ_k is equivalent to the confidence interval $\left(g(\widehat{\theta}_{k,\alpha/2}^*), g(\widehat{\theta}_{k,1-\alpha/2}^*)\right)$ for ζ. It is enough that approximate symmetry holds for any monotone transformation, $\zeta = g(\theta_k)$. Note that it is not necessary to know g.

Box 4.6

4.5.3 Bootstrap estimate of variance

The approximate equivalence in (4.16) suggests using the variance of $\widehat{\theta}_k^* - \widehat{\theta}_k$ as an estimator of the variance of $\widehat{\theta}_k - \theta_{0k}$. The variance of $\widehat{\theta}_k^* - \widehat{\theta}_k$ is with respect to the bootstrap sampling, and will be denoted $\text{var}^*(\widehat{\theta}_k^* - \widehat{\theta}_k)$. Now, $\widehat{\theta}_k$ is a constant with respect to the bootstrap sampling and so it follows that $\text{var}^*(\widehat{\theta}_k^* - \widehat{\theta}_k) = \text{var}^*(\widehat{\theta}_k^*)$. Similarly, the variance of $\widehat{\theta}_k - \theta_{0k}$ is over repeat experimentation under the fixed value of the unknown parameter $\boldsymbol{\theta}_0$, and so $\text{var}(\widehat{\theta}_k - \theta_{0k}) = \text{var}(\widehat{\theta}_k)$.

The bootstrap values $\widehat{\theta}_{k,(j)}^*$, $j = 1, ..., B$ are iid, and hence the bootstrap variance can be estimated using the sample variance estimator. This gives

$$\begin{aligned}\text{var}(\widehat{\theta}_k) = \text{var}(\widehat{\theta}_k - \theta_{0k}) &\approx \text{var}^*(\widehat{\theta}_k^* - \widehat{\theta}_k)\\ &= \text{var}^*(\widehat{\theta}_k^*)\\ &\approx \sum_{j=1}^{B} \frac{(\widehat{\theta}_{k,(j)}^* - \overline{\theta}_k^*)^2}{B-1},\end{aligned} \tag{4.19}$$

where $\overline{\theta}_k^*$ is the average of $\widehat{\theta}_{k,(j)}^*$, $j = 1, ..., B$. That is, the bootstrap estimate of variance is simply the sample variance of the bootstrap.

4.5.4 Bootstrapping test statistics

Let the continuous random variable T denote a test statistic. In some special cases (e.g. classical test statistics for linear normal models) the distribution of T under H_0 can be determined exactly. More generally, one relies on an approximation, such as the asymptotic χ^2 distribution of Wald and likelihood ratio test statistics.

The bootstrap provides a further means of approximating the distribution of T (under H_0), by simulating values $t_{(j)}^*$, $j = 1, ..., B$ under H_0. If H_0 is a simple hypothesis then $\boldsymbol{\theta}_0$ is completely specified and the true sampling distribution of t^* (under H_0) can be simulated by generating new data $\boldsymbol{y}^*$ according to the density $f(\boldsymbol{y}, \boldsymbol{\theta}_0)$ and recalculating the test statistic each time. More generally, in the composite hypothesis case where $\boldsymbol{\theta} \in \Theta_0$, t^* is simulated from generating $\boldsymbol{y}^*$ according to $f(\boldsymbol{y}, \widehat{\boldsymbol{\theta}}_0)$ where $\widehat{\boldsymbol{\theta}}_0 \in \Theta_0$ is the MLE under H_0.

The bootstrap distribution of the test statistic can be used to obtain bootstrap p-values. If t denotes the realized value of the test statistic, and assuming that large values of T represent evidence against H_0, the p-value is the probability of observing a test statistic that is at least as extreme as t,

$$p = P_{H_0}(T \geq t)\,. \tag{4.20}$$

The bootstrap p-value is then

$$\frac{\text{Number of } t_{(j)}^* \geq t}{B+1}\,. \tag{4.21}$$

Note that $B + 1$ is used in the denominator because the observed t is included in the total number of test statistics calculated.

By way of example, Exercise 4.9 provides step-by-step instructions for bootstrapping the likelihood ratio test statistic in the context of testing the null hypothesis that the Old-Faithful waiting times are distributed according to a single normal distribution.

4.5.5 Bootstrap pragmatics

In practice, it can happen (more often then you would like) that the bootstrap simulation will not finish due to a computational error. This typically occurs when one of the $\boldsymbol{y}^*$ is sufficiently 'unusual' that the optimizer struggles to find the MLE. Such errors can often be fixed by better specification of starting parameter values for the optimizer, and by placing explicit bounds on the parameter space to prevent negative variance parameters or probabilities outside of the unit interval (where relevant). It may also be necessary to impose bounds to prevent the optimizer from venturing into regions of the parameter space where calculation of the log-likelihood would result in numerical underflow or overflow. Such bounds must not be allowed to affect the ability of the optimizer to find $\widehat{\boldsymbol{\theta}}^*$.

> Experience has taught that a SAS macro will usually continue to run a bootstrap simulation even if an error message is produced from a failed attempt to fit the model to one or more $\boldsymbol{y}^*$. These errant fits can later be investigated on a case-by-case basis. In R, the bootstrap simulation will often come to a premature end unless the error is trapped. This can be done using the `try` function.

Box 4.7

4.5.6 Bootstrapping Old Faithful

The waiting times of the Old Faithful geyser (Figure 2.5) were modelled as iid from a binormal mixture distribution in Section 3.3.4, where Wald CIs for the five parameters were computed. Likelihood ratio CIs were subsequently obtained in Section 3.4.1. Here, the parametric and nonparametric bootstraps are applied to these data.

4.5.6.1 Using R

In the code below, the convergence code from the `optim` function was checked at each iteration to ensure that the optimizing algorithm had successfully converged. This code is integer-valued and takes the value 0 if successful convergence occurred, the value 1 if the iteration limit is reached, and other positive values corresponding to a variety of potential problems with the optimization. Several per cent of the optimizations returned a convergence code of 1 using the default iteration limit of 500. With the iteration limit increased to 5 000, all ten thousand bootstrap optimizations returned a convergence code of 0. The ten thousand bootstraps took a few minutes on a 3GHz computer.

This code is for a parametric bootstrap, and the variable `b=rbinom(n,1,MLE[1])` is the unobserved Bernoulli random variable that is used in the description of the binormal mixture model in Section 2.4.5. At each iteration, the value of $\widehat{\boldsymbol{\theta}}^*_{(j)}$ is saved in row j of the $10\,000 \times 5$ matrix `PBootstrapMLEs`.

```
> ###Parametric bootstrap of binormal model for Old Faithful
> nboots=10000
> PBootstrapMLEs=matrix(NA,nrow=nboots,ncol=length(MLE))
> colnames(PBootstrapMLEs)=parnames
> ConvergenceCode=NULL
> n=length(waiting)
> for(i in 1:nboots) {
+   b=rbinom(n,1,MLE[1])
+   ystar=b*rnorm(n,MLE[2],MLE[3])+(1-b)*rnorm(n,MLE[4],MLE[5])
+   fitstar=optim(c(0.5,55,5,80,5),nllhood,y=ystar,hessian=T,
+                 control=list(maxit=5000))
+   PBootstrapMLEs[i,]=fitstar$par
+   ConvergenceCode[i]=fitstar$conv
+ }

> #Check convergence codes
> table(ConvergenceCode)
ConvergenceCode
    0
10000
```

The bootstrap standard errors and percentile method 95 % CIs are given by the following code.

```
> std.err=sd(PBootstrapMLEs)
> BStrap.CI=apply(PBootstrapMLEs,2,quantile,prob=c(0.025,0.975))
> round(cbind(MLE,std.err,t(BStrap.CI)),3)
         MLE std.err   2.5%  97.5%
p      0.361   0.031  0.300  0.422
mu    54.615   0.710 53.277 56.060
sigma  5.870   0.544  4.821  6.948
nu    80.092   0.492 79.109 81.043
tau    5.869   0.388  5.107  6.636
```

It is informative to look at histograms of the bootstrapped values (Figure 4.6) because, assuming (4.16) is a reasonable approximation, the sampling distribution of $\widehat{\theta}^*_k$ provides insight into the sampling distribution of $\widehat{\theta}_k$.

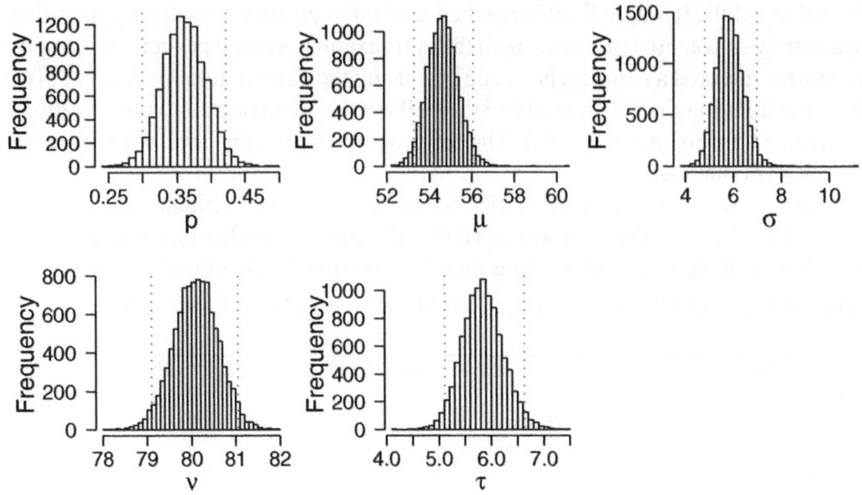

Figure 4.6 Bootstrap MLEs from ten thousand parametric bootstraps of the Old Faithful waiting time data. Vertical dashed lines show the 0.025 and 0.975 quantiles, and hence show the approximate 95 % confidence intervals obtained using the percentile method.

The nonparametric bootstrap is implemented as above, but with $\boldsymbol{y}^*$ generated by sampling with replacement from $\boldsymbol{y}$ using the code

```
ystar=sample(waiting,replace=T)
```

The nonparametric bootstrap estimates of the standard errors and confidence intervals are little different from those obtained using the parametric bootstrap, and are not shown.

4.5.6.2 Using SAS

Bootstrapping in SAS can be implemented using the SAS macro language (and bootstrap macros are available for download from SAS support at http://support.sas.com, however the code will be simpler and quicker to execute if macros can be avoided. In the SAS implementation below, the data step produces ten thousand replicate parametric bootstrap samples $\boldsymbol{y}^*$ and these are saved in a dataset containing 2 720 000 rows. The dataset has two variables, `Replicate`, and `waiting` (the simulated waiting times). `PROC NLMIXED` then processes this dataset, replicate by replicate, by virtue of the `BY Replicate` statement. The output is shown in Figure 4.7.

Parameter	the standard deviation, Estimate	CI lower limit	CI upper limit
p	0.03093	0.2999	0.4227
mu	0.69676	53.3026	56.0636
sigma	0.54705	4.8098	6.9624
nu	0.50134	79.0676	81.0620
tau	0.38796	5.1190	6.6532

Figure 4.7 Results from ten thousand parametric bootstraps of the binormal model fitted to the Old Faithful waiting times.

```
*Generate the parametric bootstrap data;
DATA ParBootstraps (KEEP=Replicate waiting);
  ARRAY MLE {5} p mu sigma nu tau (0.361 54.615 5.871 80.091 5.868);
  DO Replicate=1 TO 10000;
    DO i=1 TO 272;
      b=RANBIN(0,1,p);
      waiting=b*(mu+sigma*RANNOR(0))+(1-b)*(nu+tau*RANNOR(0));
      OUTPUT;
    END;
  END;
RUN;

ODS OUTPUT ParameterEstimates=PBootPars;
PROC NLMIXED DATA=ParBootstraps;
  BY Replicate;
  PARMS p=0.5 mu=55 sigma=5 nu=80 tau=5;
  BOUNDS 0<p<1, 0<sigma, 0<tau;
  ll=log( p*PDF("NORMAL",waiting,mu,sigma)+
          (1-p)*PDF("NORMAL",waiting,nu,tau) );
  MODEL waiting ~ GENERAL(ll);
RUN;

PROC UNIVARIATE DATA=PBootPars NOPRINT;
  CLASS parameter (ORDER=DATA);
  VAR estimate;
  OUTPUT OUT=BootStrapCIs STD=Std_Error PCTLPTS=2.5 97.5 PCTLPRE=Q;
RUN;

PROC PRINT DATA=BootStrapCIs NOOBS LABEL;
  LABEL Q2_5="CI lower limit" Q97_5="CI upper limit";
RUN;
```

The nonparametric bootstrap is accomplished by replacing the `DATA` step in the above code by the following call to the `SURVEYSELECT` procedure.

```
*Generate the nonparametric bootstrap data;
PROC SURVEYSELECT DATA=OldFaithful OUT=NonParBootstraps REP=10000
  METHOD = URS /*Unrestricted random sampling (with replacement)*/
  SAMPRATE = 1 /*Keep size of orginal data, n=272*/
  OUTHITS; /*Output repeated values*/
RUN;
```

4.5.7 How many bootstrap simulations is enough?

The number of bootstrap simulations, B, must be large enough that the estimated standard errors or quantiles are not materially affected by the inherent randomness of the bootstrap. That is, they should be negligibly different if the bootstrap were to be repeated.

In particular, quantiles in the extreme tails of a distribution require a large sample size to be estimated well. For example, for a sample size of 100, the 0.025 quantile can be calculated as the average of the 2nd and 3rd smallest observations and hence will be very sensitive to the random occurrence of just two or three extremely small values. Indeed, Davison and Hinkley (1997) recommend performing at least 1 000 bootstrap simulations when calculating 95 % bootstrap confidence intervals.

In practice it is a good idea to perform several tentative bootstrap runs with a modest value of B and to examine the variability in the observed quantiles. The standard error of the quantiles will decrease at a rate of $\sqrt{B}$, and so it will be possible to determine a suitable value of B for the full bootstrap simulation. This approach can be applied post-bootstrap. For example, a bootstrap consisting of B simulations can be partitioned into K subsets of size B/K, and the quantiles calculated for each subset. The sample standard deviations of the quantiles over the K subsets gives an unbiased estimate of the standard deviation of bootstrap quantiles from a bootstrap run of length B/K. Dividing these standard deviations by $\sqrt{K}$ then gives an estimate of the standard error of the quantiles from the full bootstrap of length B. The R code below applies this approach to the results of the parametric bootstrap of Old Faithful, by splitting the 10 000 bootstraps into ten subsets of 1 000. Equivalent SAS code is available at `http://www.stat.auckland.ac.nz/~millar`.

```
> K=10
> index=rep(1:K,nboots/K)
> BStrapQuantileSEs=matrix(NA,nrow=5,ncol=2,
+         dimnames=list(parnames,c("2.5% s.e.","97.5% s.e.")))

> for(i in 1:5) {
+   splits=split(PBootstrapMLEs[,i],index)
+   x=sapply(splits,quantile,prob=c(0.025,0.975))
+   BStrapQuantileSEs[i,]=apply(x,1,sd)/sqrt(K) }

> round(BStrapQuantileSEs,4)
      2.5% s.e. 97.5% s.e.
```

p	0.0007	0.0010
mu	0.0115	0.0122
sigma	0.0146	0.0168
nu	0.0124	0.0105
tau	0.0097	0.0141

The parametric bootstrap using R estimated an approximate 95 % confidence interval of $(53.277, 56.060)$ for μ. Under repetition of the bootstrap, the lower and upper confidence limits are estimated to have standard errors of 0.0115 and 0.0122, respectively. To avoid quoting excessive and potentially unreliable significant digits, it would be prudent to state the bootstrap confidence interval as $(53.3, 56.1)$.

4.6 Prediction

So far, interest has focused on estimation and inference about parameters $\boldsymbol{\theta}$ and functions of those parameters $g(\boldsymbol{\theta})$, where these are fixed (but unknown) quantities. However, there may be occasions when the question of interest requires inference about a random quantity. It is usual to refer to this as a *prediction* problem, to distinguish it from estimation of a fixed unknown quantity.

In many cases, the random quantity to be predicted will be a future observation of the response variable. For example, it could be tomorrow's stock market price of a company share, or the number of animals in a population next year under a particular management scheme, or the number of customers that will arrive to be served. Prediction also arises naturally in the context of models having unobserved random effects (Chapter 10) when it is desired to predict the value of those effects. Unlike a future observation, these so-called latent variables are never directly observed.

Let Z denote the random quantity (possibly vector valued) to be predicted. Our goal is to make statements about Z that have sound frequentist properties. In particular, if $Z \in \mathbb{R}$ then it is natural to calculate a $(1 - \alpha)100\,\%$ *prediction interval* for Z. This is defined analogously to a confidence interval, that is, under hypothetical repetition of the experiment that generates both the observed data $\boldsymbol{y}$ and a realization z of the random variable Z, in the long run approximately $(1 - \alpha)100\,\%$ of the prediction intervals will contain the value of z that was realized.

In special cases, such as prediction of a new observation from a linear regression model, it is possible to obtain exact prediction intervals. This can be done in R using the `predict.lm` function, and in SAS via the `PREDICT` option in the `OUTPUT` statement of `PROC GLM` or `PROC REG`. Example 4.9 shows an example of an exact prediction interval for the iid normal model.

Example 4.9. Prediction in the IID normal model. Let $Y_i, i = 1, ..., n$ be distributed iid $N(\mu, \sigma^2)$, and suppose that a $(1 - \alpha)100\,\%$ prediction interval is required for an additional iid observation Y_{n+1}. Now, the sample mean $\overline{Y}$, and

Y_{n+1} are independent, and so

$$\overline{Y} - Y_{n+1} \sim N\big(0, (1 + 1/n)\sigma^2\big) ,$$

and hence

$$\frac{\overline{Y} - Y_{n+1}}{S\sqrt{1 + \frac{1}{n}}} \sim t_{n-1} , \tag{4.22}$$

where $S^2 = \sum_{i=1}^{n}(Y_i - \overline{Y})^2/(n-1)$ is the usual sample variance estimator. Then

$$P\left(-t_{n-1,1-\alpha/2} \le \frac{\overline{Y} - Y_{n+1}}{S\sqrt{1 + \frac{1}{n}}} \le t_{n-1,1-\alpha/2}\right) = 1 - \alpha ,$$

and hence $\overline{Y} \pm t_{n-1,1-\alpha/2} S\sqrt{1 + \frac{1}{n}}$ gives a $(1-\alpha)100\,\%$ prediction interval for Y_{n+1}. □

In Example 4.9 the calculation of an exact predictive interval is possible because the expression on the left side of (4.22) is a pivotal statistic, that is, its distribution does not depend on any unknown parameters. However, in the general case such constructions are often not possible. In general, Z may not be independent of $\boldsymbol{Y}$, and since $\boldsymbol{Y} = \boldsymbol{y}$ has been observed, the density of interest is the conditional density $f_Z(z\,|\,\boldsymbol{y};\boldsymbol{\theta})$. When Z is independent of $\boldsymbol{Y}$, as with prediction of a future independent observation (e.g. Example 4.9), then of course $f_Z(z\,|\,\boldsymbol{y};\boldsymbol{\theta}) = f_Z(z;\boldsymbol{\theta})$.

The challenge for prediction is that $f_Z(z\,|\,\boldsymbol{y};\boldsymbol{\theta})$ depends on the unknown $\boldsymbol{\theta}$. It is tempting to use the so-called *plug-in* approach, which uses the predictive density $f_Z(z\,|\,\boldsymbol{y};\widehat{\boldsymbol{\theta}})$, but this fails to take into account the additional variability caused by using the MLE in place of $\boldsymbol{\theta}$. This approach, and four others, are presented below. However, of these five approaches, it is only the Bayesian, pseudo-Bayesian, and bootstrap prediction methods that can be recommended in practice.

4.6.1 The plug-in approach

The plug-in approach simply replaces $\boldsymbol{\theta}$ with the MLE, so that predictive inference about Z is obtained using the density $f_Z(z\,|\,\boldsymbol{y};\widehat{\boldsymbol{\theta}})$. In some cases the desired inference may be available in closed form. For example, if the plug-in approach was applied to the iid normal model, then the 95 % prediction interval for a future observation would simply be $\overline{y} \pm 1.96\,\widehat{\sigma}$. More generally, the plug-in distribution of Z can be examined by randomly generating values from $f_Z(z\,|\,\boldsymbol{y};\widehat{\boldsymbol{\theta}})$.

The plug-in approach is often also called the *estimative* or the *naive* approach. The latter appellation is due to the fact that it takes no account of the sampling variability in $\widehat{\boldsymbol{\theta}}$. Consequently, the predictions will tend to have too little variability and so prediction intervals will tend to have coverage that is too small.

One dramatic example of the danger of the plug-in approach was provided by Ludwig (1996) in a comparison of the plug-in approach versus fully Bayesian methodology for predicting the extinction probability of several bird species. In the case of Palila (*Loxioides balleui*), an endangered finch-billed honeycreeper, the plug-in approach predicted an extinction probability of less than 0.3 %, compared to more than 17 % from the Bayesian forecasts. This difference was largely attributed to the plug-in approach not including the uncertainty in the estimated annual variability of the population dynamics equation. However, Ludwig (1996) did not examine the sensitivity of the Bayesian forecasts to specification of alternative prior distributions for the model parameters, and did not include pseudo-Bayes or bootstrap prediction in his comparison.

4.6.2 Predictive likelihood

The general idea behind predictive likelihood is to consider the realized (but unobserved) value z as a fixed quantity, and to include it as an additional parameter via an extended definition of likelihood (Bjørnstad 1996, Lee *et al.* 2006). However, this approach is computationally challenging and not widely used, and is not without controversy. Indeed, Hall, Peng and Tajvidi (1999) argued that predictive likelihood did not improve on the naive approach, and instead recommended using a bootstrap approach to prediction.

4.6.3 Bayesian prediction

Under the Bayesian paradigm, both Z and $\boldsymbol{\theta}$ are random unobserved quantities. Realizations of Z can be generated in two steps:

1. Generate $\boldsymbol{\theta}$ from its posterior density, $\pi(\boldsymbol{\theta}|\boldsymbol{y})$.
2. Generate z from $f_Z(z|\boldsymbol{\theta}, \boldsymbol{y})$ (i.e. the predictive density given $\boldsymbol{y}$ and $\boldsymbol{\theta}$), where $\boldsymbol{\theta}$ is obtained from step 1.

An increasing number of software packages are providing Bayesian enhancements to functions and procedures that perform maximum likelihood estimation and inference. For example, the generalized linear model procedure in SAS (`PROC GENMOD`) includes a `BAYES` statement, and the R package `lme4` for mixed-effects models includes an `mcmcsamp` function which uses the Markov chain Monte-Carlo (MCMC) sampling method for generating $\boldsymbol{\theta}$ from its posterior distribution. Also, an ADMB executable can be switched into 'Bayesian mode' using the `-mcmc` command line option.

The user of these Bayesian utilities has the responsibility to read the accompanying documentation to determine what priors distributions are implicitly being placed on the parameters. By default, these will be some choice of 'non-informative' or 'reference' priors. In the case of ADMB, a flat prior is always assumed (unless the user explicitly specifies otherwise), and $\widehat{\boldsymbol{\theta}}$ therefore has the interpretation of being the mode of the posterior density $\pi(\boldsymbol{\theta}|\mathbf{y})$. It then behoves the ADMB user to ensure that the model is parameterized such that having flat prior distributions on all parameters is sensible.

4.6.4 Pseudo-Bayesian prediction

Pseudo-Bayesian prediction uses the same steps as Bayesian prediction, but in step 1 an approximation to the posterior density $\pi(\boldsymbol{\theta}|\mathbf{y})$ is used. Subject to appropriate regularity conditions and sample size, the conditional distribution of $\boldsymbol{\theta}$ given $\mathbf{y}$ can be approximated by a (multivariate) normal distribution centred at the MLE and with variance equal to the approximate variance of the MLE (Berger 1985, p. 224). That is,

$$\boldsymbol{\theta}|\mathbf{y} \stackrel{\cdot}{\sim} N_s(\widehat{\boldsymbol{\theta}}, \widehat{\mathbf{V}}) . \tag{4.23}$$

In practice, for the approximation in (4.23) to be as accurate as possible, the model should be parameterized so that likelihood profiles are close to quadratic (see Example 4.10). A formal evaluation of the accuracy of (4.23) in the context of approximating Bayesian credible intervals can be found in Severini (1994).

In the case of linear regression and generalized linear models, simulation of $\boldsymbol{\theta}$ according to (4.23) can be performed by the `sim` function in the R package `arm`. Gelman and Hill (2007) give examples of its use. More generally, simulated values from a multivariate normal can be obtained using the `mvrnorm` function in the `MASS` package of R, or using the `MVN` SAS macro available from http://support.sas.com

It should be noted that pseudo-Bayes prediction using (4.23) does not get around the issue of the dependence of Bayesian-type approaches to the assumed prior distributions on all model parameters. Rather, it sweeps it under the carpet. The approximation underlying (4.23) is able to hide the dependence on prior assumptions under the requirement of sufficiently large sample size.

4.6.5 Bootstrap prediction

Bootstrap prediction is analogous to Bayesian and pseudo-Bayesian prediction, except that the bootstrap distribution of $\widehat{\boldsymbol{\theta}}^*$ is used in place of the posterior or pseudo-posterior distribution. Harris (1989) showed that this approach performs well and is superior to the naive plug-in approach when making prediction for the class of exponential family models (defined in Section 7.2.1). The large-sample equivalence of bootstrap and Bayesian prediction is examined in Fushiki, Komaki and Aihara (2004).

Example 4.10. Binomial prediction. In Example 1.1, $y = 10$ was observed from a Bin(100, p) experiment. Suppose that it is desired to predict a future observation from this distribution.

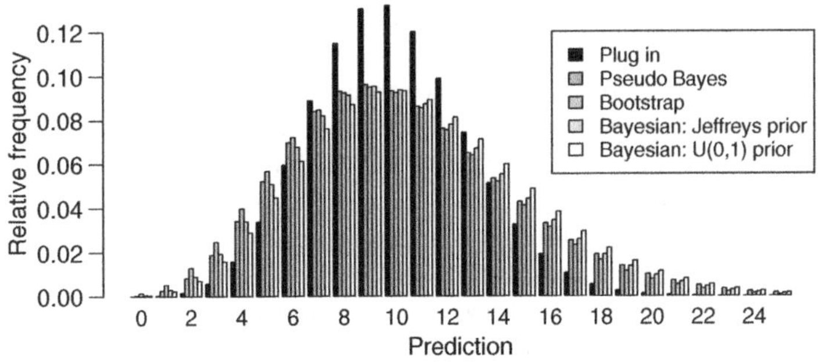

Figure 4.8 Side-by-side relative frequency histograms of ten million simulated predicted values of $Y \sim Bin(100, p)$, for five choices of prediction method.

Figure 4.8 shows a relative frequency histogram of ten million simulated predicted values of $Y \sim \text{Bin}(100, p)$, for each of five different predictors. The first three are the plug-in, pseudo-Bayes, and bootstrap, and the last two are both Bayesian predictors. The first of these uses the Jeffreys reference prior, $\pi(p) \propto p^{-1/2}(1 - p)^{-1/2}$, and the second uses the uniform prior $p \sim \text{Unif}(0, 1)$. The Jeffreys prior is the most commonly used for p, but there are plausible arguments in support of the uniform prior. Indeed, Thomas Bayes himself preferred the uniform prior on p for the purpose of predicting Y, and this choice of prior has gained further support from the work of Tuyl, Gerlach and Mengersen (2009).

The naive approach simply uses the plug-in predictive distribution, $Y \sim \text{Bin}(100, \widehat{p})$ with $\widehat{p} = 0.1$. It has much less variability compared to the other predictors, due to the failure to incorporate uncertainty in $\widehat{p}$. In the pseudo-Bayes approach, step 1 simulated $\zeta = \text{logit}(p) \sim N(\text{logit}(\widehat{p}), 1/9)$ to achieve a better normal approximation in (4.23) than would be obtained by simulating $p \sim N(\widehat{p}, 0.0009)$ (see Example 4.3 and Section 4.3.1). In this example, the parametric and nonparametric bootstrap are identical, since both are generating bootstrap samples y^* as 100 iid Bernoulli(0.1) observations, and then generating the prediction as $Y \sim \text{Bin}(100, \widehat{p}^*)$. These last four methods of prediction give similar results, with the greatest difference being that the bootstrap predicts relatively more small values of Y and relative fewer large values, in comparison to the uniform-prior Bayesian predictor. □

4.7 Things that can mess you up

4.7.1 Multiple maxima of the likelihood

For some familiar families of models (including linear regression) a unique MLE can be obtained as the unique explicit solution to the likelihood equations. In the case of generalized linear models, the MLE can not in general be obtained explicitly, however it has been proved that the MLE is unique under very general conditions (Wedderburn 1976).

In general, when implementing a novel model there may be no guarantee of the existence or uniqueness of an MLE. Thus, it may be prudent to explore the shape of the likelihood function using profile likelihood and contour plots. When using numeric optimizers to find an MLE, it is always good practice to repeat the optimization using several different starting values to see whether the optimizer will converge to different local maxima.

Multiple maxima of the likelihood is not necessarily problematic. For example, in the binormal model in Section 2.4.5 it was seen that multiple maxima arose due to non-identifiability, but that this could be negated by clearer specification of the parameter space. This model was also seen to have unbounded likelihood, but this did not preclude a sensible local maximum being obtained when this model was fitted to the Old Faithful data in Section 3.3.4.

4.7.2 Lack of convergence

If the chosen optimizing algorithm does not appear to converge then the easiest fix is to use better starting parameter values $\boldsymbol{\theta}^{(0)}$ for the optimizing algorithm, perhaps by inspecting the data using suitable exploratory plots. In high-dimensional models it can be difficult to ascertain good initial values for all parameters, but it may be possible to obtain good starting values for key parameters from fitting a cruder or simpler form of the desired model. Automatic functionality for this type of approach is included within the ADMB software (see Section 5.4.2).

Another possibility is to implement a more robust optimizer that is less sensitive to the starting values. The EM algorithm of Section 5.3 is such an example (where applicable), though it could require some effort to implement.

On occasion, the optimizer may not attain convergence due to reaching the default maximum number of iterations, and increasing this default may be all that is required. It can also be the case that the default convergence criterion is too precise (especially if calculation of the log-likelihood has low numerical precision), and needs to be set to a more appropriate level.

4.7.3 Parameters on the boundary of the parameter space

In Chapter 12, several regularity conditions are required in order to proceed with the proofs of the asymptotic distribution of MLEs and the likelihood ratio test statistic. One of these conditions, R4, prevents $\boldsymbol{\theta}_0$ from being on the boundary of the parameter space.

To see the problem that arises with parameters on the boundary of Θ, consider the case where the parameter space is the non-negative reals, $\Theta = [0, \infty)$, and suppose that $\theta_0 = 0$. Since $\widehat{\theta} \in \Theta$, then $\widehat{\theta}$ can never be less than θ_0, and hence its distribution can not be approximated by a normal distribution centred at $\theta_0 = 0$. It is also the case that the LRT statistic for $H_0 : \theta = \theta_0$ can no longer be assumed to have an asymptotic χ^2_1 distribution.

The boundary phenomenon is something to watch out for when working with mixture models (Example 4.11) and models that include multiple variance components (e.g. Section 10.4). In the latter, the boundary issue arises because variances must, of course, be non-negative (Example 4.12).

Example 4.11. Mixture model. The binormal mixture model that was used to describe waiting times of the Old Faithful geyser (Example 2.9) has a five-dimensional parameter vector $\boldsymbol{\theta} = (p, \mu, \sigma, \nu, \tau)$. Consider the null hypothesis $H_0 : p = 0$.

At first glance the null hypothesis appears to place a one-dimensional restriction on the parameter space. However, under H_0 the waiting times are from a single normal distribution, and so it could be argued that the restriction is of three dimensions. In fact, due to the unusual geometry of the parameter space under H_0, the distribution of the LRT statistic is complex, and has been shown to be close to that of a χ^2_6 random variable (McLachlan 1987, Lo 2005). See also Exercise 4.9. □

Example 4.12. Variance components model. The linear random-effects model in Section 10.4 has a variance component, σ_u^2, in addition to the residual variance σ^2. The LRT statistic of $H_0 : \sigma_u^2 = 0$ has an asymptotic distribution that takes the value 0 with probability 0.5, and otherwise is distributed as a χ^2_1 random variable (Stram and Lee 1994). The p-value for any positive LRT statistic is therefore one half of that obtained from the usual comparison with a χ^2_1 distribution, and corresponds to the usual adjustment for a one-sided test.

An intuitive explanation of this asymptotic distribution is that the LRT statistic is (asymptotically) χ^2_1 distributed whenever the MLE $\widehat{\sigma}_u^2$ is positive, because then the boundary constraint is immaterial. This corresponds to overestimation of σ_u^2 ($= 0$ under H_0) and is expected to happen about half the time under repetition of the experiment. The other half of the time the MLE is $\widehat{\sigma}_u^2 = 0$, and the LRT statistic is necessarily zero. □

As demonstrated in Examples 4.11 and 4.12, the theoretical behaviour of the LRT statistic when $\boldsymbol{\theta}_0$ is on the boundary of Θ has been established for some classes of model (e.g. Self and Liang 1987, Stram and Lee 1994, Lo 2005), However, except in special classes of models, these theoretical results can be difficult to apply, and the asymptotic distributions can be poor approximations to the true sampling distribution for moderate sample sizes (Pinheiro and Bates 2000, Crainiceanu and Ruppert 2004). Crainiceanu and Ruppert (2004) obtained an exact representation of the finite-sample LRT statistic for linear mixed models with a single variance component (in addition to the residual variance). This has been implemented in R, including an extension to a wider class of linear mixed models, by Scheipl *et al.* (2008). Application of this test is demonstrated in Section 10.4.2.

Bootstrap simulation (Section 4.5) provides a general strategy for assessing the sampling distribution of $\widehat{\boldsymbol{\theta}}$, and other statistics such as the likelihood ratio test statistic. Exercise 4.9 applies the bootstrap to the LRT statistic in the context of the boundary problem used in Example 4.11. Section 10.6.1 includes a bootstrap examination of the LRT statistic for testing the null hypothesis that a variance component is zero in the context of a generalized linear mixed model.

Burnham and Anderson (2002, Section 6.9.6 therein) consider an example of the use of AIC for comparison of models in the presence of boundary constraints, and argue that AIC does not require modification. Essentially, they note that AIC is not a formal testing procedure, and is not affected by ambiguities regarding the number of dimensions restricted by a parameter on the boundary (e.g. Example 4.11). So, for example, an iid binormal mixture model would use $s = 5$ parameters in calculation of the AIC, compared to $s = 2$ for the iid normal model.

4.7.4 Insufficient sample size

The sample size is something that should be given consideration at the experimental design stage, but in practice it is often dictated by available resources.

Some protection against 'insufficient' sample size can be gained by using a sensible model parameterization (see Section 4.3.2) if relying on the approximate normality assumed by Wald-based inference. A more formal alternative would be to explore the field of small-sample asymptotic theory (Ronchetti 1990). This oxymoron alludes to the property that these methods attempt to reduce 'sufficiently large sample size' to a relatively small number. However, the complexity of these methods has limited their practical use.

Simulation can be useful to assess the sampling distribution of MLEs and other statistics. This could be based on a bootstrap, say. This can be effective for quite small sample sizes, but there are limits because an underlying assumption of the bootstrap is that the distribution of the simulated data is emulating the sampling distribution of the real observed data. For example, in the case of the nonparametric bootstrap, because of sampling with replacement, the simulated data are effectively being sampled from a population that consists of an infinite number of copies of the observed data. If in doubt, it may be better to perform parametric simulation under a range of plausible values for $\boldsymbol{\theta}$.

Be aware that a seemingly large sample size may not be sufficient if there are relatively many parameters. This can be especially dangerous in situations where the number of parameters in the model increases with increasing sample size. Methodology for coping with this nonstandard situation is found in Chapters 9 and 10.

4.8 Exercises

4.1 Show that $X = \sin^{-1}\sqrt{\widehat{p}} = \sin^{-1}\sqrt{Y/n}$ is the variance stabilizing transformation for $Y \sim \text{Bin}(n, p)$. That is, show that the approximate variance of X does not depend on p.

4.2 In Example 2.3, the MLE of $P(Y \leq 6)$ was seen to be $g(\widehat{\theta}) = \Phi((6 - \widehat{\mu})/\widehat{\sigma})$ where Φ is the distribution function of the standard normal and $\widehat{\theta} = (\widehat{\mu}, \widehat{\sigma}) =$

(5, 2). Suppose now that the approximate variance matrix of $\widehat{\theta} = (\widehat{\mu}, \widehat{\sigma})$ is

$$\widehat{\mathbf{V}} = \begin{pmatrix} 1 & 0 \\ 0 & 0.25 \end{pmatrix}$$

Use the delta method to obtain the approximate standard deviation of $g(\widehat{\theta})$. *Hint:* Note that Φ is implemented as the `pnorm` function in R, and `PROBNORM` function in SAS.

4.3 For the two-parameter model used in Section 3.5.1 the MLE is $(\widehat{\beta}_1, \widehat{\beta}_2) = (-10.632, 0.304)$ and its approximate variance matrix is

$$\widehat{\mathbf{V}} = \begin{pmatrix} 0.7477 & -0.02022 \\ -0.02022 & 0.0005584 \end{pmatrix}.$$

In the fisheries trawling context of this example, the parameters of interest are $l_{50\,\%} = -\beta_1/\beta_2$ and $SR = 2\log(3)/\beta_2$. These correspond to the length of 50 % retention probability, and the difference between the lengths of 75 % and 25 % retention (the so-called 'selection range'), respectively. Use the delta method to determine the approximate distribution of the MLE $g(\widehat{\beta}_1, \widehat{\beta}_2) = (-\widehat{\beta}_1/\widehat{\beta}_2, 2\log(3)/\widehat{\beta}_2)$, and hence calculate approximate 95 % Wald CIs for $l_{50\,\%}$ and SR. (This model is the subject of the case study in Section 7.5.)

4.4 Use the delta theorem to determine the approximate variance of $\widehat{\theta}_1/\widehat{\theta}_2$ (as a function of $\widehat{\theta}_1$, $\widehat{\theta}_2$, $\text{var}(\widehat{\theta}_1)$, $\text{var}(\widehat{\theta}_2)$ and $\text{cov}(\widehat{\theta}_1, \widehat{\theta}_2)$).

4.5 1. Verify (4.9) using one (or both) of the following approaches:

   (a) Using (13.3), or

   (b) by an exact Taylor series expansion (it includes one second order term) of $g(\widehat{\theta}_1, \widehat{\theta}_2)$ about $g(E[\widehat{\theta}_1], E[\widehat{\theta}_2])$.

   2. Verify (4.10).

4.6 In the Old Faithful example (Section 3.3.4), calculate the 95 % Wald CI for $\zeta = \mu - \nu$.

4.7 **Batch sampling.** Randomly selected soil samples are combined into batches and then the batches are tested for the presence or absence of a toxin. Note that individual soil samples are not tested, and the measured observation is whether or not the batch contained toxin (hence the name batch sampling). From 100 batches of 10 samples each, suppose that 12 tested positive.

   1. Calculate the MLE of p, where p is the probability that toxin is present in an individual sample of soil.

2. Determine the approximating normal distribution of $\widehat{p}$ and calculate an approximate 95 % Wald CI for p.

3. Determine the approximate 95 % likelihood ratio CI for p.

Hint: You may find it easiest to parameterize the likelihood using the parameter $\theta = 1 - (1 - p)^{10}$.

4.8 Assuming $\theta \in \mathbb{R}$, show that the Wald and LR test statistics for $H_0 : \theta = \theta_0$ are identical when the log-likelihood is quadratic.

4.9 For the binormal mixture model fitted to the Old Faithful waiting time data used in Section 4.5.6, implement a parametric bootstrap simulation to assess the behaviour of the likelihood ratio test statistic under the null hypothesis $H_0 : p = 0$. That is,

1. Simulate data $\boldsymbol{y}^*_{(j)}$, $j = 1, ..., 1\,000$ from the fitted model under H_0. Note that the data are iid normal under H_0.

2. For each set of simulated data, fit an iid normal model and an iid binormal model, and hence calculate the LRT statistic for H_0.

3. Compare the actual value of the LRT to the simulated values.

4. Plot the empirical distribution function of your simulated LRT statistics, and overlay the distribution function of a χ^2_6 random variable (by way of example, see Figure 2 of McLachlan (1987)).

Note: for some simulated data, $\boldsymbol{y}^*_{(j)}$, it may be that the optimizer will not find a sensible local MLE. Your simulation will need to include the flexibility to manage this possibility should it arise.

5

Maximizing the likelihood

Arriving at one point is the starting point to another – John Dewey

5.1 Introduction

In the general case, finding an MLE will require numerical maximization of the log-likelihood. For classes of model that are well established, there will be a reliable and well documented R function or SAS procedure to perform the maximization (and provide additional inference). For less standard models, it is often fruitful to do a quick search of the CRAN (Comprehensive R Archive Network, `http://cran.r-project.org`) to see if its implementation has been made available in an R package. However, the user should be aware that R packages are not subject to formal quality control and it is not unknown to get substantively different results from using two different R packages that purport to fit the same model. The R mailing lists (`http://www.r-project.org`) can be informative for disclosing the experiences that users have voiced about various packages. For example, the `R-sig-mixed-models` mailing list is devoted to discussion of R packages for fitting mixed-effects models (Chapter 10).

If no existing software tools are available then model-specific optimization of the log-likelihood may be required, perhaps via SAS procedures `NLMIXED` or `NLP`, or the R function `optim`. In tougher cases, including very high-dimensional models or models having likelihood that can not be expressed in closed-form (e.g. Chapter 10), the capabilities of ADMB may be needed (e.g. Section 5.4.2).

For some classes of models there are specialized optimization approaches that are able to take advantage of certain structure that is known to be inherent

Maximum Likelihood Estimation and Inference: With Examples in R, SAS and ADMB, First Edition. Russell B. Millar.

in the likelihood. For example, generalized linear models are often fitted using iteratively re-weighted least squares, as is the case with the `glm` function in R. The `VGAM` package in R has extended the application of iteratively re-weighted least squares to a much wider range of models, including an assortment of zero-inflated models for count data. It has been seen that simple mixture models, such as the binormal mixture first encountered in Example 2.9, can quickly be fitted using `NLMIXED` or `optim` (Section 3.3.4). However, these general-purpose optimizers are unlikely to be successful when applied to mixture models with many component distributions, especially if the component distributions are not well distinguished. For this purpose the EM (expectation maximization) algorithm (Section 5.3) is better suited.

In SAS 9.2, the `NLMIXED` procedure uses a form of the quasi-Newton optimization algorithm (a variant of the Newton-Raphson algorithm, Section 5.2) by default. Such algorithms work well and find the MLE quickly provided that the log-likelihood is sufficiently smooth. They attain this performance by utilizing the first and second derivatives of the log-likelihood. `NLMIXED` also provides several alternatives including conjugate gradient, double dogleg, Nelder-Mead, standard Newton-Raphson, and ridge-stabilized Newton-Raphson, and the choice can be specified using the `TECHNIQUE=` procedure option. The `NLP` procedure provides an even wider choice of algorithms.

In R, function `optim` uses the Nelder-Mead method by default. The Nelder-Mead algorithm is derivative free and uses an algorithm based on local extrapolation based on the log-likelihood evaluated at a set of neighbouring points. Other possibilities include conjugate gradient, quasi-Newton and simulated annealing. These are chosen using the `method=` argument. See Nocedal and Wright (2006) for more detailed description of optimization algorithms.

When using Newton-Raphson type algorithms, most optimizers are able to approximate the required first and second derivatives numerically. Some, including `optim` and `NLP`, provide the facility for the user to provide one or both of these derivatives as analytical formulae. In `optim` this is via the `gr=` option, and in `NLP` through use of the `GRADIENT` and `HESSIAN` statements. ADMB uses a quasi-Newton optimizer, and exact (to computer precision) derivatives are algorithmically generated using automatic differentiation (see Section 15.6).

Section 5.4 looks at maximizing the likelihood in stages, and it is seen that the use of profile likelihood can sometimes be an extremely efficient method for reducing the dimensionality of a numerical optimization (Section 5.4.1). In high-dimensional problems it can be a challenge to obtain good starting values for an optimizer, and it is seen that one way to overcome this is to build up to maximization of the log-likelihood in steps (Section 5.4.2). Each step, other than the last, maximizes a reduced form of the log-likelihood for the purpose of proving good starting parameter values for the next step. The final step maximizes the full log-likelihood using the starting parameter values obtained from the penultimate step. This chapter concludes with a simple example showing the functionality of ADMB for implementing such multi-stage optimization.

5.2 The Newton-Raphson algorithm

The Newton-Raphson algorithm is based on quadratic approximation of the objective function (the log-likelihood), via linear approximation of its derivative. If $\boldsymbol{\theta} \in \mathbb{R}^s$ then this first-order partial derivative is the s-dimensional vector $l'(\boldsymbol{\theta}) = \frac{\partial l(\boldsymbol{\theta};\mathbf{y})}{\partial \boldsymbol{\theta}}$. Given the parameter value $\boldsymbol{\theta}^{(k)}$ at iteration k, the Newton-Raphson algorithm approximates $l'(\boldsymbol{\theta})$ using a Taylor series expansion about $\boldsymbol{\theta}^{(k)}$,

$$l'(\boldsymbol{\theta}) \approx l'(\boldsymbol{\theta}^{(k)}) + \mathbf{H}(\boldsymbol{\theta}^{(k)})(\boldsymbol{\theta} - \boldsymbol{\theta}^{(k)}) , \tag{5.1}$$

where $\mathbf{H}(\boldsymbol{\theta}^{(k)})$ is the $s \times s$ Hessian matrix of second-order partial derivatives of $l(\boldsymbol{\theta})$ evaluated at $\boldsymbol{\theta}^{(k)}$.

An MLE is typically obtained as a solution of the likelihood equation, $l'(\widehat{\boldsymbol{\theta}}) = \mathbf{0}$. Thus, evaluating (5.1) at $\boldsymbol{\theta} = \widehat{\boldsymbol{\theta}}$ yields

$$\mathbf{0} \approx l'(\boldsymbol{\theta}^{(k)}) + \mathbf{H}(\boldsymbol{\theta}^{(k)})(\widehat{\boldsymbol{\theta}} - \boldsymbol{\theta}^{(k)}) . \tag{5.2}$$

The above equation is linear with respect to $\widehat{\boldsymbol{\theta}}$ and, assuming that the Hessian is invertible, has the explicit solution

$$\widehat{\boldsymbol{\theta}} \approx \boldsymbol{\theta}^{(k)} - \mathbf{H}(\boldsymbol{\theta}^{(k)})^{-1} l'(\boldsymbol{\theta}^{(k)}) . \tag{5.3}$$

The right-hand side of (5.3) can be used to update $\boldsymbol{\theta}^{(k)}$, and this leads to the Newton-Raphson algorithm

$$\boldsymbol{\theta}^{(k+1)} = \boldsymbol{\theta}^{(k)} - \mathbf{H}(\boldsymbol{\theta}^{(k)})^{-1} l'(\boldsymbol{\theta}^{(k)}) . \tag{5.4}$$

It should be noted that the Newton-Raphson algorithm in (5.4) provides no guarantees that $l(\boldsymbol{\theta}^{(k+1)})$ will be greater than $l(\boldsymbol{\theta}^{(k)})$, or even that $\boldsymbol{\theta}^{(k+1)}$ will be in the parameter space Θ. So, in practice the algorithm is often used in a modified form where the increment term $-\mathbf{H}(\boldsymbol{\theta}^{(k)})^{-1} l'(\boldsymbol{\theta}^{(k)})$ is regarded as the *direction* in which to move from $\boldsymbol{\theta}^{(k)}$, but where the magnitude of the movement is optimized. This modified algorithm (see Chap 3 of Nocedal and Wright 2006) can be expressed in the form

$$\boldsymbol{\theta}^{(k+1)} = \boldsymbol{\theta}^{(k)} - \lambda_k \mathbf{H}(\boldsymbol{\theta}^{(k)})^{-1} l'(\boldsymbol{\theta}^{(k)}) ,$$

where $\lambda_k \in \mathbb{R}$ is the value that maximizes $l(\boldsymbol{\theta}^{(k+1)})$ over all possible values of $\lambda \in \mathbb{R}$. That is,

$$l(\boldsymbol{\theta}^{(k+1)}) = \max_{\lambda} \, l\big(\boldsymbol{\theta}^{(k)} - \lambda \mathbf{H}(\boldsymbol{\theta}^{(k)})^{-1} l'(\boldsymbol{\theta}^{(k)})\big) .$$

This is a one-dimensional optimization and hence is computationally fast.

The Newton-Raphson algorithm requires the computation and inversion of the $s \times s$ hessian matrix $\mathbf{H}(\boldsymbol{\theta}^{(k)})$ at each iteration, and this can be computationally demanding in high-dimensional models. For the most widely used matrix inversion algorithms, the computational demand of inverting an $s \times s$ matrix increases with the cube of s (Press *et al.* 2007). The quasi-Newton algorithm reduces the computational demand by using first-derivative updates to the Hessian and its inverse at each iteration (Nocedal and Wright 2006).

Another variation of the Newton-Raphson algorithm is given by replacing $\mathbf{H}(\boldsymbol{\theta})$ with its expected value for classes of models in which the formulae for the expected value of this matrix are known. This variation is called Fisher's method of scoring, and takes its name from the fact that the expected value of $\mathbf{H}(\boldsymbol{\theta})$ is the negative of the expected Fisher information matrix (see Equation (11.21) in Chapter 11). Finally, note that the usual approximate variance matrix of $\widehat{\boldsymbol{\theta}}$, $\widehat{\mathbf{V}}$, is readily available from application of the Newton-Raphson algorithm, because it is just the negative of the inverse of $\mathbf{H}(\widehat{\boldsymbol{\theta}})$ (Section 3.2.1).

5.3 The EM (Expectation–Maximization) algorithm

The EM algorithm can be used in situations where the statistical model can be posed as one where the observed data $\boldsymbol{y}$ are, in some sense, 'incomplete'. That is, where it can be regarded that the experimental process generates two sets of random variables, $\boldsymbol{Y}$ and $\boldsymbol{B}$, but it is only $\boldsymbol{y}$ (the realized value of $\boldsymbol{Y}$) that is observed. A surprisingly diverse variety of models can be expressed in this way, some of which are listed in Section 5.3.2.

In the form in which it is presented in Section 5.3.1, the EM algorithm is suitable when the likelihood $L(\boldsymbol{\theta}; \boldsymbol{y})$ is unwieldy to work with (perhaps not even being expressible in closed form), but the likelihood from the 'completed' data $L(\boldsymbol{\theta}; \boldsymbol{y}, \boldsymbol{b})$ is relatively easy to maximize. Here, $L(\boldsymbol{\theta}; \boldsymbol{y}, \boldsymbol{b})$ denotes the likelihood obtained when both $\boldsymbol{Y} = \boldsymbol{y}$ and $\boldsymbol{B} = \boldsymbol{b}$ are observed. The binormal mixture model of Section 2.4.5 provides a good example (and is used throughout this section). Its likelihood is moderately complex because it is obtained as a weighted sum of the densities of the $N(\mu, \sigma^2)$ and $N(\nu, \tau^2)$ distributions. For each observation, there is conceptually an unobserved Bernoulli random variable that identifies which normal distribution generated that particular observation (see Equation (2.7)). If these Bernoulli random variables were to be observed then the model for $\boldsymbol{Y}$ would cease to be a mixture model, but rather would just be an iid $N(\mu, \sigma^2)$ model for all Y_i such that $B_i = 1$, and an iid $N(\nu, \tau^2)$ model for all Y_i such that $B_i = 0$.

The EM algorithm was brought to the widespread attention of the statistical community by Dempster, Laird and Rubin (1977), and Wu (1983) subsequently corrected some errors in that work regarding convergence of the algorithm. An extensive treatment of the EM algorithm and its subsequent extensions is provided by McLachlan and Krishnan (2008), while Navidi (1997) provides a gentler presentation using a graphical representation of the EM algorithm in the context of maximizing the likelihood of a one-parameter model.

5.3.1 The simple form of the EM algorithm

The first step of the EM algorithm requires calculating the expected value of the log-likelihood from the completed data, but it takes a much more intuitive form in many applications and it is this form that is presented here. This simplified form is appropriate when the log-likelihood of the completed data, $l(\boldsymbol{\theta}; \boldsymbol{y}, \boldsymbol{b})$, is linear in $\boldsymbol{b}$ (see Box 5.1 for generalizations).

Let $\boldsymbol{\theta}^{(k)}$ denote the value of $\boldsymbol{\theta}$ at iteration k. The simplified form of the EM algorithm consists of the following two steps:

Step 1 (Expectation). Under the completed-data model with joint density function $f(\boldsymbol{y}, \boldsymbol{b}; \boldsymbol{\theta}^{(k)})$ for $(\boldsymbol{Y}, \boldsymbol{B})$, calculate $\boldsymbol{b}^*$ as the expected value of $\boldsymbol{B}$ given $\boldsymbol{y}$. That is,

$$\boldsymbol{b}^* = E_{\boldsymbol{\theta}^{(k)}}[\boldsymbol{B}|\boldsymbol{y}] .$$

Step 2 (Maximization). Using the value of $\boldsymbol{b}^*$ calculated in Step 1, maximize the completed-data likelihood $L(\boldsymbol{\theta}; \boldsymbol{y}, \boldsymbol{b}^*) = f(\boldsymbol{y}, \boldsymbol{b}^*; \boldsymbol{\theta})$ to obtain a new parameter estimate $\boldsymbol{\theta}^{(k+1)}$. Return to Step 1.

> If the completed data $\boldsymbol{y}^{(c)} = (\boldsymbol{y}, \boldsymbol{b})$ belong to the class of exponential family distributions of the form given by (14.5) then the log-likelihood is linear with respect to sufficient statistics $T_j(\boldsymbol{y}^{(c)})$ (where T_j is defined in Exercise 14.3 for the iid case). Consequently, the above simplified EM algorithm remains valid if it is modified such that it is the expected values of $T_j(\boldsymbol{y}^{(c)})$ that are calculated in the expectation step.
>
> The general version of the EM algorithm obtains $\boldsymbol{\theta}^{(k+1)}$ by maximizing $E_{\boldsymbol{\theta}^{(k)}}[\log L(\boldsymbol{\theta}; \boldsymbol{y}, \boldsymbol{B})|\boldsymbol{y}]$. See Dempster *et al.* (1977) or Pawitan (2001) for more details.

Box 5.1

Example 5.1. Mixture of two known distributions. Recall that the model for the waiting times of the Old Faithful geyser (Example 2.9) was a mixture whereby Y_i was observed from either a $N(\mu, \sigma^2)$ or $N(\nu, \tau^2)$ distribution according to an unobserved Bernoulli random variable, B_i. Since B_i is unobserved, the data can be viewed as incomplete, and the complete data for each observation is the bivariate pair (Y_i, B_i). For simplicity, it will be assumed here that the two distributions in the mixture are of arbitrary, but known form, leaving only p to be estimated. Their density functions will be denoted f_1 and f_0, where Y_i is sampled from f_1 if $B_i = 1$, or from f_0 if $B_i = 0$.

From (2.8), the completed-data pair (Y_i, B_i) has density function

$$f(y_i, b_i; p) = \begin{cases} pf_1(y_i), & b_i = 1 \\ (1-p)f_0(y_i), & b_i = 0\,, \end{cases} \tag{5.5}$$

and hence the marginal density for Y_i is

$$f(y_i; p) = pf_1(y_i) + (1-p)f_0(y_i)\,. \tag{5.6}$$

In Step 1 of the EM algorithm, the unobserved $B_i, i = 1, ..., n$ are replaced by their expected values. These expected values are real-valued quantities in the interval [0,1], and so to implement Step 2 it is necessary to extend the completed-data likelihood in (5.5) so that it is also defined for any value of b between 0 and 1. A natural way to do this is to write the completed-data likelihood in the form

$$f(y_i, b_i; p) = \left(pf_1(y_i)\right)^{b_i}\left((1-p)f_0(y_i)\right)^{(1-b_i)}\,. \tag{5.7}$$

Note that the log of this likelihood is linear in b_i, and hence the simple form of the EM algorithm is applicable here.

EM step 1. Given $p^{(k)}$, it is necessary to calculate $b_i^* = E[B_i|y_i], i = 1, ..., n$. Since B_i is Bernoulli,

$$\begin{aligned} b_i^* = E_{p^{(k)}}[B_i|y_i] &= P_{p^{(k)}}(B_i = 1|y_i) \\ &= \frac{f(y_i, 1; p^{(k)})}{f(y_i; p^{(k)})} \\ &= \frac{p^{(k)} f_1(y_i)}{f(y_i; p^{(k)})}\,. \end{aligned}$$

EM step 2. Denoting $b^* = \sum b_i^*$, it follows from Equation (5.7) that the likelihood from the completed data is

$$\begin{aligned} L(p; \boldsymbol{y}, \boldsymbol{b}^*) &= \prod_{i=1}^{n} f(y_i, b_i^*; p) \\ &= \prod_{i=1}^{n} \left(pf_1(y_i)\right)^{b_i^*}\left((1-p)f_0(y_i)\right)^{(1-b_i^*)} \\ &= p^{b^*}(1-p)^{(n-b^*)}\prod_{i=1}^{n} f_1(y_i)^{b_i^*} f_0(y_i)^{(1-b_i^*)}\,. \end{aligned} \tag{5.8}$$

In (5.8), parameter p occurs only in the term $p^{b^*}(1-p)^{(n-b^*)}$. To within a constant, this term is equivalent to the likelihood arising from a Bin(n, p) experiment in which the value b^* is 'observed'. Here, b^* is not necessarily integer valued, but nonetheless, by analogy with the binomial likelihood it is immediate that (5.8) is maximized by $p^{(k+1)} = b^*/n$ That is, the EM algorithm reduces to the simple updating equation

$$p^{(k+1)} = \frac{b^*}{n} = \frac{1}{n}\sum_{i=1}^{n} \frac{p^{(k)} f_1(y_i)}{f(y_i; p^{(k)})} \,. \tag{5.9}$$

□

In the above example the Newton-Raphson algorithm would also be quite easy to implement. However, if the number of known components in the mixture distribution were increased to $m \geq 3$, then the EM algorithm would simply consist of $m - 1$ simple update formulae for the probabilities $p_1, p_2, ..., p_{m-1}$, all of similar form to (5.9). In comparison, the computational demand of a conventional implementation of the Newton-Raphson algorithm would increase roughly at a rate proportional to the cube of $m - 1$. Also, the Newton-Raphson algorithm may give infeasible solutions since it does not take into account that the p_j, $j = 1, ..., m - 1$ are bounded between 0 and 1. This could be circumvented by a suitable re-parameterization, such as using the multinomial logit transformation $\eta_j = \log(p_j/p_m), i = 1, ..., m - 1$, However, this parameterization can be prone to numerical instability due to probabilities close to zero, and this would be particularly problematic for moderately high m. In contrast, by construct, the EM algorithm cannot give infeasible parameter values. For example, in (5.9), if follows immediately from (5.6) that each term in the summation is between 0 and 1, and hence so too is $p^{(k+1)}$.

Example 5.2. Mixture of two unknown normals. In the binormal mixture model where μ, σ^2, ν and τ^2 are also unknown, the maximization step of the EM algorithm must be extended to also include maximization over these four parameters (Exercise 5.1). This results in the additional updating equations

$$\mu^{(k+1)} = \frac{\sum_{i=1}^n b_i^* y_i}{np^{(k+1)}} \,, \qquad \nu^{(k+1)} = \frac{\sum_{i=1}^n (1 - b_i^*) y_i}{n(1 - p^{(k+1)})} \,,$$

and

$$(\sigma^2)^{(k+1)} = \frac{\sum_{i=1}^n b_i^*(y_i - \mu^{(k+1)})^2}{np^{(k+1)}} \,, \qquad (\tau^2)^{(k+1)} = \frac{\sum_{i=1}^n (1 - b_i^*)(y_i - \nu^{(k+1)})^2}{n(1 - p^{(k+1)})} \,.$$

□

5.3.2 Properties of the EM algorithm

The main advantage of the EM algorithm is its relative ease of implementation (where applicable), especially in high-dimensional models. Under weak regularity conditions it can be shown that the likelihood increases at each iteration, that is, $L(\boldsymbol{\theta}^{(k+1)}; \boldsymbol{y}) \geq L(\boldsymbol{\theta}^{(k)}; \boldsymbol{y})$, with the inequality being strict unless $\boldsymbol{\theta}^{(k)}$ is a local maximum of the likelihood (Dempster *et al.* 1977). Under additional weak conditions, including the requirement that the likelihood is bounded on Θ, it can be shown that $\boldsymbol{\theta}^{(k)}$ converges to a local maximum of $L(\boldsymbol{\theta}; \boldsymbol{y})$ (Wu 1983)[1].

There are a diverse range of applications of the EM algorithm, including:

- Situations where the data are grouped, truncated, or censored (e.g. see Example 5.3 and Exercise 5.6).
- In estimation of animal abundance (Section 6.4), where the missing data is the number of unobserved animals (e.g. Van Deusen 2002).
- In unbalanced experimental designs, where completing the design by filling in the missing cells (the E-step) results in a simpler completed-data likelihood.
- In mixed-effects models, the data are completed by inclusion of the unobserved random effects, and the completed-data likelihood is then that of a fixed effects model.
- In state-space models, the data are completed by inclusion of the unobserved process errors, and the completed-data likelihood is then that of a deterministic process model.
- When multivariate data contain missing values, these can be imputed (the E-step), so that the completed data contain the full multivariate vector of observations on every individual (Schafer 1997). See R package `mirf`, and SAS procedure `MI`.
- In maximum likelihood modelling of evolutionary trees, the likelihood of observed DNA sequences is routine to compute if the evolutionary tree has been completed by specification of ancestry. For example, see Felsenstein (1981).

The chief drawback of the EM algorithm is that convergence is typically linear and can be very slow (Box 5.2), especially once $\boldsymbol{\theta}^{(k)}$ gets close to $\widehat{\boldsymbol{\theta}}$. A further drawback is that the EM algorithm provides no immediate way to obtain approximate standard errors of the MLE (but see Section 5.3.4).

For the case $\theta \in \mathbb{R}$ (or for a single element of parameter vector $\boldsymbol{\theta}$), and for sufficiently large k, the linear convergence of the EM algorithm is typically such that

$$\theta^{(k+1)} - \widehat{\theta} \approx a(\theta^{(k)} - \widehat{\theta}) , \tag{5.10}$$

where $|a| < 1$. Substituting $\theta^{(k+1)} = \theta^{(k)} + (\theta^{(k+1)} - \theta^{(k)})$ into (5.10) and rearranging gives

$$\theta^{(k+1)} - \theta^{(k)} \approx (a - 1)(\theta^{(k)} - \widehat{\theta}) .$$

[1] Wu (1983) corrects some invalid convergence statements made in Dempster *et al.* (1977).

Convergence rates of algorithms. Suppose that the algorithmic sequence $\boldsymbol{\theta}^{(k)}$, $k = 0, 1, 2, ...,$ converges to $\widehat{\boldsymbol{\theta}}$. The algorithm has convergence rate $q \geq 1$ if, for all k sufficiently large,

$$||\boldsymbol{\theta}^{(k+1)} - \widehat{\boldsymbol{\theta}}|| \leq C||\boldsymbol{\theta}^{(k)} - \widehat{\boldsymbol{\theta}}||^q \tag{5.11}$$

where $0 < C$ and $|| \cdot ||$ denotes Euclidean distance. If the value of $\widehat{\boldsymbol{\theta}}$ is to be determined to high accuracy then an algorithm with larger value of q will require fewer iterations. For sufficiently smooth log-likelihoods the Newton-Raphson algorithm converges at a quadratic rate, $q = 2$. The EM algorithm typically converges at a linear rate, $q = 1$, and the smallest value of C for which the above inequality holds may be only slightly less than unity (see Example 5.3), making it extremely slow to converge. In some situations the EM algorithm converges at a sublinear rate, which also corresponds to $q = 1$, but with $C = 1$ being the smallest value of C for which (5.11) holds.

Box 5.2

It follows that the increments, $\theta^{(k+1)} - \theta^{(k)}$, also decrease linearly with constant a. That is,

$$\theta^{(k+1)} - \theta^{(k)} \approx a(\theta^{(k)} - \theta^{(k-1)}) , \tag{5.12}$$

and hence a can be estimated from the ratio of increments. The linear rate of decrease in the increments is exploited in Section 5.3.3 for the purpose of accelerating the EM algorithm.

In the example below, the completed-data model is multinomial and hence has log-likelihood that satisfies the linearity requirement of the simple form of the EM algorithm.

Example 5.3. EM algorithm for a contingency table with grouped data. The objective is to estimate the MLEs of the row and column probabilities in a two-by-three contingency table in which rows and columns are independent. The unknown parameters are the probability of row 1, and columns 1 and 2, denoted $\boldsymbol{\theta} = (r_1, c_1, c_2)$, respectively. For the sake of this example, 100 subjects are put into a two-by-three table, but the experimenter confused the counts in the [1,3] and [2,2] cells. Sixty five subjects were put into the other cells and so 35 subjects are known to have been assigned to the [1,3] and [2,2] cells.

The observed table was

10	5	
20		30

Here, the observed data are the observed cells counts $\boldsymbol{y} = (y_{11}, y_{12}, y_{21}, y_{23})$. The values in the unobserved cells will be denoted Y_{13} and Y_{22}, where the capitalization is used as a reminder that these remain random quantities by virtue of not being observed.

The expectation step of the EM algorithm requires determination of the expected values in the unobserved cells, with the expectation calculated using the current value of $\boldsymbol{\theta}$. Since the total count is 100, and the four observed counts sum to 65, it must be the case that $Y_{13} + Y_{22} = 35$. It can be shown (directly from the definition of conditional probability) that

$$Y_{13}|\boldsymbol{y} \sim \text{Bin}(35, \rho)\,, \tag{5.13}$$

where

$$\rho = \frac{p_{13}}{(p_{13} + p_{22})}\,,$$

and p_{13} and p_{22} denote cell probabilities. These cell probabilities are $p_{13} = r_1 c_3$ and $p_{22} = r_2 c_2$, due to the assumption of row and column independence.

So, at iteration k with parameter values $\boldsymbol{\theta}^{(k)} = (r_1^{(k)}, c_1^{(k)}, c_2^{(k)})$, step 1 of the EM algorithm estimates the missing values by taking expectation under the binomial model in (5.13). That is,

$$\begin{aligned} y_{13}^* = E_{\boldsymbol{\theta}^{(k)}}[Y_{13}|\boldsymbol{y}] &= 35\rho^{(k)} \\ &= 35\frac{p_{13}^{(k)}}{p_{13}^{(k)} + p_{22}^{(k)}} \\ &= 35\frac{r_1^{(k)}(1 - c_1^{(k)} - c_2^{(k)})}{r_1^{(k)}(1 - c_1^{(k)} - c_2^{(k)}) + r_2^{(k)} c_2^{(k)}}\,, \end{aligned}$$

and of course $y_{22}^* = 35 - y_{13}^*$.

The maximization step simply calculates new values $r_1^{(k+1)}$, $c_1^{(k+1)}$, and $c_2^{(k+1)}$ as the row and column proportions from the completed contingency table. That is,

$$r_1^{(k+1)} = \frac{15 + y_{13}^*}{100}\,, \quad c_1^{(k+1)} = 0.3\,, \quad c_2^{(k+1)} = \frac{5 + y_{22}^*}{100}\,. \tag{5.14}$$

The following R code implements 200 iterations of this simple algorithm, with starting value $\boldsymbol{\theta}^{(0)} = (1/3, 1/3, 1/6)$, and prints out $\boldsymbol{\theta}^{(k)}$ for the first and last five of these iterations.

```
> #Observed data
> y11=10; y12=5; y21=20; y23=30
> #Initial parameter guesses
> r1=1/3; r2=1-r1; c1=1/3; c2=1/6; c3=1-c1-c2
> for(k in 1:200) {
+ # E step: apportion the 35 missing obs to cells [1,3] and [2,2]
+   p13=r1*c3; p22=r2*c2
+   y13=35*p13/(p13+p22); y22=35-y13
+ #M step
```

```
+     r1=(y11+y12+y13)/100; r2=1-r1
+     c1=(y11+y21)/100; c2=(y12+y22)/100; c3=1-c1-c2
+     if(k<6|k>195) cat("\n","Iter",k,": P(Row 1)=",r1,
+                        ", P(Col 1)=",c1,", P(Col 2)=",c2,sep="") }

Iter1: P(Row 1)=0.36, P(Col 1)=0.3, P(Col 2)=0.19
Iter2: P(Row 1)=0.3605505, P(Col 1)=0.3, P(Col 2)=0.1894495
Iter3: P(Row 1)=0.3610844, P(Col 1)=0.3, P(Col 2)=0.1889156
Iter4: P(Row 1)=0.361602, P(Col 1)=0.3, P(Col 2)=0.1883980
Iter5: P(Row 1)=0.3621036, P(Col 1)=0.3, P(Col 2)=0.1878964
Iter196: P(Row 1)=0.3749976, P(Col 1)=0.3, P(Col 2)=0.1750024
Iter197: P(Row 1)=0.3749977, P(Col 1)=0.3, P(Col 2)=0.1750023
Iter198: P(Row 1)=0.3749978, P(Col 1)=0.3, P(Col 2)=0.1750022
Iter199: P(Row 1)=0.3749979, P(Col 1)=0.3, P(Col 2)=0.1750021
Iter200: P(Row 1)=0.374998, P(Col 1)=0.3, P(Col 2)=0.1750020
```

Note that the $c_1^{(k)}$ values converged to the value 0.3 at the first iteration (since column 1 is fully observed), and $r_1^{(k)}$ and $c_2^{(k)}$ appear to be converging towards 0.375 and 0.175, respectively. It is left as an exercise (Exercise 5.4) to verify that the MLE is indeed $\widehat{\boldsymbol{\theta}} = (\widehat{r}_1, \widehat{c}_1, \widehat{c}_2) = (0.375, 0.3, 0.175)$ exactly. □

5.3.3 Accelerating the EM algorithm

The linear rate of convergence of the EM algorithm can be used to accelerate the algorithm by extrapolation. To see how this works, note that the convergence of $\boldsymbol{\theta}^{(k)}$ to $\widehat{\boldsymbol{\theta}}$ as $k \to \infty$ allows $\widehat{\boldsymbol{\theta}}$ to be expressed as

$$\widehat{\boldsymbol{\theta}} = \boldsymbol{\theta}^{(k)} + \sum_{j=k}^{\infty}(\boldsymbol{\theta}^{(j+1)} - \boldsymbol{\theta}^{(j)}) .$$

Restricting attention to $\theta \in \mathbb{R}$, it follows from the linear convergence of the increments in (5.12), that

$$\widehat{\theta} \approx \theta^{(k)} + (\theta^{(k+1)} - \theta^{(k)}) \sum_{j=0}^{\infty} a^j , \tag{5.15}$$

where $|a| < 1$. The summation in (5.15) is a geometric series with limit of $(1 - a)^{-1}$. That is,

$$\widehat{\theta} \approx \theta^{(k)} + \frac{(\theta^{(k+1)} - \theta^{(k)})}{1 - a} . \tag{5.16}$$

Formula (5.16) is a form of Aitken acceleration (Aitken 1926). Extension to the multi-parameter case is described in McLachlan and Krishnan (2008, Chap. 4 therein).

Example 5.3 continued. Restricting attention to just the single parameter r_1, a plot of k versus $r_1^{(k)}$ shows the slow convergence to $\widehat{r}_1 = 0.375$ (Figure 5.1).

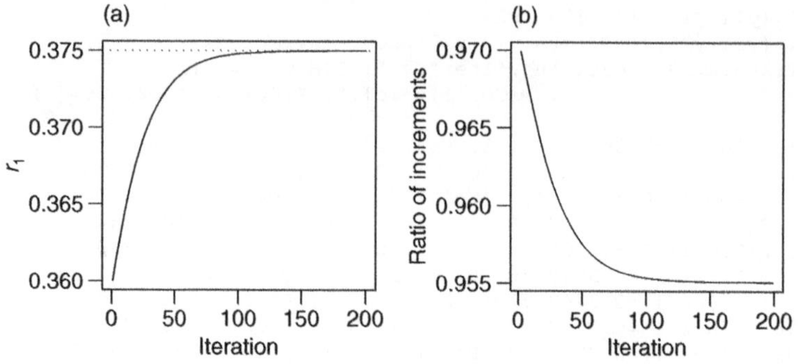

Figure 5.1 (a) The sequence $r_1^{(k)}, k = 1, ..., 200$. *(b) Ratio of increments,* $(r_1^{(k+1)} - r_1^{(k)})/(r_1^{(k)} - r_1^{(k-1)})\ , k = 1, ..., 199$.

Furthermore, for large k it appears that the ratio of successive increments approaches approximately 0.955. That is,

$$r_1^{(k+1)} - r_1^{(k)} \approx 0.955(r_1^{(k)} - r_1^{(k-1)})\ . \tag{5.17}$$

Suppose that it is decided to apply acceleration after 50 iterations of the EM algorithm. The acceleration calculation requires the values of $(r_1^{(48)}, r_1^{(49)}, r_1^{(50)}) = (0.372890, 0.372982, 0.373070)$. An estimate of the constant a in (5.12) is required, and this can be obtained as

$$a = \frac{r_1^{(50)} - r_1^{(49)}}{r_1^{(49)} - r_1^{(48)}} \approx 0.957\ .$$

Using $k = 49$ in (5.16), the accelerated estimate is

$$r_1^{(49)} + \frac{r_1^{(50)} - r_1^{(49)}}{1 - a} = 0.372982 + \frac{0.000088}{0.043} = 0.3750285,$$

which differs from $\widehat{r}_1$ by just 0.0000285. By comparison, the EM algorithm takes until iteration 144 to get this close to $\widehat{r}_1$. □

5.3.4 Inference

The EM algorithm is particularly useful in situations where the log-likelihood $l(\boldsymbol{\theta}; \boldsymbol{y})$ is challenging to maximize using general-purpose optimizers such as those available in `optim` and `NLMIXED`. Indeed, the EM algorithm is able to find $\widehat{\boldsymbol{\theta}}$ without explicit calculation of $l(\boldsymbol{\theta}; \boldsymbol{y})$. Thus, inference based on likelihood ratio or approximate normality (i.e. obtaining standard errors utilizing the curvature of $l(\boldsymbol{\theta}; \boldsymbol{y})$) could be problematic to implement.

It has been shown that, via partial derivative calculations on the completed-data log-likelihood, the EM algorithm can be extended to compute the curvature of $l(\boldsymbol{\theta}; \boldsymbol{y})$, without calculation of $l(\boldsymbol{\theta}; \boldsymbol{y})$ itself (Louis 1982). Alternatively, in situations where $l(\boldsymbol{\theta}; \boldsymbol{y})$ can be calculated but `optim` and `NLMIXED` are slow or unstable, it may be the case that they will reach convergence if they are given a starting value that is very close to $\widehat{\boldsymbol{\theta}}$ – this starting value could be the value of the MLE as provided by the EM-algorithm.

Another alternative is use bootstrap simulation (Section 4.5). The robustness of the EM algorithm makes it well-suited to this form of inference. Bootstrapping requires the estimation of the parameter value $\widehat{\boldsymbol{\theta}}^*$ that maximizes $l(\boldsymbol{\theta}; \boldsymbol{y}^*)$ for each of many different simulated sets of data $\boldsymbol{y}^*$. It is therefore very advantageous to use an algorithm that can be relied upon to converge (under very weak conditions) for any $\boldsymbol{y}^*$ that could be simulated.

5.4 Multi-stage maximization

The general idea of multi-stage maximization is to break the maximization into two (or more) maximizations of lower dimension. That is, the s-dimensional vector of parameters is partitioned as $\boldsymbol{\theta} = (\boldsymbol{\psi}, \boldsymbol{\lambda})$, and separate maximizations are done with respect to $\boldsymbol{\psi} \in \mathbb{R}^r$ and $\boldsymbol{\lambda} \in \mathbb{R}^{s-r}$. The reader may recognize profile likelihood as an instance of this. Indeed, profile likelihood is revisited in Section 5.4.1 but from the viewpoint of being a potential device for effective optimization rather than a convenient tool for inference on parameters of interest (as presented in Section 3.6).

Example 5.5 presents a variant of profile likelihood that maximizes an approximation to the likelihood, and hence the maximizing value of this approximate likelihood, $\widetilde{\boldsymbol{\theta}}$, will not in general equal $\widehat{\boldsymbol{\theta}}$. This approach is used in practice when it can be argued that the approximate likelihood effectively contains the relevant information contained in the true likelihood, so that it can be argued that $\widetilde{\boldsymbol{\theta}}$ will be negligibly different from $\widehat{\boldsymbol{\theta}}$. This approach wears a lot of different names within the published literature because it can be justified using arguments based on elimination of nuisance parameters using conditional or marginal likelihood (Section 9.2). Here, it will simply be called profile approximate-likelihood to reinforce the fact that it is used as an optimization technique rather than a technique for mitigating the effects of nuisance parameters.

Section 5.4.2 introduces the idea of performing the optimization in phases. In the first phase a simplified version of the likelihood is maximized with respect to a subset of the parameters. In a two-phase optimization, the second phase would then maximize the full likelihood, using the values estimated from phase one as the starting values for that subset of parameters.

5.4.1 Efficient maximization via profile likelihood

With the parameter vector partitioned as $\boldsymbol{\theta} = (\boldsymbol{\psi}, \boldsymbol{\lambda})$, recall (Section 3.6) that the profile likelihood for $\boldsymbol{\psi}$ is defined as

$$l^*(\boldsymbol{\psi}; \boldsymbol{y}) = \max_{\boldsymbol{\lambda}} \; l(\boldsymbol{\psi}, \boldsymbol{\lambda}; \boldsymbol{y}) .$$

Maximization of $l(\boldsymbol{\theta}; \boldsymbol{y})$ is obtained in two stages, by the maximization of $l^*(\boldsymbol{\psi}; \boldsymbol{y})$, which in turn requires maximization over $\boldsymbol{\lambda}$ for a given value of $\boldsymbol{\psi}$. This provides a particularly effective means of obtaining $\widehat{\boldsymbol{\theta}}$ if $\boldsymbol{\psi}$ is of relatively low dimension, and the maximization with respect to $\boldsymbol{\lambda}$ is fast.

Profile likelihood allows for the speedy fitting of so-called transformation-regression models in which both the observations $\boldsymbol{y}$ and covariates may be subject to nonlinear transformation. A example of this is the linear regression model applied to a Box-Cox transformation of $\boldsymbol{y}$, and is presented in Section 6.2. Profile likelihood effectively reduces this to a one-dimensional numerical optimization. Similarly, fitting of negative binomial models within the generalized linear model framework is accomplished by the fact that, for a fixed value of the shape parameter m, the negative binomial distribution is of the exponential dispersion family form assumed by GLM software (see Equation (7.1) and Exercise 7.3).

Example 5.4. Nonlinear (least-squares) regression model. Let each $Y_i, i = 1, ..., n$ be independently distributed as $N(\mu_i(\boldsymbol{\beta}), \sigma^2)$, where each μ_i is a smooth (i.e. differentiable) function of parameters $\boldsymbol{\beta} = (\beta_1, ..., \beta_p)$.

The MLE of $\boldsymbol{\beta}$ is found using nonlinear least squares, which is implemented in SAS procedure `NLIN` or the R function `nls`. In particular, `nls` has the facility to take advantage of any partial linearity (see Golub and Pereyra 1973, for details) in the specification of $\mu_i(\boldsymbol{\beta})$.

By way of example, the 3-parameter von-Bertalanffy curve (von Bertalanffy 1938) is a nonlinear curve that is commonly used to model the length of fish. If a_i is the age[2] of fish i, then the von-Bertalanffy curve for the expected length of fish i can be expressed in the form

$$\mu_i(\boldsymbol{\beta}) = E[Y_i] = \beta_1 - \beta_2 \exp(-\beta_3 a_i) ,$$

where β_1, β_2 and β_3 are all assumed to be positive. This is an increasing curve with upper asymptote of β_1 (Figure 5.2).

If the value of β_3 is fixed then the von-Bertalanffy curve has the form of a simple linear regression with respect to β_1 and β_2. Specifically, the expected length of fish i is then

$$E[Y_i] = a + bx_i , \tag{5.18}$$

where $a = \beta_1$, $b = -\beta_2$, and $x_i = \exp(-\beta_3 a_i)$ is the explanatory variable. That is, the maximizing values of β_1 and β_2 can be obtained explicitly for any given value of

[2] Fish are typically aged by counting rings on hard body parts, such at otoliths (an ear-bone) or scales.

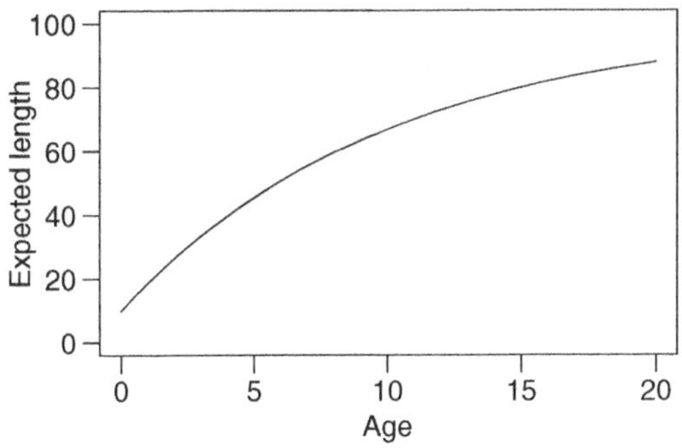

Figure 5.2 Von-Bertalanffy curve for parameter values $\boldsymbol{\beta} = (100, 90, 0.1)$.

β_3. So, instead of fitting the von-Bertalanffy curve using a 3-dimensional numerical maximization over $(\beta_1, \beta_2, \beta_3)$, it is enough to perform 1-dimensional numerical maximization of

$$\begin{aligned} l^*(\beta_3; \mathbf{y}) &= \max_{\beta_1, \beta_2} l(\beta_1, \beta_2, \beta_3; \mathbf{y}) \\ &= l(\widehat{\beta}_{1,\beta_3}, \widehat{\beta}_{2,\beta_3}, \beta_3; \mathbf{y}) , \end{aligned} \quad (5.19)$$

where $\widehat{\beta}_{1,\beta_3}$ and $\widehat{\beta}_{2,\beta_3}$ denote the least-squares estimates of β_1 and β_2 obtained from the linear regression model (5.18) for the fixed value β_3.

The code snippet below demonstrates the ability of the `nls` function to exploit partial linearity. The first call of `nls` in this code demonstrates its standard use. The second call takes advantage of the partial linearity of $\mu_i(\boldsymbol{\beta})$ and performs numerical optimization with respect to only β_3. The code segment

```
y~cbind(1,exp(-beta3*age))
```

is used to specify (5.18). That is, given β_3, the linear model to be fitted includes an intercept term and the explanatory variable $\exp(-\beta_3 a_i)$.

```
#Data frame vonB contains variables age and y (fish length)
#Standard use of nls
nls(y~beta1-beta2*exp(-beta3*age),data=vonB,
       start=list(beta1=90,beta2=80,beta3=0.2))

#Taking advantage of partial linearity
nls(y~cbind(1,exp(-beta3*age)),data=vonB,
       start=list(beta3=0.2),algorithm="plinear")
```

□

In the next example, it is an approximation to the likelihood that is profiled.

Example 5.5. Copulas. Copula models provide a general methodology for specifying dependency structure in multivariate data where the measurement on each individual is the m-dimensional observation $\boldsymbol{y}_i = (y_{i1}, ..., y_{im}) \in \mathbb{R}^m, i = 1, ..., n$. They are especially relevant to non-normal data where traditional multivariate-normal theory does not apply. See Trivedi and Zimmer (2007) for a comprehensive introduction to copulas.

A copula is a m-dimensional joint distribution function defined on the m-dimensional unit cube, and such that its marginal distributions are all Unif(0, 1). Specifically, let $\boldsymbol{U} = (U_1, ..., U_m) \in [0, 1]^m$ have distribution function given by the copula $C(u_1, ..., u_m)$. Then, $C(u_1, ..., u_m) = \text{Prob}(U_1 \leq u_1, ..., U_m \leq u_m)$, and $\text{Prob}(U_j \leq u_j) = u_j, j = 1, ..., m, 0 \leq u_j \leq 1$. The versatility of copulas is due to the property that any multivariate distribution for $\boldsymbol{Y}_i = (Y_{i1}, ..., Y_{im})$ can be represented by its m marginal distributions, $F(y_{ij}) = \text{Prob}(Y_{ij} \leq y_{ij})$, and an appropriate copula.

Now, introducing parameters into the model, let $\boldsymbol{\theta} = (\boldsymbol{\psi}, \boldsymbol{\lambda})$ where $\boldsymbol{\lambda}$ denote the parameters of the marginal distributions $F_{ij}(y_{ij}; \boldsymbol{\lambda}), i = 1, ..., n, \ j = 1, ..., m$, and let $\boldsymbol{\psi}$ denote the parameters of the copula. Assuming independence of the $\boldsymbol{Y}_i$, it can be shown (Trivedi and Zimmer 2007) that the log-likelihood has the form

$$l(\boldsymbol{\theta}; \boldsymbol{y}_1, ..., \boldsymbol{y}_n) = \sum_{i=1}^{n}\sum_{j=1}^{m} \log f_{ij}(y_{ij}; \boldsymbol{\lambda}) \tag{5.20}$$

$$+ \sum_{i=1}^{n} \log c(F_{i1}(y_{i1}; \boldsymbol{\lambda}), ..., F_{im}(y_{im}; \boldsymbol{\lambda}); \boldsymbol{\psi}) \, . \tag{5.21}$$

Note that the term in (5.21) is a specification of the correlation structure on the unit cube, and that each y_{ij} value is mapped from its domain into the unit interval via the transformation $F_{ij}(y_{ij}; \boldsymbol{\lambda})$.

Direct maximization of the above log-likelihood generally requires an s-dimensional numerical optimization to determine the MLE $\widehat{\boldsymbol{\theta}} = (\widehat{\boldsymbol{\psi}}, \widehat{\boldsymbol{\lambda}}) \in \mathbb{R}^s$. However, an approximate MLE, $\widetilde{\boldsymbol{\theta}} = (\widetilde{\boldsymbol{\psi}}, \widetilde{\boldsymbol{\lambda}})$ can more readily be obtained using an approximation to the profile log-likelihood. This approximation obtains $\widetilde{\boldsymbol{\lambda}} \in \mathbb{R}^{s-r}$ by maximization over the term in (5.20) only. This maximization is typically relatively straightforward because (5.20) is equivalent to the log-likelihood from assuming that all Y_{ij} are independent. The estimate $\widetilde{\boldsymbol{\psi}} \in \mathbb{R}^r$ is then obtained from maximization of (5.21), given the (fixed) value of $\widetilde{\boldsymbol{\lambda}}$. This is commonly known as the inference function for margins (IFM) approach. It can be shown that the estimator $\widetilde{\boldsymbol{\theta}}$ is a consistent estimator, but that it is less efficient than the MLE (Joe and Xu 1996). See Yan (2007) for demonstration of fitting copulas using the `copula` package in R. In SAS, copulas can be fitted using `PROC MODEL` in the ETS module. □

5.4.2 Multi-stage optimization

The successful global maximization of a likelihood often requires specification of a good starting value for the parameter vector $\boldsymbol{\theta}^{(0)}$. In some situations it may be possible to deduce good starting values (for at least a subset of the parameters) from graphical inspection or exploratory analysis of the data. However, in more complex models this may not be enough. This section presents a simple form of multi-stage optimization that obtains a value of $\boldsymbol{\theta}^{(0)}$ from a sequence of partial optimizations, and concludes with a brief description of the automatic implementation of multi-stage optimization within the ADMB software.

For convenience, two-stage optimization is presented, since this permits use of the existing notation whereby the parameter vector is partitioned as $\boldsymbol{\theta} = (\boldsymbol{\psi}, \boldsymbol{\lambda})$. At the first stage, a restricted form of the log-likelihood is optimized with respect to $\boldsymbol{\lambda}$ only. Let $\widetilde{\boldsymbol{\lambda}}$ denote the resulting maximizing value of $\boldsymbol{\lambda}$. The second stage performs a full maximization of the log-likelihood, using the starting value $\boldsymbol{\theta}^{(0)} = (\boldsymbol{\psi}^{(0)}, \widetilde{\boldsymbol{\lambda}})$, where $\boldsymbol{\psi}^{(0)}$ denotes user-specified starting values for this subset of parameters.

It is often convenient to implement the first-stage maximization by simply maximizing the log-likelihood $l(\boldsymbol{\theta})$ with $\boldsymbol{\psi}$ fixed at $\boldsymbol{\psi}^{(0)}$. Then the first stage optimization has the profile likelihood form

$$\max_{\boldsymbol{\lambda}} l(\boldsymbol{\psi}^{(0)}, \boldsymbol{\lambda}) ,$$

from which $\widetilde{\boldsymbol{\lambda}}$ is obtained. However, any sensible modification of $l(\boldsymbol{\theta})$ can be used. For example, it may be the case that some component of the log-likelihood contains most of the relevant information about $\boldsymbol{\lambda}$, and this would be a suitable candidate if it were convenient to maximize with respect to $\boldsymbol{\lambda}$ (with other parameters held fixed if present).

Example 5.5 continued. Two-stage optimization of the copula log-likelihood. The copula log-likelihood in Example 5.5 is the sum of terms (5.20) and (5.21). The first term involves only $\boldsymbol{\lambda}$, and maximization of this term gives the so-called IFM estimate, of $\widetilde{\boldsymbol{\lambda}}$. In a two-stage optimization of the full log-likelihood, $\widetilde{\boldsymbol{\lambda}}$ is a natural starting value for $\boldsymbol{\lambda}$. □

The extension to three or more stages is immediate. Specifically, the parameter vector $\boldsymbol{\theta}$ is partitioned into a collection of parameter subsets and these are sequentially added in a series of restricted optimizations. At each stage, the maximizing values of those parameters from the previous stage are used as initial values.

5.4.2.1 Using ADMB

Multi-stage optimization is incorporated within ADMB, where the terminology 'multi-phase' is used instead of 'multi-stage'. In the `PARAMETER_SECTION`, an optional integer value can be associated with the declaration of a parameter to indicate the stage at which the parameter enters the optimization. Within the `PROCEDURE_SECTION`, program code is provided to return the value of the objective function to be minimized at each stage. This is enabled by use of the `current_phase()` function, which returns the integer value of the stage currently being processed. See the ADMB user guide (ADMB-project 2008a, or later version) for full details.

The simple example below demonstrates multi-stage maximization of the log-likelihood of the iid $Y_i \sim N(\mu, \sigma^2)$ model using ADMB. This model is parameterized using $\boldsymbol{\theta} = (\mu, \log \sigma)$ to be consistent with use of ADMB in Chapter 10 (and see also Figure 10.4.), Note that $\widehat{\sigma}^2$ and its approximate variance (via the delta method) are obtained by using the `sdreport_number` declaration. The declaration `init_number mu(1)` specifies that parameter μ is to be included at the first stage, and the objective function at this stage is just the sum of squared residuals. The declaration `init_number log_sigma(2)` specifies that parameter $\log \sigma$ is to be made active at the second stage. The second-stage objective function is the full negative log-likelihood.

```
DATA_SECTION
  init_int n                          //Number of observations
  init_vector y(1,n)                  //The data

PARAMETER_SECTION
  init_number mu(1)                   //Phase 1 parameter
  init_number log_sigma(2)            //Phase 2 parameter
  sdreport_number sigmasq
  objective_function_value f          //Negative log-likelihood

PROCEDURE_SECTION
  sigmasq=exp(2*log_sigma);
  f=f+norm2(y-mu);                    //norm2() is the sum-of-squares
  if(current_phase()==2)
    f=n*log_sigma+f/(2*sigmasq);
  f=f+0.5*n*log(2*3.141593);          //Constant term
```

5.5 Exercises

5.1 Obtain the updating formulae in Example 5.2 by maximizing the complete data likelihood $\prod_{i=1}^{n} f(y_i, b_i^*; p, \mu, \sigma, \nu, \tau)$. (This likelihood has the form given by Equation (5.8), but is now a function of all five parameters.)

5.2 Using R or SAS, apply the EM algorithm of Example 5.2 to the Old-Faithful waiting time data (Figure 2.5), and for each parameter determine (approximately) the value of a in (5.12). The data are available as R dataframe `faithful`.

5.3 The following question is based on the lead example used in Dempster *et al.* (1977), and is derived from a model for recombination in gene mapping studies (e.g. see Lange 2002).

1. Let $(X_1, X_2, X_3, X_4, X_5)$ be a multinomial random vector with cell probabilities $(\frac{1}{2}, \frac{\theta}{4}, \frac{1-\theta}{4}, \frac{1-\theta}{4}, \frac{\theta}{4})$. Given $(x_1, x_2, x_3, x_4, x_5)$ show that the maximum likelihood estimate of θ is

$$\widehat{\theta} = \frac{x_2 + x_5}{x_2 + x_3 + x_4 + x_5}.$$

2. Suppose now that the observations in cells 1 and 2 are pooled, so that it is only the incomplete data $(y_1, y_2, y_3, y_4) = (x_1 + x_2, x_3, x_4, x_5)$ that are observed. Given the values $(y_1, y_2, y_3, y_4) = (125, 18, 20, 34)$, use the result of part 1 to apply the EM algorithm for calculation of $\widehat{\theta}$. Perform three iterations of the EM algorithm by hand, starting with the initial value $\theta^{(0)} = 0.5$.

5.4 For Example 5.3, verify that $\widehat{\boldsymbol{\theta}} = (\widehat{r}_1, \widehat{c}_1, \widehat{c}_2) = (0.375, 0.3, 0.175)$ is the MLE using one of the following methods.

1. Show that $\widehat{\boldsymbol{\theta}}$ is the fixed-point solution of the EM algorithm. That is, show that if $\boldsymbol{\theta}^{(k)} = (0.375, 0.3, 0.175)$ then $\boldsymbol{\theta}^{(k+1)} = \boldsymbol{\theta}^{(k)}$. (From the convergence results in Wu (1983), it follows that this is the unique MLE.)

2. Determine the likelihood for the observed (incomplete) data and maximize it using the R function `optim` or the SAS procedure `NLMIXED`.

5.5 In Example 5.3 the observed (incomplete) data $\boldsymbol{y} = c(10, 5, 20, 30, 35)$ are multinomial, denoted Mult$(5, p_1, p_2, ..., p_5)$ where $p_5 = 1 - \sum_{i=1}^{4} p_i$, and hence the likelihood from the observed data can be explicitly calculated.

1. Under the null hypothesis of row and column independence, the MLEs of $(p_1, p_2, ..., p_5)$ are obtained from $(\widehat{r}_1, \widehat{c}_1, \widehat{c}_2) = (0.375, 0.3, 0.175)$. Calculate the log-likelihood evaluated at this MLE.

2. Calculate the maximized log-likelihood without restriction on the parameter space. That is, assuming $\boldsymbol{y}$ is an observation from a Mult$(5, p_1, p_2, ..., p_5)$ distribution.

3. Use the G-test to obtain a p-value for the null hypothesis of row and column independence.

5.6 Suppose that $Y_i, i = 1, 2, 3, 4$ are iid from an exponential distribution with mean μ. However, only $y_1 = 5$, $y_3 = 12$ and $y_4 = 17$ are observed. The value of y_2 is not observed, but it is known that $y_2 > 10$.

The data are incomplete because y_2 is not observed. The data can be completed by estimating y_2 in the E-step, in which case the completed data model is simply that $y_i, i = 1, ..., 4$ are iid observations from the Exp(μ) distribution. Note that the completed data likelihood is linear in y_2, and hence the simple version of the EM algorithm is applicable.

1. Use the lack-of-memory property of the exponential distribution to construct the E-step. This property states that, if Y has an exponential distribution then, for any $s \geq 0$,

$$E[Y \mid Y > s] = s + E[Y].$$

(See also Exercise 6.1).

2. Choose a starting value $\mu^{(0)}$ and apply two iterations of the EM algorithm.

3. Implement Aitken acceleration using formula (5.16) for $k = 1$.

4. Regardless of your choice of $\mu^{(0)}$, the value of μ obtained after Aitken acceleration should be $\mu = 14\frac{2}{3}$. Verify that this is the MLE by showing that it is the fixed-point solution to the EM algorithm. That is, if $\mu^{(k)} = 14\frac{2}{3}$, then $\mu^{(k+1)} = 14\frac{2}{3}$ also.

5.7 Let $\boldsymbol{Y} = (Y_1, ..., Y_n)$ be iid observations from a zero-inflated Poisson distribution (see Exercise 2.11). That is, with probability p, Y necessarily takes the value 0, and with probability $(1 - p)$, Y is observed from a Pois(λ) distribution.

1. Determine the updating formulae of the EM algorithm for calculation of $(\widehat{p}, \widehat{\lambda})$.

2. Use R or SAS to apply the updating formulae to the apple micropropagation data in Exercise 3.9.

6

Some widely used applications of maximum likelihood

The best thing about being a statistician is that you get to play in everyone's backyard.
– John Tukey

6.1 Introduction

This Chapter presents three examples chosen from the diverse myriad of applications of maximum likelihood inference. The aim is to give a flavour for the nature of the likelihoods used in each of these applications, and some very simple examples of their use.

Section 6.2 presents the Box-Cox transformation as a very elegant application of profile likelihood. The total number of model parameters may be large, but numerical optimization is required over only the sole parameter that determines the optimal power transformation. Section 6.3 looks very briefly at survival analysis. A notable feature of likelihoods constructed from survival data is that they contain terms arising from both continuous and discrete forms of data. Section 6.4 presents some simple forms of mark–recapture models. These models have the unusual feature that the parameter of primary interest is the unknown number of animals in a population, and therefore is integer valued.

Maximum Likelihood Estimation and Inference: With Examples in R, SAS and ADMB, First Edition. Russell B. Millar.

6.2 Box-Cox transformations

The methodology developed by Box and Cox (1964) can be useful for the modelling of non-negative continuous data that violate the normality and homogeneous variance assumptions required by normal linear models such as linear regression models and ANOVA. Rather, it is assumed that a suitable transformation of the data can be found such that the transformed data will satisfy these assumptions.

> As a general rule, one should always attempt to use a model that is appropriate for the measured data without any need for their manipulation or transformation. Indeed, the poison data that are used in Section 6.2.1 are from an example in Box and Cox (1964), but these data (Table 6.1) can now be analyzed directly (Exercise 6.3) using the survival analysis models seen in Section 6.3. Nonetheless, there are many situations in which continuous data do not conform to any standard model and for which the Box-Cox transformation may be necessary.

Box 6.1

The standard Box-Cox transformation is a power transformation of the form

$$y_i^{(\lambda)} = \begin{cases} \frac{(y_i^\lambda - 1)}{\lambda}, & \lambda \neq 0 \\ \log y_i, & \lambda = 0 \end{cases},$$

where $\lambda \in \mathbb{R}$ is to be estimated. For any nonzero value of λ, note that it would be equivalent to use the simpler power transformation y_i^λ for the purpose of linear modelling of the transformed data. However, the Box-Cox power transformation has the advantage that it can be considered a smooth family of transformations with respect to λ, since the log function is the limit of the transformation $\frac{(y_i^\lambda - 1)}{\lambda}$ as λ tends to zero.[1]

Under the Box-Cox model, the transformed data $\boldsymbol{y}^{(\lambda)} = (y_1^{(\lambda)}, ..., y_n^{(\lambda)})$ are assumed to follow a normal linear model. That is, the transformed observation $y_i^{(\lambda)}$ is assumed to be observed from a $N(\boldsymbol{x}_i^T\boldsymbol{\beta}, \sigma^2)$ distribution where $\boldsymbol{x}_i$ is a vector of known covariates and $\boldsymbol{\beta} \in \mathbb{R}^p$ and σ^2 are to be estimated. This formulation includes ANOVA and ANCOVA (analysis of covariance), by appropriate specification of dummy variables as elements of the covariate vector $\boldsymbol{x}_i$ (e.g. see Seber and Lee 2003). The parameters to be estimated are $\boldsymbol{\theta} = (\boldsymbol{\beta}, \sigma^2, \lambda)$.

In Section 13.3 it was noted (see Box 13.1) that likelihood is invariant to any fixed transformation of the data. However, this does not apply here because the transformation is not fixed due to the fact that λ is a parameter to be estimated. It is therefore essential to work with the likelihood of the observed (untransformed)

[1] This limit result can be established by application of L'Hôpital's rule. See, for example, Stewart (1999).

data. The calculations below derive the likelihood for y_i from the assumed normal linear model for $y_i^{(\lambda)}$.

From the transformation of variables formula in Section 13.3, the density function for y_i is

$$\begin{aligned} f(y_i;\boldsymbol{\theta}) &= f(y_i^{(\lambda)};\boldsymbol{\beta},\sigma^2)\left|\frac{dy_i^{(\lambda)}}{dy_i}\right| \\ &= f(y_i^{(\lambda)};\boldsymbol{\beta},\sigma^2)y_i^{(\lambda-1)} . \end{aligned} \tag{6.1}$$

The transformed data are assumed to follow the linear model, and so

$$\log f(y_i^{(\lambda)};\boldsymbol{\beta},\sigma^2) = -\log\left(\sqrt{2\pi}\sigma\right) - \frac{\left(y_i^{(\lambda)} - \boldsymbol{x}_i^T\boldsymbol{\beta}\right)^2}{2\sigma^2} .$$

From (6.1) it follows that the log-likelihood for the observed data $\boldsymbol{y}$ is

$$l(\boldsymbol{\beta},\sigma^2,\lambda;\boldsymbol{y}) = \log f(\boldsymbol{y}^{(\lambda)};\boldsymbol{\beta},\sigma^2) + (\lambda-1)\sum_{i=1}^{n}\log y_i , \tag{6.2}$$

where

$$\log f(\boldsymbol{y}^{(\lambda)};\boldsymbol{\beta},\sigma^2) = -\frac{n}{2}\log(2\pi\sigma^2) - \sum_{i=1}^{n}\frac{\left(y_i^{(\lambda)} - \boldsymbol{x}_i^T\boldsymbol{\beta}\right)^2}{2\sigma^2} .$$

Note that for any fixed value of λ, $l(\boldsymbol{\beta},\sigma^2,\lambda;\boldsymbol{y})$ is maximized with respect to $\boldsymbol{\beta}$ and σ^2 by maximizing $\log f(\boldsymbol{y}^{(\lambda)};\boldsymbol{\beta},\sigma^2)$. This latter maximization is explicitly obtained from the least-squares fit of the linear model on the transformed data. Thus, it is computational quick and easy to produce a likelihood profile plot for λ.[2] In practice, this plot is used to suggest a 'convenient' value of λ. For example, if $\widehat{\lambda}$ is close to zero (e.g. the 95 % CI for λ includes zero) then a log transformation might be chosen.

6.2.1 Example: the Box and Cox poison data

Example 1 of Box and Cox (1964) used a two-way ANOVA to model the survival time of animals[3] after being administered with a poison. Each of 48 animals received one of three poisons and one of four treatments according to a replicated two-way design. The data are the hours until death of the animals (Table 6.1).

[2] The parameter of the Box-Cox transformation is almost always denoted λ, because this was the Greek symbol used by Box and Cox (1964), and this usage is followed here. Unfortunately, this causes a parameter naming conflict with the definition of profile likelihood in Section 3.6. The Box-Cox λ takes the role of ψ in that definition.

[3] Box and Cox do not disclose the source of the data, or the species of animal involved.

Table 6.1 Survival times (hr) after poisoning of an unidentified species of animal. The data are from Box and Cox (1964).

Poison 1				Poison 2				Poison 3			
Treatment				Treatment				Treatment			
A	B	C	D	A	B	C	D	A	B	C	D
3.1	8.2	4.3	4.5	3.6	9.2	4.4	5.6	2.2	3.0	2.3	3.0
4.5	11.0	4.5	7.1	2.9	6.1	3.5	10.2	2.1	3.7	2.5	3.6
4.6	8.8	6.3	6.6	4.0	4.9	3.1	7.1	1.8	3.8	2.4	3.1
4.3	7.2	7.6	6.2	2.3	12.4	4.0	3.8	2.3	2.9	2.2	3.3

The data were entered into a tab-delimited text file, `SurvTimes`, in standard rectangular form with each row corresponding to one observation. The first six lines of this file are:

```
poison  trmt    time
1       A        3.1
1       A        4.5
1       A        4.6
1       A        4.3
1       B        8.2
```

The R and SAS programs below both specify that the profile log-likelihood, $l^*(\lambda; \boldsymbol{y})$, is to be calculated for a sequence of λ values between -2 and 2, in increments of 0.01.

6.2.1.1 Using SAS

`PROC TRANSREG` (for transformation regression) is capable of many forms of regression on both raw and transformed data. In the `MODEL` statement used below, the `CLASS` keyword is used to specified that `poison` and `trmt` are factors. This causes `TRANSREG` to construct the appropriate dummy variables to form the design matrix corresponding to the two-way ANOVA.

```
PROC TRANSREG DATA=SurvTimes;
  MODEL BOXCOX(time / LAMBDAS=-2 to 2 by 0.01)=CLASS(poison|trmt);
```

6.2.1.2 Using R

The `MASS` library provides the `boxcox` function. This function automatically draws the likelihood profile plot (Figure 6.1) over the specified range of λ values.

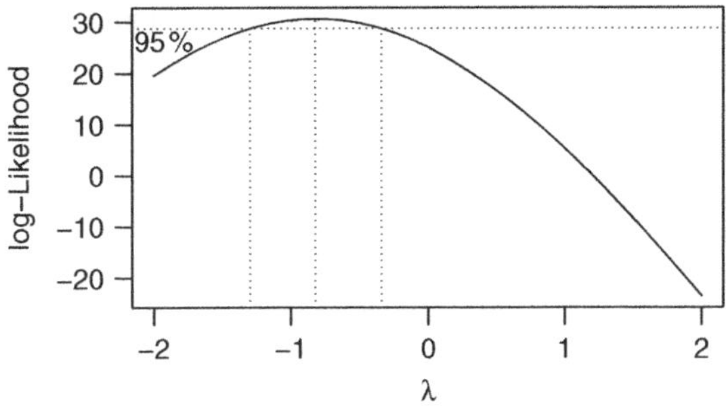

Figure 6.1 The profile log-likelihood plot for λ produced by the `boxcox` *function.*

```
> library(MASS)
> bc.fit=boxcox(time~as.factor(poison)*trmt,
+              lambda=seq(-2,2,0.01),data=SurvTimes)
> #Find maximizing value of lambda over the sequence
> lambdahat=bc.fit$x[order(bc.fit$y,decreasing=T)][1]
> cat("\n MLE of lambda is",lambdahat)

 MLE of lambda is -0.82
```

The 95 % likelihood ratio confidence interval suggests that a reciprocal transformation would be appropriate. Box and Cox (1964) note that the reciprocal transformation 'has a natural appeal for the analysis of survival times since it is open to the simple interpretation that it is the rate of dying which is to be considered'. This leads to the next application of maximum likelihood inference.

6.3 Models for survival-time data

Here, it is desired to model the time until 'failure' of a component or individual. This time could be the lifetime of a piece of equipment (e.g. lightbulb, bearing, microchip etc.), or the time until death or time that a disease remains in remission after treatment. These times will (interchangeably) be called survival times or failure times. Inference will typically include questions about the effect of covariates $\boldsymbol{x}_i$ associated with individual i, such as whether a new drug treatment is superior to the current treatment, say.

To be generally applicable, methods of analysis for survival-time data must be able to accommodate censored data. Such data routinely arise due to an individual being lost from the study prior to failure. In a clinical trial this could happen if a patient moves address and is no longer available for observation, or drops out for some other reason (such as being hit by a bus!). Also, the study will typically be of fixed duration and there may be individuals who have not failed by the time the study is concluded. In such cases, the actual survival time of a censored individual is not directly observed, but it is known that the individual had not failed at the time of censoring. This is known as right censoring. Other forms are possible, including left censoring where it is known only that an individual failed prior to a certain time, and interval censoring where the failure is known to have occurred within some interval of time. For simplicity, only right censoring is considered here. Furthermore, it is assumed that individuals are independent, and that the censoring process is independent of the survival times.

Sections 6.3.2 and 6.3.3 present the accelerated failure time and parametric proportional hazards models, respectively. These models use the likelihood that is developed in Section 6.3.1. In contrast, Section 6.3.4 presents Cox's semi-parametric version of the proportional hazards model, and shows that it is derived using a modified likelihood that is motivated from arguments that are presented in Section 9.2. The accelerated failure-time model and Cox's proportional hazards model are fitted to leukaemia data in Section 6.3.5.

6.3.1 Notation

Let t_i, $i = 1, ..., n$ be the time at which individual i was either observed to fail or was right censored, and let w_i be the indicator variable for the type of observation,

$$w_i = \begin{cases} 1, & \text{failure at time } t_i \\ 0, & \text{right censoring at time } t_i \,. \end{cases} \tag{6.3}$$

Note that the data observed on individual i is the pair (t_i, w_i). The corresponding contribution to the likelihood function is therefore the joint density function of this pair, regarded as a function of the model parameters. However, under the assumption that censoring times are independent of survival times, it follows[4] that, for the purpose of modelling survival times, no relevant information is lost by treating the w_i as fixed. That is, the fact that the w_i may actually be realizations of a Bernoulli random variable can be ignored, and they can be treated as fixed. The likelihood function is therefore obtained solely from the density function of the observed times, $\boldsymbol{t} = (t_1, ..., t_n)$, at which failure or censoring occurred.

If the event that is observed on individual i is a failure ($w_i = 1$), then t_i is the failure time and this observation therefore contributes $f_i(t_i; \boldsymbol{\theta})$ to the likelihood, where $f_i(t; \boldsymbol{\theta})$ denotes the density function for the failure time of individual i. However, if the event observed at time t_i is a censoring event ($w_i = 0$), then it is the case that individual i had not yet failed at time t_i. Then, all that is known is that the actual value of the failure time is in excess of time t_i at which the censoring occurred. In effect, the right censoring at time t_i corresponds to observing the discrete event $T_i > t_i$, where T_i denotes the failure-time random variable for individual i. The contribution to the likelihood is the probability of this event. Letting $F_i(t; \boldsymbol{\theta})$ denote the distribution function for the survival time of individual i, this probability

[4] From application of the conditionality principle in Section 14.3.

is $1 - F_i(t_i; \boldsymbol{\theta})$. The function $S_i(t; \boldsymbol{\theta}) = 1 - F_i(t; \boldsymbol{\theta})$ is called the survivor function because it gives the probability of surviving beyond time t.

As noted above, if t_i is an observed time of failure then the contribution to the likelihood is $f_i(t_i; \boldsymbol{\theta})$, and otherwise the contribution is $S_i(t_i; \boldsymbol{\theta})$ when t_i is a right-censoring time. Thus, the log-likelihood function arising from survival data that are subject to possible right censoring is

$$l(\boldsymbol{\theta}; \boldsymbol{t}) = \sum_{i=1}^{n} (w_i \log f_i(t_i; \boldsymbol{\theta}) + (1 - w_i) \log S_i(t_i; \boldsymbol{\theta})) . \tag{6.4}$$

This likelihood is used by the parametric survival-time models developed in Sections 6.3.2 and 6.3.3.

The description of survival models in the next three sections is necessarily brief, and does not cover model assessment or comparison. Also, it is assumed that the covariates $\boldsymbol{x}_i$ associated with individual i are constant over time. This would not be a valid assumption in a long-term health study in which, say, body-mass index (BMI) was used as a covariate. For greater coverage of the modelling of survival data, the reader is referred to Kleinbaum and Klein (2005). Survival analysis using R is a topic covered in Everitt and Hothorn (2006), while Der and Everitt (2009) demonstrate its implementation in SAS.

6.3.2 Accelerated failure-time model

The accelerated failure-time model (AFT) can intuitively be described using the popular convention that one dog year is equivalent to seven human years. That is, if $F_h(t)$ and $F_d(t)$ are the distribution functions for the age-at-death of humans and dogs, respectively, then this convention assumes

$$F_d(t) = F_h(7t) .$$

For example, the probability that a dog dies by the age of 10, $F_d(10)$, is equal to the probability that a human dies by the age of 70, $F_h(70)$.

The AFT model specifies the failure-time distribution for individual i as

$$F_i(t) = F_0(\gamma_i t; \boldsymbol{\psi}) ,$$

where γ_i depends on covariate vector $\boldsymbol{x}_i$. The distribution function F_0 denotes the 'baseline' distribution function, and is indexed by parameters $\boldsymbol{\psi}$. The time-multiplier parameter γ_i is positive, and a natural choice is to model $\log \gamma_i$ as a linear function of $\boldsymbol{x}_i$. Then the AFT model can be written

$$F_i(t; \boldsymbol{\theta}, \boldsymbol{x}_i) = F_0(\exp(\boldsymbol{x}_i^T \boldsymbol{\beta}) t; \boldsymbol{\psi}) ,$$

where $\boldsymbol{\theta} = (\boldsymbol{\beta}, \boldsymbol{\psi})$. Using this notation, the AFT model to compare dog longevity to human longevity could be expressed

$$F_d(t) = F_0(\exp(\beta) t) ,$$

where $F_0 = F_h$ is taken to be the baseline distribution. The popular convention corresponds to $\beta = \log 7 \approx 1.95$ (but see Box 6.2).

The Weibull distribution (see Exercise 6.4) is a popular choice for the form of the baseline failure-time distribution, F_0. Other popular choices include exponential, lognormal, log-logistic, and generalized gamma. Application of the AFT model is demonstrated in Section 6.3.5.

> The AFT model that arises from the popular convention that a dog year is equivalent to seven human years is not a good model. With respect to physiology, the first year of a dog's life is equivalent to about 15 human years, and later years are equivalent to about 4 or 5 human years, though this varies considerably with breed and size (see e.g. Patronek, Waters and Glickman 1997).

Box 6.2

6.3.3 Parametric proportional hazards model

Rather than using a model that specifies the effect of covariates directly through the failure-time distribution (as with the AFT model), the model presented in this section specifies the effect through the hazard function. The hazard function at time t is defined to be $h(t) = f(t)/S(t)$, and is the instantaneous rate (or risk) of failure at that time (Box 6.3).

In many ways, it is much more intuitive to pose a model for the hazard function than for the failure-time distribution. By way of example, the hazard function of humans is reasonably well modelled by the bathtub shaped hazard function (Figure 6.2). This is of course a gross simplification, and one of its omissions is that it does not show the temporary increase in hazard that occurs when the legal age of driving is first reached. In medical studies it may be natural to assume that a successful treatment results in an immediate reduction in hazard. This section shows how the log-likelihood for survival data can be constructed from a specified hazard function.

> To see that the hazard function, $h(t) = f(t)/S(t)$, is the instantaneous rate of failure at time t, let Δt denote some small interval of time, and note that
>
> $$h(t)\Delta t = \frac{f(t)\Delta t}{1 - F(t)} = \frac{F'(t)\Delta t}{1 - F(t)} \approx \frac{F(t + \Delta t) - F(t)}{1 - F(t)} .$$
>
> That is, for small Δt, $h(t)\Delta t$ gives the probability of failing in the interval $(t, t + \Delta t)$, given survival to time t.

Box 6.3

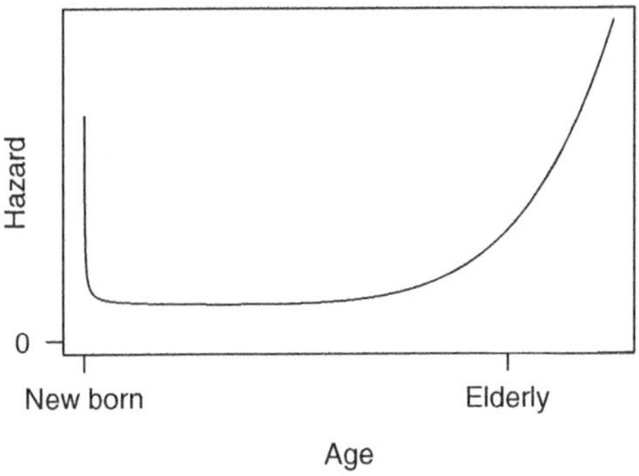

Figure 6.2 A typical bathtub hazard function for humans.

In the proportional hazards model the hazard functions of any two individuals are assumed to be proportional, with the constant of proportionality being constant over time and depending only on the covariates. The usual formulation for the hazard function of individual i is

$$h_i(t; \boldsymbol{\lambda}, \boldsymbol{\beta}) = h_0(t; \boldsymbol{\lambda}) \exp(\boldsymbol{x}_i^T \boldsymbol{\beta}) , \tag{6.5}$$

where $h_0(t; \boldsymbol{\lambda})$ is the baseline hazard function and depends on parameters $\boldsymbol{\lambda}$. Note that the hazard function depends on time only through $h_0(t; \boldsymbol{\lambda})$, and hence the ratio of hazards for any two individuals is a constant.

To compute the log-likelihood function in (6.4) it is necessary to determine both the failure-time density function, $f_i(t)$, and the survivor function, $S_i(t) = 1 - F_i(t)$, from specification of the hazard function. The survivor function can be obtained from the hazard function by noting that

$$\begin{aligned} h_i(t) = \frac{f_i(t)}{S_i(t)} &= \frac{\frac{\partial F_i(t)}{\partial t}}{S_i(t)} \\ &= \frac{-\partial \log S_i(t)}{\partial t} . \end{aligned}$$

This differential equation can be solved for $S_i(t)$ by the following re-arrangement and integration,

$$\begin{aligned} S_i(t) &= \exp\left(-\int_0^t h_i(u)du\right) \\ &= \exp\left(-\int_0^t h_0(u)\exp(\boldsymbol{x}_i^T \boldsymbol{\beta})du\right) \\ &= \exp\left(-H_0(t)\exp(\boldsymbol{x}_i^T \boldsymbol{\beta})\right) , \end{aligned} \tag{6.6}$$

where $H_0(t) = \int_0^t h_0(u)du$ is the cumulative baseline hazard function. With $S_i(t)$ obtained as above, and $f_i(t)$ given by $f_i(t) = h_i(t)S_i(t)$, the log-likelihood function in (6.4) can be computed and maximized with respect to model parameters $\boldsymbol{\theta} = (\boldsymbol{\lambda}, \boldsymbol{\beta})$.

Example 6.1. Exponential survival times. Suppose that an individual has a hazard function that is constant over time, $h(t) = \lambda$. From (6.6), the survivor function is

$$S(t) = \exp\left(-\int_0^t \lambda du\right) = \exp(-\lambda t),$$

and the density function of the failure times is therefore

$$f(t) = h(t)S(t) = \lambda \exp(-\lambda t).$$

That is, a constant hazard rate of λ corresponds to exponentially distributed failure times with mean $\mu = \lambda^{-1}$. Note that constant hazard corresponds to the lack-of-memory property of the exponential distribution (see Exercise 6.1). □

The above example shows that a constant baseline hazard corresponds to exponentially distributed failure times. Other parametric forms of the baseline hazard are possible. However, the parametric proportional hazards model is not widely used in practice because the AFT is generally more robust to model misspecification (Hutton and Monaghan 2002). This can be mitigated to some extent by employing flexible classes of models, such as that of Royston and Parmar (2002) who used cubic splines to fit a smooth cumulative baseline hazard function. In practice, proportional hazard models are most commonly fitted using the semi-parametric approach developed by Cox (1972) and presented in Section 6.3.4. This was the original application of the proportional hazards model, and the reader should be aware that the term 'proportional hazards model' usually refers to Cox's implementation of this model.

6.3.4 Cox's proportional hazards model

Cox's implementation (Cox 1972) of the proportional hazards model has the advantage that it requires no specification of the baseline hazard function $h_0(t; \boldsymbol{\lambda})$, and hence can be considered a semi-parametric model.

The essence of Cox's approach is the argument that, in the absence of knowledge about the baseline hazard function, there is no relevant information in the actual times of failure. For example, it could be argued that times at which failure can occur can effectively be pre-determined by the baseline hazard being zero outside of those times. Rather, the relevant information is provided solely by consideration of which individual failed at each failure time. The likelihood so obtained is called the partial likelihood – see Example 9.4 and Cox (1975) for more details.

Each observed failure event contributes a term to the partial likelihood, but censored observations do not, and so only the times of failure are used in the notation below. For simplicity, it is assumed that there are no tied failures times, and that the m individuals that failed have been re-ordered such that $t_1 < t_2 < ... < t_m$ are the distinct failure times. Let $R(t_j)$ denote the risk set at failure time t_j, that is, the set of individuals still in the experiment immediately prior to t_j. The observed censoring times enter the partial likelihood indirectly, since they are needed in order to determine membership of the risk set at each failure time.

Given that one failure is to occur at time t_j, the relative probabilities of failure for the individuals in $R(t_j)$ are proportional to the values of their hazard functions at t_j. To see this, note that the probability that individual j fails in time interval $(t_j, t_j + \Delta t)$, given that a member of $R(t_j)$ fails in this interval, is (from Box 6.3)

$$\begin{aligned} P(\text{indiv } j \text{ fails in } (t_j, t_j + \Delta t)|\text{one failure in } (t_j, t_j + \Delta t)) \\ \approx \frac{h_j(t_j; \boldsymbol{\lambda}, \boldsymbol{\beta})\Delta t}{\sum_{i \in R(t_j)} h_i(t_j; \boldsymbol{\lambda}, \boldsymbol{\beta})\Delta t} \\ = \frac{h_j(t_j; \boldsymbol{\lambda}, \boldsymbol{\beta})}{\sum_{i \in R(t_j)} h_i(t_j; \boldsymbol{\lambda}, \boldsymbol{\beta})} . \end{aligned}$$

Under the proportional hazards formulation in (6.5), with $\boldsymbol{x}_j$ denoting the covariate vector of the individual failing at time t_j, the above probability is

$$P_j(\boldsymbol{\beta}) = \frac{h_0(t_j; \boldsymbol{\lambda}) \exp(\boldsymbol{x}_j^T \boldsymbol{\beta})}{\sum_{i \in R(t_j)} h_0(t_j; \boldsymbol{\lambda}) \exp(\boldsymbol{x}_i^T \boldsymbol{\beta})} = \frac{\exp(\boldsymbol{x}_j^T \boldsymbol{\beta})}{\sum_{i \in R(t_j)} \exp(\boldsymbol{x}_i^T \boldsymbol{\beta})} , \tag{6.7}$$

which depends only on $\boldsymbol{\beta}$ and not on $h_0(t; \boldsymbol{\lambda})$.

The MLE of $\boldsymbol{\beta}$ is obtained by maximizing the 'partial likelihood' given by the product over the m failure times of the probabilities given in (6.7).

Example 6.2. The data here have been contrived for illustrative purposes. The scenario is that two groups of four patients each were used in a study of the effectiveness of a drug. The first group received a placebo and the second group received the drug. The times until relapse were

Placebo group:	3 7 8 (15)
Drug group:	5 (10) 12 27

where values in parentheses denote censored observations.

A natural parameterization would be to set the hazard function of the placebo group equal to the baseline hazard function, $h_0(t)$. The hazard of the drug group is then $h_0(t)e^{\beta}$, where β is the sole parameter to be estimated.

Letting $t_1 < t_2 < .... < t_6$ denote the six observed relapse times, the partial likelihood is

$$L_p(\beta) = \prod_{j=1}^{6} P_j(\beta) .$$

At the first relapse time, $t_1 = 3$, the risk set includes all eight individuals, and it is an individual from the placebo group who relapses. Thus,

$$P_1(\beta) = \frac{h_0(t_1)}{4h_0(t_1) + 4h_0(t_1)e^{\beta}} = \frac{1}{4 + 4e^{\beta}} .$$

At the second relapse time, $t_2 = 5$, the risk set includes the four drug group individuals and the remaining three from the placebo group. It is an individual from the drug group who relapses, and so $P_2(\beta) = e^{\beta}/(3 + 4e^{\beta})$. Continuing with this logic over the remaining four relapses results in the partial likelihood

$$L_p(\beta) = \frac{1}{4+4e^{\beta}} \times \frac{e^{\beta}}{3+4e^{\beta}} \times \frac{1}{3+3e^{\beta}} \times \frac{1}{2+3e^{\beta}} \times \frac{e^{\beta}}{1+2e^{\beta}} \times \frac{e^{\beta}}{e^{\beta}} . \quad (6.8)$$

The partial log-likelihood function $\log L_p(\beta)$ can be treated as a conventional log-likelihood. That is, it can be maximized to obtain $\widehat{\beta}$, and used to perform subsequent inference. Here $\widehat{\beta} = -0.689$, which has the interpretation that the hazard of patients receiving the drug is estimated to be $\exp(-0.689) = 0.502$ that of untreated patients. However, the 95 % likelihood ratio confidence interval (Figure 6.3) for β is $(-2.73, 1.12)$, which includes zero. It is perhaps not surprising that there is little statistical evidence of a difference between placebo and drug – this is a consequence of the very small sample sizes used in this example. □

6.3.5 Example in R and SAS: Leukaemia data

The data in Table 6.2 are the times in remission of 21 patients in each of two groups. One group received an experimental drug treatment, and the control group was given a placebo treatment. It will be assumed that these 42 observations are in a dataset containing the three variables `trmt`, `time`, and the censoring indicator variable `w`.

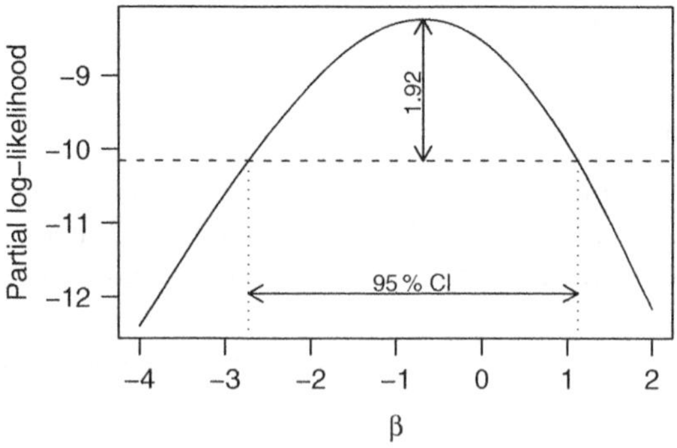

Figure 6.3 The partial log-likelihood in (6.8), from applying Cox's proportional hazards model to the contrived data.

Table 6.2 Duration of remission, t (weeks), for two groups of patients with acute leukaemia, from Table 10 of Freireich *et al.* (1963). Variable w is the censoring indicator variable and takes the value 1 if relapse was observed to occur at time t, and otherwise is zero if censoring occurred at time t.

Drug						Placebo					
t	w	t	w	t	w	t	w	t	w	t	w
6	0	10	1	22	1	1	1	5	1	11	1
6	1	11	0	23	1	1	1	5	1	12	1
6	1	13	1	25	0	2	1	8	1	12	1
6	1	16	1	32	0	2	1	8	1	15	1
7	1	17	0	32	0	3	1	8	1	17	1
9	0	19	0	34	0	4	1	8	1	22	1
10	0	20	0	35	0	4	1	11	1	23	1

The accelerated failure-time and Cox proportional hazards models can be fitted using the `survreg` and `coxph` functions in the `survival` package, or using SAS procedures `LIFEREG` and `PHREG`, respectively. The code below fits the AFT model with Weibull failure-time distribution using `PROC LIFEREG`. Note that it includes the model statement `MODEL time*w(0)=trmt;` to specify that right censoring of `time` corresponds to `w` = 0.

```
*AFT model with Weibull failure-time distribution;
*Variables trmt, time and w are in dataset Leukemia;
PROC LIFEREG DATA=Leukemia;
  CLASS trmt;
  MODEL time*w(0)=trmt;
RUN;
```

From the output in Figure 6.4 it is seen that the AFT model fits a baseline Weibull distribution $F(t; \widehat{\phi}, \widehat{k})$ to the placebo group with MLEs $\widehat{\phi} = \exp(2.2484) = 9.4726$

Analysis of Maximum Likelihood Parameter Estimates								
Parameter		DF	Estimate	Standard Error	95 % Confidence Limits		Chi-Square	Pr > ChiSq
Intercept		1	2.2484	0.1660	1.9231	2.5737	183.51	<.0001
trmt	Drug	1	1.2673	0.3106	0.6585	1.8762	16.64	<.0001
trmt	Plac	0	0.0000	.	.	.	.	.
Scale		1	0.7322	0.1078	0.5486	0.9772		
Weibull Shape		1	1.3658	0.2012	1.0233	1.8228		

Figure 6.4 Table of parameter estimates from using `PROC LIFEREG` *to fit a Weibull accelerated failure-time model to the leukaemia data.*

and $\widehat{k} = 1.3658$ (See Exercise 6.4 for the definition of $F(t;\phi,k)$.) The Wald test statistic of 16.64 shows that the drug effect is highly significant, and the drug group has fitted Weibull distribution $F(t;\exp(1.2673)\widehat{\phi},\widehat{k}) = F(t;3.5514\widehat{\phi},\widehat{k})$. This is equal to the baseline distribution evaluated at time $3.5514t$, $F(3.5514t;\widehat{\phi},\widehat{k})$. That is, one week of remission in the placebo group equates to about 3.55 weeks of remission in the group that received the drug.

The R code below shows the Cox proportional hazards fit using the `coxph` function, and AFT fit from the `survreg` function. For the Cox model, the likelihood ratio test statistic is 16.4 and indicates strong evidence of a treatment effect. The placebo treatment is the baseline, and the hazard function of the drug treatment is estimated to be about $100e^{-1.57} \approx 21\ \%$ that of the placebo. For the AFT model, the LR test statistic is seen to be 19.65, compared to the Wald test statistic of 16.64 shown in Figure 6.4.

```
> library(survival)
> Leukemia=Surv(times,w) #Create a survival object
> #Cox's proportional hazards model fitted to leukemia data
> coxph(Leukemia~trmt)
            coef exp(coef) se(coef)      z       p
trmtDrug -1.57       0.208    0.412 -3.81 0.00014

Likelihood ratio test=16.4  on 1 df, p=5.26e-05  n= 42

> #Accelerated failure time fit
> survreg(Leukemia~trmt)
Coefficients:
(Intercept)     trmtDrug
   2.248352     1.267335

Scale= 0.7321944

Loglik(model)= -106.6   Loglik(intercept only)= -116.4
    Chisq= 19.65 on 1 degrees of freedom, p= 9.3e-06  n= 42
```

For a more in-depth analysis of these data, including model assessment and consideration of other covariates, see Kleinbaum and Klein (2005).

6.4 Mark–recapture models

Mark-recapture models are one of a multitude of methods that can be used to estimate the number of animals, N, in a population (Seber 1982, Borchers, Buckland and Zucchini 2002). It will be assumed here that the population is *closed*, that is, the number of animals is not changed by births, deaths, immigration or emigration over the duration of the study, and so N can be regarded as constant throughout the course of the study, and is the parameter of primary interest.

In a simple mark–recapture experiment, a random sample of animals of size n_1 is captured, marked[5], and returned to the population. Some time later (having allowed enough time for the marked animals to randomly disperse within the population) a second random sample of size n_2 is taken. The number of marked animals in the second sample is recorded, and will be denoted m_2. Intuitively, one could argue that the proportion of marked animals in the second sample, $\widetilde{p} = m_2/n_2$, is an obvious estimator of the proportion of marked animals in the population $p = n_1/N$. Therefore, since $N = n_1/p$, an intuitive estimate of N is

$$\widetilde{N} = \frac{n_1}{\widetilde{p}} = \frac{n_1 n_2}{m_2}.$$

This is commonly known as the Petersen estimate (Petersen 1896) or Lincoln index.

The Petersen estimator is positively biased and the relative bias can be large for small sample sizes, if not infinite (which is the case when m_2 is able to take the value 0 with positive probability). In practice it is common to use Chapman's bias-adjusted variant of the Petersen estimate,

$$N^* = \frac{(n_1+1)(n_2+1)}{m_2+1} - 1\,.$$

Approximate standard errors of $\widetilde{N}$ and N^* can be obtained from application of the delta method (see Exercise 6.5) and related methods (Seber 1982).

The above mark–recapture experiment presents some interesting challenges for application of likelihood-based inference. First, the unknown parameter, N, is integer valued and the log-likelihood is undefined for non-integer values of N (but see Section 6.4.2). Thus, it is not possible to differentiate the log-likelihood (i.e. regularity conditions R4–R7 in Chapter 12 are not applicable). Moreover, the range of possible values that n_1, n_2 and m_2 can take varies with N (i.e. condition R3 is also violated). Nonetheless, subject to due caution, likelihood-based inference can be successfully applied here, and can also be used as a general purpose tool for estimation of animal abundance in more complex situations. In particular, in Section 6.4.2 the definition of the likelihood is extended to non-integer values of N, thereby enabling standard likelihood-based estimation and inference to be utilized.

Different likelihood formulations for the above mark–recapture experiment can be postulated. Section 6.4.1 uses a likelihood formulated from the hypergeometric distribution. This likelihood is arguably a truer description of the statistical model than the multinomial likelihood demonstrated in Section 6.4.3. However, the latter is more readily extendable to generalizations of the simple mark–recapture

[5] Options for 'marking' an animal include the use of dye, tagging or banding, and clipping (salmon fry from hatcheries are routinely marked by the removal of their adipose fin), or inspection for distinguishing marks (such as individually unique scar patterns on whales).

experiment, and is widely used in ML estimation of animal abundance from these types of experiments.

6.4.1 Hypergeometric likelihood for integer valued N

In the hypergeometric model the values of n_1 and n_2 are regarded as fixed. This is certainly appropriate if n_1 and n_2 were indeed specified in advance of the experiment.[6] With n_1 and n_2 fixed, only m_2 is random and the likelihood is determined by the probability of the observed value of m_2 as a function of N. Now, m_2 is the number of marked animals recorded when a sample of size n_2 is taken from a population with n_1 marked animals and $N - n_1$ unmarked animals. Assuming all possible samples of n_2 animals are equally probable, m_2 is distributed according to the hypergeometric distribution $\mathrm{H}(n_2, n_1, N - n_1)$. The probability function of this distribution is given by (15.3), and results in the likelihood

$$L(N; m_2) = f(m_2; n_2, n_1, N - n_1) \tag{6.9}$$

$$= \frac{\binom{n_1}{m_2}\binom{N-n_1}{n_2-m_2}}{\binom{N}{n_2}}, \quad N \geq \max(n_1, n_2)\,. \tag{6.10}$$

Note that the terms in (6.10) that do not involve N are constant terms since N is the sole parameter to be estimated. Thus, to within an additive constant, the log-likelihood function for N is

$$\begin{aligned} l(N; m_2) = {} & \log(N - n_1)! + \log(N - n_2)! \\ & - \log(N - n_1 - (n_2 - m_2))! - \log N!\,. \end{aligned} \tag{6.11}$$

The argument used below obtains an explicit formula for a MLE of N. It proceeds by determining the change in log-likelihood resulting from a unit change in N.

For any integer $N > \max(n_1, n_2)$, let

$$\Delta l(N; m_2) = l(N; m_2) - l(N - 1; m_2)$$

denote the difference in the log-likelihood when evaluated at N versus $N - 1$. Noting that

$$\log n! - \log(n - 1)! = \log n\,, \tag{6.12}$$

[6] More generally, this conditioning is appropriate if it can be argued that n_1 and n_2 contain no useful information about N. This would be reasonable in the absence of reliable knowledge about the catchability of the species, say.

it follows that

$$\Delta l(N; m_2) = \log\left(\frac{(N-n_1)(N-n_2)}{N(N-n_1-(n_2-m_2))}\right)$$

$$= \log\left(\frac{N^2 - Nn_1 - Nn_2 + n_1n_2}{N^2 - Nn_1 - Nn_2 + Nm_2}\right) . \tag{6.13}$$

The numerator and denominator in (6.13) differ only in their last terms, n_1n_2 and Nm_2, respectively. It follows that $\Delta l(N; m_2)$ is negative if and only if $n_1n_2 < Nm_2$. That is,

$$l(N; m_2) < l(N-1; m_2) \quad \text{if and only if} \quad \frac{n_1n_2}{m_2} < N , \tag{6.14}$$

with equality if $N = n_1n_2/m_2$. So, if one starts with an initial choice of integer N that is arbitrarily large, then it is the case that the log-likelihood is higher at $N - 1$, providing that N is greater than the Petersen estimate $\widetilde{N} = n_1n_2/m_2$. In other words, sequentially decreasing N by unity results in an increase in likelihood until N takes a value less than $\widetilde{N}$. Thus, a MLE of N is given by the biggest integer less than or equal to $\widetilde{N}$. This can be denoted

$$\widehat{N} = [\widetilde{N}] = \left[\frac{n_1n_2}{m_2}\right] ,$$

where $[x]$ denotes the integer part of x. If the Petersen estimate is integer valued then equality holds in (6.14) with $N = \widetilde{N}$. It is then the case that the MLE is not unique and $\widetilde{N}$ and $\widetilde{N} - 1$ are both MLEs.

6.4.2 Hypergeometric likelihood for $N \in \mathbb{R}^+$

The unknown number of animals is necessarily a positive integer value. However, if one is prepared to think laterally, it can be very convenient to extend the log-likelihood to positive real values. This can be achieved using the gamma function, $\Gamma(N)$, for $N > 0$. The gamma function is a smooth function that has the property that $\Gamma(N+1) = N!$ when N is integer valued. In that sense, it provides an extension of the factorial function to the positive real numbers (Figure 6.5).

The hypergeometric log-likelihood in (6.11) can be defined on $\mathbb{R}^+$ by replacing the factorial terms by their gamma equivalents. That is,

$$l(N; m_2) = \log\Gamma(N - n_1 + 1) + \log\Gamma(N - n_2 + 1)$$
$$- \log\Gamma(N - n_1 - (n_2 - m_2) + 1) - \log\Gamma(N+1) . \tag{6.15}$$

This log-likelihood can easily be programmed in both R and SAS because they both have a log-gamma function `lgamma` for evaluation of $\log\Gamma(\cdot)$.

Example 6.3. Hypergeometric likelihood for mark–recapture experiment. Borchers *et al.* (2002, p. 107) use the example of a mark–recapture experiment

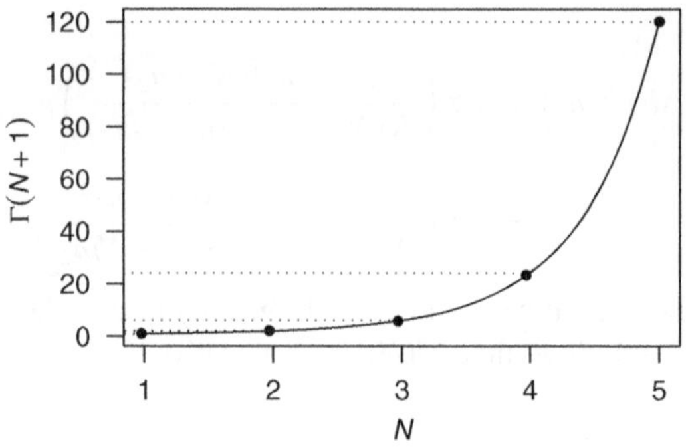

Figure 6.5 Plot of $\Gamma(N+1)$ *for real valued N between 1 and 5. Note that* $\Gamma(N+1) = N!$ *for integer valued N.*

where $n_1 = 64$, $n_2 = 67$ and $m_2 = 34$. The log-likelihood from (6.15) is unimodal with real-valued MLE $\widehat{N} = 125.615$. The integer-valued MLE is therefore either 125 or 126, and is found to be 126. The approximate standard error (obtained from the second derivative of the log-likelihood at $\widehat{N}$) is 10.4. This gives a 95 % Wald CI of (105.2, 146.0). The 95 % likelihood ratio CI is (109.7, 152.3) (Figure 6.6). For reporting purposes, one could present these CIs as (105, 146) and (109, 153), respectively. □

6.4.3 Multinomial likelihood

The hypergeometric likelihood in (6.10) is exact when the sample sizes n_1 and n_2 are fixed and the assumptions of the hypergeometric distribution are satisfied, that is, when all possible samples of size n_2 are equally probable. However, it is

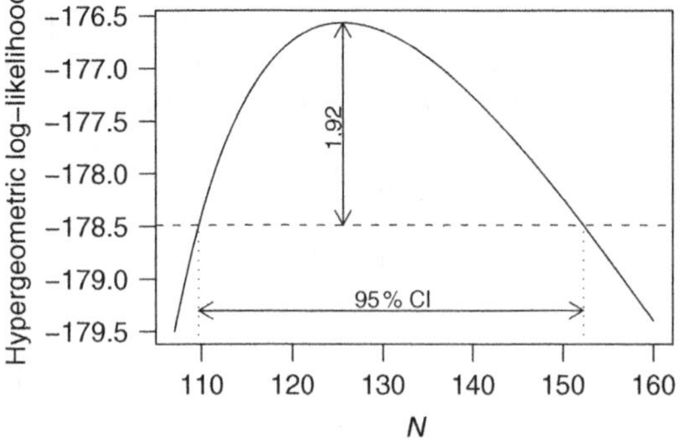

Figure 6.6 Hypergeometric log-likelihood from a mark–recapture experiment with $n_1 = 64$, $n_2 = 67$ *and* $m_2 = 34$.

cumbersome to extend the hypergeometric model to more complex mark recapture experiments, and a more flexible alternative is to employ a likelihood derived from a multinomial model for counts.

The multinomial model is approximate because it assumes that the observation on each captured animal in the second sample (i.e. whether the animal is marked or unmarked) is independent of that on all other captured animals. This is not strictly the case in most situations, because the second sample is typically most accurately described as being a sample of size n_2 taken without replacement from a finite population. Hence, the probability that a captured animal is marked depends on how many other captured animals are marked. However, the multinomial approximation works well in practice (e.g. compare Examples 6.3 and 6.4).

The multinomial likelihood necessarily contains additional parameters corresponding to capture probabilities. These additional parameters can often be explicitly estimated for any given value of N, and hence profile likelihood (Section 3.6) can be employed. For example, for any given value of N, the data from the simple mark–recapture experiment can be arranged in the following two-way table:

	Sample 2		
	Captured	Not captured	
Marked	m_2	$n_1 - m_2$	n_1
Unmarked	$n_2 - m_2$	u_2	$N - n_1$
	n_2	$N - n_2$	N

where $u_2 = N - n_1 - (n_2 - m_2)$ is the (unobserved) number of animals that are not caught in either sample. As demonstrated in Example 6.4, the likelihood to be maximized is a function of N and the cell probabilities of this table.

Example 6.4. Multinomial likelihood for a mark–recapture experiment. For the values $n_1 = 64$, $n_2 = 67$ and $m_2 = 34$ used in Example 6.3, the above table is

	Sample 2		
	Captured	Not captured	
Marked	34	30	64
Unmarked	33	u_2	$N - 64$
	67	$N - 67$	N

Assuming that the two samples (mark and recapture) are independent, this is a two-way contingency table with independent rows and columns. The parameters to be estimated are therefore (r_1, c_1, N) where r_1 and c_1 are the probabilities of the row 1 event and column 1 event, respectively. For any N, the MLEs of r_1 and c_1 are their respective row and column proportions $\widehat{r}_1 = 64/N$ and $\widehat{c}_1 = 67/N$ (see Exercise 3.5). This enables quick calculation of the profile likelihood for N (Figure 6.7).

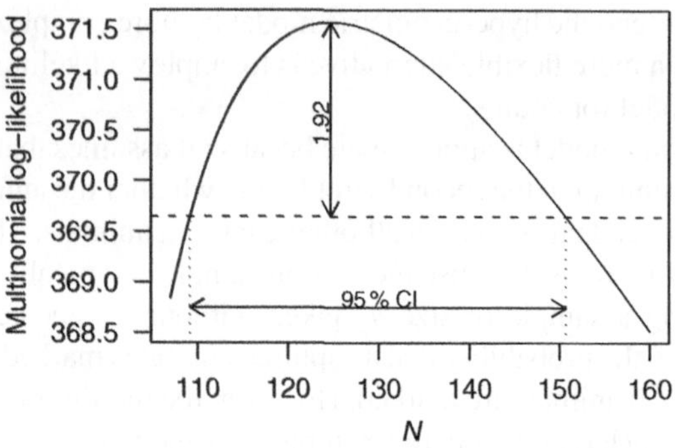

Figure 6.7 Multinomial log-likelihood from a mark–recapture experiment with $n_1 = 64, n_2 = 67$ and $m_2 = 34$.

Using an extended multinomial likelihood for $N \in \mathbb{R}^+$, the real-valued MLE is $\widehat{N} = 124.692$ and the integer-valued MLE is 125. The approximate standard error is 10.1, and the 95 % likelihood ratio CI is (109.2, 150.7). □

The multinomial likelihood has the advantage that it can be readily generalized to a multitude of variations on the mark–recapture theme, including the modelling of experiments with multiple recapture phases and experiments where distinct marks are used on every captured individual. This enables the capture history of individual animals to be observed and permits the construction of models to describe this history. Such models can, for example, allow for the possibility of trap avoidance by animals that have previously been caught (e.g. Pollock 1975, Pledger 2000). See also Borchers *et al.* (2002, p. 108) and references therein.

6.4.4 Closing remarks

The sampling distribution of MLEs of abundance can be severely right skewed for small sample sizes, and likelihood ratio or bootstrap techniques are generally to be preferred over methods that assume approximate normality (Cormack 1992, Evans, Kim and O'Brien 1996).

The R package `Rcapture` provides a general suite of functions for the estimation of abundance from mark–recapture experiments based on the multinomial model. The implementation in `Rcapture` is based on the equivalence between multinomial and Poisson models (see Example 9.2), and hence is able to express the multinomial as a log-linear model (Chapter 7). The R package `wisp` is a comprehensive wildlife simulation package and contains functions for estimation of abundance from a wide variety of experimental designs, including mark–recapture, removal (e.g. Exercise 6.6) and distance-based methods.

6.5 Exercises

6.1 **Lack-of-memory property of the exponential distribution:** Let T be distributed exponentially with hazard rate λ, that is, $T \sim \text{Exp}(\lambda^{-1})$ with density function $f(\lambda) = \lambda \exp(-\lambda t)$. For any $s \geq 0$, let f_s denote the density of the remaining time to failure given survival to time s. That is, f_s is the density of $T - s$ given $T > s$. Show that

$$f_s(t) = f(t) \ .$$

That is, the remaining time to failure is also distributed $\text{Exp}(\lambda^{-1})$.

6.2 Suppose that observations (t_i, w_i), $i = 1, ..., n$ are generated from a survival experiment with exponentially distributed survival times with hazard rate λ. The indicator variable w_i takes the value 1 if t_i is an observed failure time, or the value 0 if observation i was censored at time t_i. Maximize the log-likelihood for λ to show that

$$\widehat{\lambda} = \frac{\sum_{i=1}^n w_i}{\sum_{i=1}^n t_i} \ .$$

6.3 Fit a Cox's proportional hazards model to the poison data in Table 6.1 assuming additive effects of poison and treatment. Repeat using the full model with interaction, and compare these two models using AIC.

6.4 Weibull(ϕ, k) distributed survival times have density that can be written in the form

$$f(t; \phi, k) = k\phi^{-k}t^{k-1} \exp\left(-(t/\phi)^k\right) , \qquad y > 0,\ \phi > 0,\ k > 0 \ .$$

Show that the distribution function is

$$F(t; \phi, k) = 1 - \exp\left(-(t/\phi)^k\right) ,$$

and hence that the hazard function is

$$h(t; \phi, k) = k\phi^{-k}t^{k-1} \ .$$

6.5 In a mark–recapture experiment, if N is large compared to n_2 then it will be reasonable to assume $m_2 \sim \text{Bin}(n_2, p)$, where p is the proportion of animals that have been marked. Under this model for m_2, and assuming that n_1 and n_2 are fixed, apply the delta method to obtain the following approximate variance for the Petersen estimator,

$$\widehat{\text{var}}(\widetilde{N}) = \frac{n_1^2 n_2 (n_2 - m_2)}{m_2^3} \ .$$

6.6 The unknown number, N, of animals in a closed population can be estimated by the method of removal. In this type of experiment, n_1 animals are removed on the first removal occasion, leaving $N - n_1$ animals. A further

n_2 are removed on a second removal occasion. It will be assumed that n_1 is Bin(N, p) and that n_2 is Bin($N - n_1$, p), where p is common to both removal occasions.

Assume that p is known, leaving only N to be estimated. The likelihood for N is given by the product of the binomial likelihoods from the two removal occasions.

1. Show that an integer-valued MLE of N is

$$\widehat{N} = \left[\frac{n_1 + n_2}{2p - p^2}\right],$$

where $[x]$ denotes the integer part of x.

2. Is the above MLE the unique integer-valued MLE?

6.7 Implement and maximize the hypergeometric log-likelihood for the mark–recapture experiment in Example 6.3, and obtain the 95 % LR confidence interval.

6.8 Implement and maximize the multinomial log-likelihood for the mark–recapture experiment in Example 6.4, and obtain the 95 % LR confidence interval.

7

Generalized linear models and extensions

To generalize means to think. – Georg Hegel

7.1 Introduction

From the likelihood point of view, generalized linear models (GLMs) can simply be regarded as a flexible class of models, all of which share a convenient form of likelihood function. GLMs are in wide use because this class includes models that are natural for the modelling of count and binomial data. The likelihood of a GLM, and its derivatives, have a simple explicit form and optimization of the likelihood is fast. The existence and uniqueness of the MLE are guaranteed under very weak general conditions (e.g. Wedderburn 1976). Furthermore, by virtue of the generalized linear model structure, the regression parameters of a typical[1] GLM are not constrained, in the sense that they can take any value on the real line. This avoids the complications of constrained optimization, and it also helps to improve the approximate normality of the MLEs, by alleviating concerns about parameter values being close to the boundary of the parameter space (Section 4.7.3).

Software for fitting generalized linear models is well developed and widely available. GLMs can be fitted using the `glm` function in R, or `GENMOD` procedure in SAS. `PROC GENMOD` includes numerous options for model evaluation and

[1] But, see Box 7.2.

Maximum Likelihood Estimation and Inference: With Examples in R, SAS and ADMB, First Edition. Russell B. Millar.

inference, including automatic calculation of likelihood ratio confidence intervals. Similar functionality is provided in R, using functions that can apply appropriate methods to the fitted `glm` object.

A generalized linear model has two key features. The first is that the data have a distribution belonging to the exponential family (Section 7.2.1). This includes the normal, Poisson, binomial, gamma, inverse-Gaussian, and fixed-dispersion negative binomial. The second feature is that covariates $\boldsymbol{x}_i$ (associated with observation i) are included using a linear model, but this is a linear model for some appropriate transformation of $E[Y_i]$ (Section 7.2.2). For example, in the case of count data the standard model assumes that the data are Poisson distributed and that the log of $E[Y_i]$ follows a linear model. This is commonly known as log-linear regression.

Section 7.3 briefly presents the general form of the likelihood equations that underly the maximum likelihood modelling of exponential family data. Section 7.4 presents methods for model evaluation and comparison. The model that is used in the case study of Section 7.5 is a standard logistic regression on proportion data, and no obvious problems with the model fit are found. However, one interesting facet of this case study is that the regression coefficients of the logistic regression are not of direct interest. Rather, the case study is an example of an inverse prediction (i.e. calibration) problem in which it is desired to estimate the covariate value that corresponds to a specified expected value of the response variable.

In practice, it is very often the case that proportion data will not be well modelled by a binomial distribution, and that count data will not be well modelled by a Poisson distribution. Section 7.6 presents practical strategies for this situation, including the use of natural generalizations of the binomial and Poisson distributions, and the use of quasi-likelihood. Section 7.7 concludes this chapter with a second case study that employs these strategies in an analysis of count data.

7.2 Specification of a GLM

Section 7.2.1 defines the form of the likelihood function assumed by GLMs. At a first reading, it can by skipped by readers who are primarily interested in the application of GLMs.

7.2.1 Exponential family distribution

In a generalized linear model, it is required that the log-density function of $Y \in \mathbb{R}$ can be written in the form

$$\log f(y; \psi, \phi, w) = \frac{y\psi - b(\psi)}{a(\phi, w)} + c(y, \phi, w)\,, \tag{7.1}$$

where $\psi \in \mathbb{R}$ and $\phi \in \mathbb{R}^+$ are parameters, and w is a known 'weight'. Parameter ϕ is the dispersion parameter, and function $a(\phi, w)$ is required to be of the form ϕ/w. Formula (7.1) is a special case of the general class of exponential family distributions, and some texts refer to it as the exponential dispersion model. This distinction will not be made here, and in what follows it is implicitly understood that the terminology 'exponential family distribution' corresponds to the form of density function in (7.1).

To maintain consistency with previous notation, it is convenient to drop w from the above notation for the density function, since w is not a parameter to be estimated. That is, the log-density will be written

$$\log f(y; \psi, \phi) = \frac{y\psi - b(\psi)}{\phi/w} + c(y, \phi, w)\,. \tag{7.2}$$

The data $Y_i, i = 1, ..., n$, are assumed to be independent where each Y_i has exponential family distribution with density $f(y_i, \psi_i, \phi)$. Any covariates associated with Y_i are modelled through ψ_i (Section 7.2.2), and the dispersion parameter ϕ is assumed to be common to all observations. However, the known weights, w_i, may vary between observations.

The following two examples demonstrate that the normal and binomial distributions are of exponential family form. In the binomial case it is seen that the dispersion parameter ϕ is redundant, and it can be fixed at unity by setting the weight equal to the number of trials. Also, note that the binomial distribution is expressed for the binomial proportion instead of the number of 'successes'.

Example 7.1. If $Y \sim N(\mu, \sigma^2)$ then the log-density function can be written as

$$\log f(y; \mu, \sigma^2) = \frac{y\mu - \mu^2/2}{\sigma^2} - 0.5\left(\log(2\pi\sigma^2) + \frac{y^2}{\sigma^2}\right)\,.$$

This establishes that the $N(\mu, \sigma^2)$ distribution is of exponential family form under the parameterization $\psi = \mu$, $\phi = \sigma^2$, and with $b(\psi) = \psi^2/2$, and weight w set to unity. □

> The linear-regression model for normal data is included amongst the class of GLMs. However, there is no need to use maximum likelihood theory for these models because the sampling properties of the least-squares estimators of the regression coefficients (which are also the MLEs, Section 11.7.3) and associated test statistics are known exactly.

Box 7.1

Example 7.2. For the binomial model it is the proportion of 'successes', rather than the number of 'successes', that is expressed in exponential family form. So, let $Y = T/n$ where $T \sim \text{Bin}(n, p)$. Then, for $y \in \{0, \frac{1}{n}, \frac{2}{n}, ..., \frac{n-1}{n}, 1\}$, the log-density function is

$$\log f(y; p) = \log \binom{n}{ny} + ny \log p + n(1-y)\log(1-p)$$

$$= ny \log\left(\frac{p}{1-p}\right) + n\log(1-p) + \log\binom{n}{ny} .$$

Setting $\psi = \log(\frac{p}{1-p}) = \text{logit}(p)$, and noting that $\log(1-p) = -\log(1+e^{\psi})$, the log-density function can be written in a one-parameter exponential family form by writing

$$\log f(y; \psi) = \frac{y\psi - \log(1+e^{\psi})}{1/n} + \log\binom{n}{ny} .$$

Here, $b(\psi) = \log(1+e^{\psi})$. Parameter ϕ is absent, and can be set to unity. The weight is $w = n$. □

7.2.2 GLM formulation

In Section 7.2.1, the exponential family density function was expressed in terms of parameters ψ and ϕ. However, this parameterization is not a practical one for the purpose of specifying models for exponential family data. It is much more natural to consider the effect of covariates upon the mean $\mu = E[Y]$, and this is the approach taken by a GLM. (See Exercise 7.4 for the relationship between μ and ψ, and Section 11.7.2 for more detail.)

A GLM generalizes the linear regression model by fitting a linear model to a transformation of the mean. That is, if $\mu_i = E[Y_i]$ and $\boldsymbol{x}_i = (x_{i1}, ..., x_{ip})$ are the covariates associated with Y_i, then the GLM specifies

$$g(\mu_i) = \sum_{j=1}^{p} x_{ij}\beta_j , \tag{7.3}$$

where β_j, $j = 1, ..., p$ are parameters to be estimated. The linear model term on the right-hand side of (7.3) is called the linear predictor, and will be denoted by

$$\eta_i = \sum_{j=1}^{p} x_{ij}\beta_j . \tag{7.4}$$

The transformation g is called the link function because it links the mean to the linear predictor, as $g(\mu_i) = \eta_i$.

The link function is required to be invertible. Denoting $h = g^{-1}$, this gives

$$h(\eta_i) = \mu_i .$$

Parameters β_j, $j = 1, ..., p$ are regression coefficients and it is natural (although not absolutely necessary – see Box 7.2) that they be permitted to take any values in $\mathbb{R}$, and consequently that the linear predictor also ranges over $\mathbb{R}$. However, it may be that μ_i is not defined on all of $\mathbb{R}$. For example, it Y_i are Poisson distributed then it is necessarily the case that $\mu_i > 0$. It is therefore standard practice to choose h to be a function that maps from $\mathbb{R}$ into the domain of valid values of μ_i. In the Poisson case, a natural choice for h is the exponential transformation, $h(\eta_i) = \mu_i = e^{\eta_i}$, corresponding to the log-link function $g(\mu_i) = \log \mu_i = \eta_i$.

> Most GLM software will permit the use of nonstandard forms of the link function. For example, an identity link can be used to model Poisson data if it is felt that it is appropriate to model the mean directly using a linear model (see the case study in Section 7.7). However, an identity-link Poisson model must be treated with care, because it induces constraints on $\boldsymbol{\beta}$ by virtue of the fact that the likelihood will be undefined for values of the regression coefficients that result in a negative value of $\eta_i (= \mu_i)$ for any i.

Box 7.2

Example 7.2 continued. Here, Y_i are binomial proportions, $Y_i \sim \text{Bin}(n_i, p_i)/n_i$ and so $E[Y_i] = p_i$. If p_i is formulated as $p_i = h(\eta_i)$, then it is natural to require that h is an invertible function that maps onto the unit interval, that is, $h : \mathbb{R} \to [0, 1]$. These requirements are satisfied by the distribution function of any continuous random variable with sample space $\mathbb{R}$.

Logistic regression makes the choice of taking h to be the logistic function, which is the distribution function of a logistic random variable. With this choice,

$$p_i = h(\eta_i) = \frac{e^{\eta_i}}{1 + e^{\eta_i}}, \tag{7.5}$$

and its inverse is the logit link function

$$\begin{aligned} \eta_i = g(p_i) &= \log\left(\frac{p_i}{1 - p_i}\right) \\ &= \text{logit}(p_i) . \end{aligned} \tag{7.6}$$

Probit regression uses the distribution function of a normal random variable as the choice of h. The inverse of the normal distribution function is the probit function, from which this model inherits its name. Another choice that is commonly provided in many software packages is the complementary log-log link, $g(p_i) = \log(-\log(1 - p_i))$, which is obtained as the inverse of the extreme value distribution function.

The distribution functions of the logistic and normal are both symmetric and are very similar in shape (Figure 7.1). In practice, logistic and probit regression will give similar fits (Chambers and Cox 1967), but the logistic is generally to be

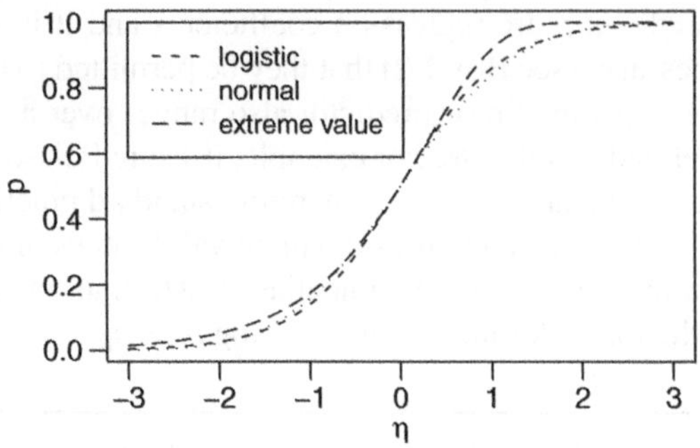

Figure 7.1 Comparison of the distribution functions of the logistic, normal, and extreme value distributions. Their inverses are the logit, probit, and complementary log-log link functions, respectively. The distribution functions have been standardized to be those of random variables with median equal to zero, and variance equal to unity.

preferred due to ready interpretability of regression coefficients in terms of log odds (see the case study in Section 7.5), and because the logit is the natural link to use for binomial data (see Box 7.3). □

7.3 Likelihood calculations

It is assumed that the data Y_i, $i = 1, ..., n$ are independently distributed from an exponential family distribution where $\mu_i = E[Y_i]$ depends on parameters $\boldsymbol{\beta} = (\beta_1, ..., \beta_p)$. In the previous section it was assumed that the relationship between μ_i and $\boldsymbol{\beta}$ was specified using the generalized linear model of Equation (7.3). However, the likelihood calculations hold more generally and it is enough to require only that μ_i is a smooth (i.e. differentiable) function of $\boldsymbol{\beta}$. For example, the likelihood calculations are also applicable to nonlinear regression models for exponential family data.

The parameters to be estimated are $\boldsymbol{\beta}$, and the dispersion parameter ϕ if it is present. The dispersion parameter is not usually of direct interest and here focus is on estimation of the MLE $\widehat{\boldsymbol{\beta}}$. It can be shown (see Section 11.7.2) that the likelihood equations to be solved for finding $\widehat{\boldsymbol{\beta}}$ are

$$\begin{aligned}\frac{\partial l(\boldsymbol{\beta}, \phi; \mathbf{y})}{\partial \beta_j} &= \sum_{i=1}^{n} \frac{\frac{\partial \mu_i}{\partial \beta_j}(y_i - \mu_i)}{\text{var}(Y_i)} \\ &= 0\,, \quad j = 1, ..., p\,. \end{aligned} \tag{7.7}$$

Note that this system of equations depends only on the specification of the mean and variance of each Y_i. This is also true of the second derivative calculations that are required to obtain the approximate variance matrix of $\widehat{\boldsymbol{\beta}}$.

> **Canonical link functions.** For GLMs, the second-order partial derivatives of the log-likelihood have an especially convenient form when the canonical choice of the link function is used (see Section 11.7.5 for more details). In particular, it is then the case that these second derivatives do not depend on the data. The canonical link functions for binomial proportion and Poisson count data are the logit and log links, respectively.

Box 7.3

7.4 Model evaluation

Many of the model evaluation tools for linear regression have analogous variants for evaluating the fit of GLMs to exponential family data. Deviance (Section 7.4.1) plays much the same role as the residual sum-of-squares, and analysis of deviance (Section 7.4.2) is the direct analogue of analysis of variance. The (raw) residual is generalized by the Pearson and deviance residuals, which are introduced in Section 7.4.3.

Section 7.4.4 uses the Pearson chi-square statistic and/or the deviance statistic to provide omnibus tests of goodness-of-fit for binomial and Poisson models fitted to proportion and count data, respectively. These tests do not have a linear regression counterpart, because they require a known value of the dispersion parameter ϕ.

7.4.1 Deviance

To within an additive constant, the deviance of an exponential family model is just the negative of twice the log-likelihood of that model. To mimic the behaviour of the residual sum-of-squares, an additive constant is chosen so that the deviance of a model with perfect fit is zero. Analysis of deviance (for hypothesis testing within nested sets of models, say) is equivalent to the use of likelihood ratio tests, and for these tests to be meaningful it is necessary to require that the value of the dispersion parameter ϕ be constant across the collection of models. This is the case for the Poisson and binomial distributions, since they implicitly fix $\phi = 1$. It also holds for the negative binomial distribution, notwithstanding that the use of deviance in the context of negative binomial models must be done with care due to the presence of an additional parameter m (see Example 7.4 and Box 7.7).

Here, it is convenient to write the log-likelihood of an exponential family model as a function of the vector of means $\boldsymbol{\mu} = (\mu_1, ..., \mu_n)$ and the dispersion parameter ϕ (rather than as a function of $\boldsymbol{\beta}$ and ϕ), and this formulation will be denoted $l(\boldsymbol{\mu}, \phi; \mathbf{y})$.

A model with perfect fit is one that sets the fitted mean of each observation equal to its observed value, and hence has log-likelihood $l(\mathbf{y}, \phi; \mathbf{y})$. This is called the saturated model, though some texts may refer to it as the full model (but more commonly, full model is used to describe a model that includes all possible explanatory terms).

The deviance of a given fitted model is simply twice the difference between the log-likelihood of the saturated model and the log-likelihood of that model. So, if the fitted means are $\widehat{\boldsymbol{\mu}}$, then the deviance of the model is

$$D = 2\big(l(\mathbf{y}, \phi; \mathbf{y}) - l(\widehat{\boldsymbol{\mu}}, \phi; \mathbf{y})\big) . \tag{7.8}$$

This may also be referred to as the residual deviance of the given model. In the case of a GLM, the fitted values in $\widehat{\boldsymbol{\mu}}$ are obtained from the fitted values of the linear predictors, that is, $\widehat{\mu}_i = h(\widehat{\eta}_i) = h(\sum_j x_{ij}\widehat{\beta}_j)$.

Since the saturated model maximizes the likelihood over all possible values of $\widehat{\boldsymbol{\mu}}$ (Box 7.4), it follows that the deviance is always non-negative. The deviance of the model that fits a common mean to all observations is commonly called the null deviance. The null deviance is the GLM analogue to the total sum-of-squares in a linear regression.

> The saturated model is obtained from fitting a model with i as a factor variable having n levels. This model can be parameterized such that $\mu_i = \beta_i, i = 1, ..., n$ under the formulation assumed in Section 7.3. The likelihood equation in (7.7) then takes the form
>
> $$\frac{\partial l}{\partial \mu_i} = \frac{(y_i - \mu_i)}{\text{var}(Y_i)} = 0 , \quad i = 1, ..., n ,$$
>
> and hence it follows that $\widehat{\boldsymbol{\mu}} = \mathbf{y}$ is a solution to these equations.

Box 7.4

Example 7.3. Poisson deviance. If $y_i, i = 1, ..., n$, are the observed values of independently distributed Pois(λ_i) random variables, then the deviance of the model with fitted means $\widehat{\lambda}_i$ is

$$D = 2\left(\sum_{i=1}^{n}\big(-y_i + y_i \log y_i - \log(y_i!)\big) - \sum_{i=1}^{n}\big(-\widehat{\lambda}_i + y_i \log \widehat{\lambda}_i - \log(y_i!)\big)\right)$$

$$= 2\sum_{i=1}^{n}\Big((\widehat{\lambda}_i - y_i) + y_i \log(y_i/\widehat{\lambda}_i)\Big) . \tag{7.9}$$

□

Example 7.4. Negative binomial deviance. The negative binomial distribution has several common parameterizations. Here, the form in Equation (15.2) is assumed, and the parameters are μ and m. For a given fixed value[2] of parameter m, the negative binomial is an exponential family distribution with $\phi = 1$ (Exercise 7.3), and the deviance of a negative binomial model with fitted means $\widehat{\mu}_i$ is

$$D = 2\sum_{i=1}^{n}\left[y_i \log\left(\frac{y_i}{\widehat{\mu}_i}\right) - (m + y_i)\log\left(\frac{y_i + m}{\widehat{\mu}_i + m}\right)\right] . \tag{7.10}$$

It is important to note that this deviance is obtained by assuming that the fitted negative binomial model and saturated negative binomial model share the same common value of m. □

7.4.2 Model selection

The general tools of model selection that were covered in Section 4.4 are equally applicable to GLMs and other exponential family models. In particular, likelihood ratio tests can be used for selecting between nested models, although a more holistic approach is to use AIC or BIC (Section 4.4.1), which also has the advantage that non-nested models can be compared.

In the case of binomial or Poisson models, using the likelihood ratio to perform hypothesis tests on nested models can conveniently be implemented by comparison of the model deviances. If model A with deviance D_A is a subset of model B with deviance D_B, then the difference in deviances is

$$\begin{aligned} D_A - D_B &= 2\big(l(y; y) - l(\widehat{\boldsymbol{\mu}}_A; y)\big) - 2\big(l(y; y) - l(\widehat{\boldsymbol{\mu}}_B; y)\big) \\ &= 2\big(l(\widehat{\boldsymbol{\mu}}_B; y) - l(\widehat{\boldsymbol{\mu}}_A; y)\big) , \end{aligned}$$

where $\widehat{\boldsymbol{\mu}}_A$ and $\widehat{\boldsymbol{\mu}}_B$ denote the fitted means under models A and B, respectively. That is, the difference in deviances is the likelihood ratio test statistic for the hypothesis that the restricted model (model A) is true. If this hypothesis is true then this test statistic has an approximate chi-square distribution with degrees of freedom equal to the difference in the number of parameters between the two models.

The deviances of a collection of nested binomial or Poisson models can be partitioned in a similar way to the partitioning of the sums-of-squares in an ANOVA table. Rather than using the decrease in the residual sum-of-squares, it is the reduction in deviance that is used to measure the improvement in fit obtained by adding a term to the model. This procedure is commonly known as analysis of deviance.

7.4.3 Residuals

Pearson and deviance residuals are commonly used in the evaluation of GLMs and other exponential family models. The Pearson residual for observation i is the raw

---

[2] The negative binomial model with fixed m has the property that $\mathrm{var}(Y_i)$ is a quadratic function of $E[Y_i]$ – see the description of the NB distribution in Section 15.4.1.

residual divided by the estimated standard deviation of Y_i. That is,

$$r_i^P = \frac{y_i - \widehat{\mu}_i}{\sqrt{\widehat{\text{var}}(Y_i)}} .$$

The deviance residual for observation i is the signed square root of that observation's contribution to the deviance. That is

$$r_i^D = \text{sign}(y_i - \widehat{\mu}_i)\sqrt{2}\big(l(y_i, \phi; y_i) - l(\widehat{\mu}_i, \phi; y_i)\big)^{\frac{1}{2}} ,$$

where sign$(\cdot)$ is the function that takes the value -1 if its argument is negative, and the value 1 otherwise. Note that the sum of the squared deviance residuals (over the n observations) is equal to the deviance, D.

Example 7.5. Poisson residuals. If y_i is the observed value of a Poisson random variable and $\widehat{\lambda}_i$ is the fitted mean, then the Pearson residual for this observation is

$$r_i^P = \frac{y_i - \widehat{\lambda}_i}{\sqrt{\widehat{\lambda}_i}} ,$$

and the deviance residual is

$$r_i^D = \text{sign}(y_i - \widehat{\lambda}_i)\sqrt{2}\big((\widehat{\lambda}_i - y_i) + y_i \log(y_i/\widehat{\lambda}_i)\big)^{\frac{1}{2}} .$$ □

7.4.4 Goodness of fit

When the dispersion parameter ϕ is known, then the log-likelihood $l(\boldsymbol{\mu}, \phi; \boldsymbol{y})$ is well defined for the saturated model having $\boldsymbol{\mu} = \boldsymbol{y}$. (This is in contrast to the case of normally distributed data with unknown variance, because $l(\boldsymbol{y}, \sigma^2; \boldsymbol{y})$ can be made arbitrarily large by making σ^2 arbitrarily small.) An omnibus goodness-of-fit test statistic for a given model can therefore be obtained as the likelihood ratio test statistic for the test of that model nested within the saturated model. This test statistic is none other than the deviance defined in (7.8). The saturated model has one parameter for each of the n observations, and so the degrees of freedom is $n - p$ where p is the number of parameters in the given model.

An alternative omnibus goodness-of-fit test statistic is provided by Pearson's chi-square. This is the sum of the squared Pearson residuals,

$$P_{\chi^2} = \sum_{i=1}^{n} \frac{(y_i - \widehat{\mu}_i)^2}{\widehat{\text{var}}(Y_i)} , \qquad (7.11)$$

and the degrees of freedom is again $n - p$.

Caution is advised when using the deviance and Pearson chi-square as goodness-of-fit statistics, because their approximate distribution under H_0 (i.e. that the given model is true) may not be well approximated by a χ^2_{n-p} distribution (see Example 7.6). For example, in the case of the deviance, this is because the usual large-sample theory of likelihood ratio statistics does not hold here. This theory requires that the models are of a particular fixed form. However, the saturated model is not of fixed form, since the number of parameters is equal to the sample size n, and hence increases with increasing sample size.

The chi-square approximation to the null distribution of the deviance and Pearson chi-square statistics is especially questionable if the data are overly sparse. There are various rules of thumb regarding what is meant by 'overly sparce'. For example, an analysis of contingency tables (a Poisson GLM in which all explanatory variables are factors) can be implemented using the FREQ procedure in SAS. This procedure gives a warning message if less than 80 % of the fitted cell means exceed the value 5. For binomial data, $Y_i \sim \text{Bin}(n_i, p_i)$, an analogous rule of thumb would be to require $\min(n_i\widehat{p}_i, n_i(1-\widehat{p}_i)) \geq 5$ for most of the observations.

In practice, the Pearson chi-square is more robust (see Example 7.6) to sparseness than deviance (McCullagh and Nelder 1989), and is generally preferred. However, unlike deviance, it can not be used for model comparison.

Example 7.6. Deviance and P_{χ^2} for sparse binomial data. As an extreme example of sparse binomial data, suppose that Y_i, $i = 1, \ldots, n$ are iid Bin$(1, p)$, that is, iid Bernoulli with Prob$(Y_i = 1) = p$. The MLE is $\widehat{p} = \sum_{i=1}^n y_i/n$, and assuming that $0 < \widehat{p} < 1$, the Pearson chi-square statistic is

$$\begin{aligned}\sum_{i=1}^n \frac{(y_i - \widehat{p})^2}{\widehat{p}(1-\widehat{p})} &= \sum_{\{i:y_i=0\}} \frac{\widehat{p}}{1-\widehat{p}} + \sum_{\{i:y_i=1\}} \frac{1-\widehat{p}}{\widehat{p}} \\ &= n(1-\widehat{p})\frac{\widehat{p}}{1-\widehat{p}} + n\widehat{p}\frac{1-\widehat{p}}{\widehat{p}} \\ &= n\,.\end{aligned}$$

The saturated model has log-likelihood of zero, since the contribution from each Bernoulli observation is $y_i \log y_i$ if $y_i = 1$, or $(1-y_i)\log(1-y_i)$ if $y_i = 0$. Thus, the deviance statistic is

$$D = -2l(\widehat{p}; \mathbf{y}) = -2n\left(\widehat{p}\log\widehat{p} + (1-\widehat{p})\log(1-\widehat{p})\right).$$

Here, the iid Bernoulli model is the true model, and it is desirable that these two statistics have approximate χ^2_{n-1} distribution, but this is not the case. The Pearson statistic is the constant n, which is close to its nominal $n-1$ degrees of freedom and, if compared to the upper quantiles of a χ^2_{n-1} distribution, provides no evidence against the true model. The sampling distribution of the deviance depends on p. In particular, if $p = 0.5$ then it can be shown that $D/(n-1)$ converges (in probability) to $2\log(2) = 1.386$ as n increases. It follows that, for large n and $p = 0.5$, the deviance statistic will invariably lead to rejection of the true model if it is assumed that the deviance possesses a χ^2_{n-1} sampling distribution. □

Example 7.7. Fish counts. The following fifteen counts of snapper (*Pagrus auratus*) are a subset of the data used in the case study in Section 7.7, and were obtained from independent replicate deployments of a baited underwater video camera.

2 9 6 2 17 2 9 8 9 11 3 7 3 2 1

It was desired to test the null hypothesis $H_0 : Y_i \sim \text{Pois}(\lambda)$, that is, whether the observations conform to the iid Poisson model with mean λ. Under H_0, $\widehat{\lambda} = \overline{y} = 6.0667$, and from (7.9), the deviance statistic for the goodness of fit under H_0 is 45.910.

With count data, the Pearson chi-square test statistic is frequently written using an alternative notation where $O_i = y_i$ and $E_i = \widehat{\lambda}_i$. That is, O_i and E_i denote the observed and expected values for observation i, respectively. This yields the commonly seen expression

$$P_{\chi^2} = \sum_{i=1}^{n} \frac{(O_i - E_i)^2}{E_i} ,$$

which evaluates to 46.967 for the above counts. Both test statistics are very extreme in comparison to the χ^2_{14} distribution, and it is concluded that the iid Poisson model is not appropriate for these data. □

7.5 Case study 1: Logistic regression and inverse prediction in R

To avoid redundancy, R is used here and SAS is used in the next case study in Section 7.7. The R code in Section 7.5.1 fits a logistic regression model to proportion data. The log-likelihood function of this model has previously been used in this text – it was presented in the form of a contour plot (Figure 3.5) in Section 3.5.1.

The research question here is to investigate the relationship between the length of a fish and the probability that it is retained by a trawl. One responsibility of fisheries managers is to legislate a mesh size for the fishing fleet that is commensurate with the minimum legal size (MLS) at which fish can be kept. It is known that undersized fish that are landed onboard and discarded can suffer substantial mortality, and so it is preferable that such fish are able to pass through the trawl meshes. In practice, the mesh size needs to be chosen as a compromise between the need to minimize retention of sub-legal sized fish and the need to maintain high catchability of legal-sized fish. A suitable compromise may be to legislate that the mesh size is such that the length at which the retention probability is 0.5, $l_{50\,\%}$, is equal to the MLS.[3]

Fish length is the sole covariate in this model, and so (including the intercept) the logistic regression model has just two regression parameters to be estimated, β_1

[3] This policy varies with fishery agency. For example, the International Council for the Exploration of the Sea recommends (ICES 1979) that MLS correspond to the length of 25 % retention.

and β_2. However, these regression parameters do not have any meaning to fisheries managers. Rather, as noted above, the managers require inference about quantities such as $l_{50\,\%}$. From the logistic equation in (7.12) it is immediate that the retention probability is 0.5 at length l for which $\beta_1 + \beta_2 l = 0$. That is, $l_{50\,\%} = -\beta_1/\beta_2$, is the quantity of primary interest to fisheries managers.

This is an example of a so-called 'inverse prediction' problem, because rather than being used to predict values of the response variable for a given covariate value, the model is being used to estimate the covariate value for a given expected value of the response. More generally, inverse prediction arises in bioassay studies where it is desired to quantify the effectiveness of a substance when used at different dosages (the covariate). In these experiments, the dosage with 50 % effectiveness is often given a name such as ED50 or LD50, the former denoting the effective dose corresponding to a 50 % response rate, and the latter denoting the lethal dose corresponding to a 50 % mortality rate in the context of studies where the dichotomous response variable is survival or death.

This case study proceeds by implementing and assessing the fit of a logistic regression model to data collected from an experimental trawl. It concludes by addressing the research question of interest. That is, by making inference about $l_{50\,\%} = -\beta_1/\beta_2$.

7.5.1 Size-selectivity modelling in R

An experiment was conducted to investigate the ability of a fishing trawl to release undersized haddock. The aft part of a trawl is called the codend, and this was constructed using 113 mm diamond-mesh netting. A fine mesh cover was placed around the codend, and held clear of the codend by the use of plastic hoops (Figure 7.2). The data (Table 7.1) are from Clark (1957), and are the length frequencies of fork lengths (i.e. length from the nose to the base of the fork in the tail) in both the codend and cover, in 1-cm length-classes. Small fish can escape

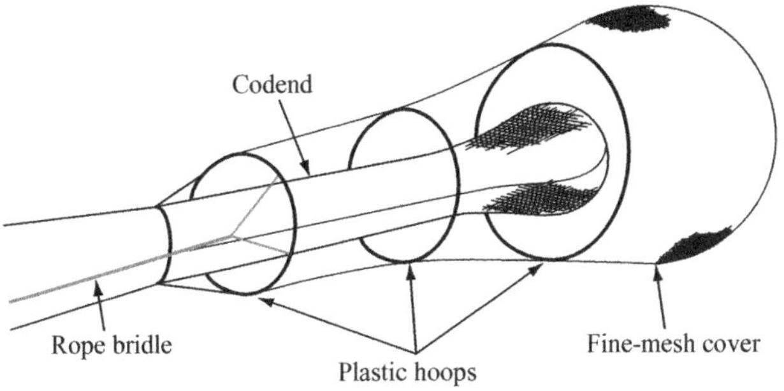

Figure 7.2 An experimental trawl with a fine-mesh cover over the codend. Small fish are able to pass through the meshes of the codend, but will be retained in the cover. Reproduced by permission of William MacBeth.

Table 7.1 Fork-length frequencies of haddock caught in the codend and its cover. The recorded length is the midpoint of the corresponding 1-cm length-class.

length	codend	cover	length	codend	cover	length	codend	cover
19.5	0	2	32.5	3	2	45.5	12	0
20.5	0	5	33.5	4	4	46.5	9	0
21.5	0	11	34.5	5	12	47.5	3	0
22.5	0	28	35.5	8	9	48.5	5	1
23.5	1	53	36.5	13	14	49.5	3	0
24.5	5	46	37.5	29	15	50.5	5	0
25.5	1	35	38.5	29	8	51.5	2	0
26.5	3	27	39.5	34	9	52.5	2	0
27.5	1	5	40.5	30	3	53.5	1	0
28.5	0	3	41.5	29	3	54.5	1	0
29.5	1	2	42.5	18	2	55.5	4	0
30.5	0	0	43.5	16	1			
31.5	0	2	44.5	11	0			

the codend by swimming through the 113 mm mesh and most will be caught in the cover, but large fish will be less able to squeeze through the codend mesh.

The R code below reads in the tab-delimited data file `haddock.dat` which contains the data in Table 7.1 arranged in three columns. The column names are `forklen`, `codend` and `cover`.

The raw data are counts and could be modelled using a Poisson log-linear regression (Exercise 7.7). However, it is equivalent (see Example 9.2), and much more intuitive, to regard the data as binomial where, for each length, `codend` is the number of 'successes' (from the fisherman's point of view) out of a total of `codend` + `cover` trials. Note that although `haddock.dat` has 37 lines of data, the codend and cover counts for fork length of 30.5 cm are both zero, and so there is no binomial observation for this fork length. Thus, when calculating the degrees of freedom, the number of binomial observations is actually 36, and the logistic regression models assumes that they are independent.

A plot of the proportion of haddock retained in the codend versus `forklen` suggests a very strong effect of fork length, and also that a logistic curve may be a reasonable choice to model these data (Figure 7.3). The equation for the logistic curve is given by Equation (7.5), and in the present context this is

$$p_i = \frac{\exp(\beta_1 + \beta_2 l_i)}{1 + \exp(\beta_1 + \beta_2 l_i)}, \quad i = 1, ..., 37, \tag{7.12}$$

where p_i is the probability that a fish of length l_i is retained. That is,

$$\text{logit}(p_i) = \log\left(\frac{p_i}{1 - p_i}\right) = \beta_1 + \beta_2 l_i . \tag{7.13}$$

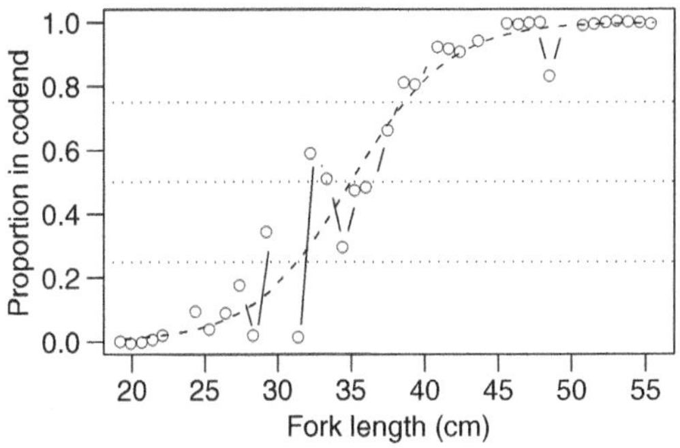

Figure 7.3 Proportion of total haddock catch (by length) that was retained in the codend. The dashed line is the fitted logistic curve from Equation (7.12).

The ML fit of this model uses the `glm` function. The fit, and Figure 7.3, are obtained using the following code.

```
> Haddock.df=read.table("Haddock.dat",head=T)
> #Rename codend as y, and calculate total catch at length
> Haddock.df=transform(Haddock.df,y=codend,n=codend+cover)
> attach(Haddock.df)
> #Take a quick look at the data
> plot(forklen,y/n,las=1,type="b",
+      xlab="Fork length (cm)",ylab="Proportion in codend")

> #Logistic model: logit link is default when family is binomial
> Haddock.glm=glm(y/n~forklen,family=binomial,weight=n)
> #Overlay fitted retention probabilities
> lines(forklen[n>0],fitted(Haddock.glm),type="l",lty=2)
> abline(h=c(0.25,0.5,0.75),lty=3)
> summary(Haddock.glm)
Coefficients:
             Estimate Std. Error z value Pr(>|z|)
(Intercept) -10.63219    0.86468  -12.30   <2e-16 ***
forklen       0.30396    0.02363   12.86   <2e-16 ***
---
Signif. codes:  0 *** 0.001 ** 0.01 * 0.05 . 0.1   1

(Dispersion parameter for binomial family taken to be 1)

    Null deviance: 432.464  on 35  degrees of freedom
Residual deviance:  23.436  on 34  degrees of freedom
  (1 observation deleted due to missingness)
AIC: 79.75
```

As anticipated, `forklen` is highly significant, with a Wald test statistic of $12.86^2 = 165.4$, on 1 d.o.f. under the null hypothesis $H_0 : \beta_2 = 0$. However, it would be preferable to employ a likelihood ratio test, and this is straightforward since the difference between the deviances of the null and logistic models is the required LR test statistic of H_0. This statistic is approximately 409.0, and with an approximate χ_1^2 distribution under H_0, the null hypothesis is again soundly rejected.

The estimated value of β_2 is 0.304, and from (7.13) it follows that each 1-cm increase in fork length results in an estimated increase of 0.304 in the log-odds of a haddock being retained in the codend.

The residual (i.e. model) and null deviances were seen to be 23.436 and 432.464, respectively. The values of these deviances can be verified from evaluation of the log-likelihoods of the relevant models. As a demonstration, the next code segment uses the `logLik` function to obtain the log-likelihoods of the null, logistic, and saturated models.

```
> #Log-likelihoods of null, logistic, and saturated models
> Null.lhood=logLik(glm(y/n~1,family=binomial,weight=n))
> Model.lhood=logLik(Haddock.glm)
> Satd.lhood=logLik(glm(y/n~as.factor(forklen),
+                       family=binomial,weight=n))
> cat("\n Null model log-likelihood=",Null.lhood,
+      "\n Logistic model log-likelihood=",Model.lhood,
+      "\n Saturated model log-likelihood=",Satd.lhood,"\n")

 Null model log-likelihood= -242.3891
 Logistic model log-likelihood= -37.87508
 Saturated model log-likelihood= -26.15727
```

From the above log-likelihoods it is verified that the residual deviance of the logistic model, 23.436, is obtained as $2(-26.157 - (-37.875))$, and similarly for the null deviance.

The above inferences are conditional on the fitted model being a good fit to the data. Since the data are sparse for some fork lengths, especially for larger fish where p_i is close to unity, it is advisable to use the Pearson chi-square statistic to test goodness of fit rather than the residual deviance. The Pearson chi-square statistic is provided by the following code.

```
> cat("\n Pearson chi-square=",
+      sum(resid(Haddock.glm,type="pearson")^2))

 Pearson chi-square= 27.09488
```

The 34 degrees of freedom for the Pearson chi-square goodness-of-fit test is likely to be inappropriately high due to the sparsity of the data. It is reassuring to see that the observed statistic of 27.1 is well below 34, and so a somewhat inflated d.o.f. will not alter the conclusion that this goodness-of-fit test finds no evidence of lack of fit. Combining the counts into wider length classes when the data are sparse would be one way of more formally mitigating this issue.

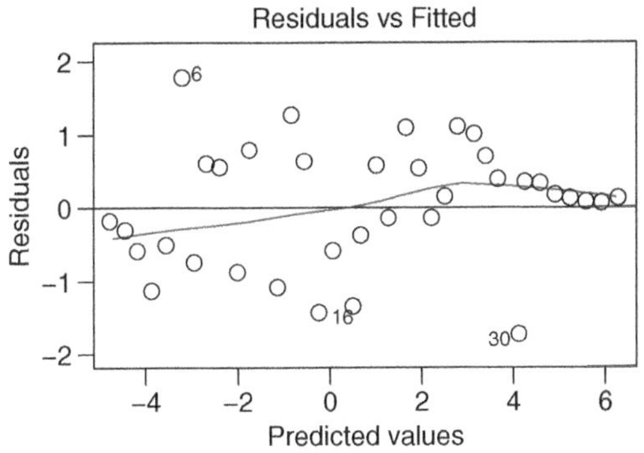

Figure 7.4 Plot of deviance residuals versus linear predictor values.

Further model evaluation can be obtained using the `plot` function, which draws several diagnostic plots. In particular, a plot of the deviance residuals versus linear predictors ($\eta_i = \beta_1 + \beta_2 l_i$) is obtained from the following code.

```
> #Use which=1 to select only the plot of resids
      vs linear predictors
> plot(Haddock.glm,which=1)
> abline(h = 0)
```

The residual plot (Figure 7.4) suggests that there are more negative residuals for small predicted values, and more positive residuals for large predicted values. However, it should be borne in mind that, unlike residuals from a linear regression, the deviance residuals from this logistic model do not have a sampling distribution with a median of zero. For example, for sufficiently large lengths, the probability of retention is so close to unity that the most likely outcome is that all fish of that length will be retained, and hence most of the very large lengths will result in a small positive residual.

The fitted logistic regression model appears to be adequate, and so it is reasonable to proceed with further inference. Likelihood ratio confidence intervals for β_1 and β_2 are preferable to Wald-based intervals (Section 4.3) and these are easily obtained using the `confint` function.

```
> #Likelihood ratio 95% confidence interval for regression coefs
> confint(Haddock.glm,level=0.95)
                  2.5 %      97.5 %
(Intercept) -12.4563256 -9.0501022
forklen       0.2604867  0.3535532
```

It remains to make inference about the quantity of primary interest, the length of 50 % retention, $l_{50\%} = -\beta_1/\beta_2$. The MLE is $\widehat{l}_{50\%} = -\widehat{\beta}_1/\widehat{\beta}_2 = 34.98$ cm.

The approximate standard error of $\widehat{l}_{50\,\%}$ can be obtained via the delta method (Section 4.2.3), as shown below.

```
> #Wald 95% confidence interval for length of 50% retention
> MLE=coef(Haddock.glm)
> Vhat=vcov(Haddock.glm)
> L50hat=-MLE[1]/MLE[2]
> library(msm)
> L50.se=deltamethod(~-x1/x2,MLE,Vhat)
> L50.se
[1] 0.4246743

> cat("\n L50% is estimated to be",L50hat,"with 95% Wald CI (",
+      round(L50hat+c(-1,1)*qnorm(0.975)*L50.se,2),") \n")

 L50% is estimated to be 34.97927 with 95% Wald CI ( 34.15 35.81 )
```

The likelihood ratio confidence interval for $l_{50\,\%}$ is preferred over the above Wald CI, but it is not as straightforward to obtain. The LRCI can be calculated by re-parameterizing the model using $(\beta_2, l_{50\,\%})$, say, but fitting this model will require use of a nonlinear optimizer. The LRCI can also be determined from the contour plot of the log-likelihood (Figure 3.5). See Exercise 3.8 for further details.

The 95 % likelihood ratio CI is found to be (34.12, 35.79), which is negligibly different from the Wald CI. In this application of logistic regression, the sample size is large (the total number of fish is 590) and the normal approximation works well.

A confidence interval for $\zeta = -\beta_1/\beta_2$ can be obtained using Fieller's method (Fieller 1954, Finney 1971) which is derived from the statistical properties of the ratios of normally distributed random variables. Calculation of the Fieller CI is relatively easy, and is obtained explicitly as the solution of a quadratic equation. It is implemented in R (`mratios` package) and in the SAS procedure `PROBIT`.

A number of authors have compared the relative performance of the CIs constructed from use of the delta method, Fieller's method, and likelihood-ratio. In general, while Fieller's method tended to be preferable to the delta method, it was found to be conservative (i.e. to have greater than the required coverage probability) and produced wider CIs. Overall, likelihood-ratio CIs are preferred (Williams 1986, Faraggi, Izikson and Reiser 2003).

Box 7.5

7.6 Beyond binomial and Poisson models

The binomial and Poisson distributions provide the starting point for modelling of proportion and count data, respectively. However, it is frequently the case that a goodness-of-fit test will provide strong evidence against these models, yet there will be little indication (from diagnostic plots, say) that the specification of $E[Y_i]$ is at fault. Indeed, in the next case study, the full model includes all three factor variables and their interactions, but nonetheless it is seen to exhibit gross 'lack of fit' when the counts are modelled using a Poisson distribution. It is the assumption that the counts are Poisson distributed that is at fault.

The inadequacy of the binomial or Poisson models is often due to lack of independence of the underlying Bernoulli trials in the case of proportion data, or of the events being enumerated in the case of count data. For example, if B_j, $j = 1, \ldots, m$ are iid Bernoulli(p) and Y is the proportion of 'successes', $Y = \frac{1}{m}\sum_{j=1}^{m} B_j$, then Y is distributed as a binomial proportion, $Y \sim \frac{1}{m}\text{Bin}(m, p)$. However, if the B_j are all positively correlated then Y will still have mean p, but variance in excess of $p(1 - p)/m$. A real-world example would be that of an insurance company, where financial risk would be grossly underestimated if it was assumed that the proportion of house-insurance policies that incurred claims was distributed according to the binomial distribution. A claim from one policy could be due to a natural disaster that was also experienced by many other policy holders.

Similarly, count data on animal abundance is notoriously non-Poisson, because individuals of many species interact and congregate. That is, the presence of an animal tends to be positively correlated with the presence of other individuals of that species, and hence counts of animals typically have variance well in excess of that assumed by the Poisson distribution. Hospitals can not assume that arrivals of emergency patients will be Poisson distributed because, in the event of an epidemic (say), the arrival of one infected patient is likely to be associated with the arrival of many.

The above mentioned violations of independence are with regard to the underlying random processes that give rise to binomial or Poisson random variables. These violations typically result in the variance of the counts (or proportions) being greater than that expected under the Poisson (or binomial) model, and the data are then said to be over-dispersed, or variance-inated, or to have extra variation. It will continue to the case that the counts (or proportions) are assumed to be independent observations.

Section 7.6.1 presents the convenient quasi-likelihood approach for fitting models to over- (or under-) dispersed count or proportion data. The dispersion parameter, ϕ, of the Poisson and binomial distributions is unity, but quasi-likelihood instead allows the value of ϕ to be estimated. However, the resulting likelihood may not correspond to the density function of a random variable (since it may not integrate to unity when integrated over the sample space), and hence this approach is called *quasi*-likelihood. One limitation of quasi-likelihood is that it is not suitable when it is required to make predictions (of future observations, say), because there is no concept of an underlying statistical model from which predictions can be postulated.

With proportion data, it can sometimes be the case that there is no clear notion of the number of trials. This can happen when the proportion is obtained from the percentage of a surface or volume that is composed of some constituent. For example, it could be the percentage of a bodily organ that is tumorous, or the percentage cover of a field by a particular grass, or the percentage of a leaf's surface that is covered in blotch. This is a situation to which quasi-likelihood (Sections 7.6.1 and Chapter 8) is well suited. Indeed, analysis of a historical set of barley-blotch data is given in Section 8.2.1.

Box 7.6

Section 7.6.2 returns to the framework of the parametric statistical model, and introduces the negative binomial model as a candidate for the modelling of over-dispersed count data using R or SAS. This section also introduces the zero-inflated Poisson (ZIP) model. Zero-inflation is another frequent cause of departure from the Poisson (or binomial) distributions, and refers to the presence of more zeros than would otherwise be plausible under those models.

The ZIP model can be represented as a mixture of a Poisson distribution and a degenerate zero distribution, the latter taking the value zero with probability of unity. For example, Bøhning, Dietz and Schlattmann (1999) used the zero-inflated Poisson to model the number of damaged teeth in seven-year-old school children. Conceptually, one might infer that some children took care of their teeth and incurred no damage to them, but for children who neglected their teeth the number that were damaged could be modelled by a Poisson distribution. The population can be regarded as a mixture of the two types of childen, those who took care of their teeth and those who neglected them.

The negative binomial distribution can also be very effective at modelling count data that have more zeros than would be reasonable under the Poisson model (Warton 2005). Moreover, unlike the ZIP model, the negative binomial also has the ability to model extra-Poisson variation in the nonzero counts.

7.6.1 Quasi-likelihood and quasi-AIC

Here, $\widehat{v}_i$ is used to denote the estimated variance of Y_i under the fitted Poisson or binomial model. The variances under these models are, of course, $v_i = \mu_i = E[Y_i]$ for counts modelled as Poisson, and $v_i = p_i(1 - p_i)/n_i = \mu_i(1 - \mu_i)/n_i$ for proportion data modelled as binomial proportions, $Y_i \sim \frac{1}{n_i}\text{Bin}(n_i, p_i)$, and are estimated by $\widehat{v}_i = \widehat{\mu}_i$ and $\widehat{v}_i = \widehat{\mu}_i(1 - \widehat{\mu}_i)/n_i$, respectively. With this notation, the Pearson chi-square statistic can be written

$$P_{\chi^2} = \sum_{i=1}^{n} \frac{(y_i - \widehat{\mu}_i)^2}{\widehat{v}_i} .$$

If the data y_i, $i = 1, .., n$ are not overly sparse, and the model is correct, then the Pearson chi-square statistic will be approximately distributed χ^2_{n-p}, where p is the number of estimated parameters in the fitted model (Section 7.4.4).

If the Pearson chi-square provides evidence of lack of fit, but specification of μ_i appears to be satisfactory, then this suggests that v_i is not correctly specifying the variance of Y_i. A convenient adjustment is obtained by assuming that the true variance of Y_i, var(Y_i), is a common multiplicative constant of the assumed variance v_i, for all i. That is,

$$\text{var}(Y_i) = kv_i \ . \tag{7.14}$$

With this adjustment the Pearson chi-square becomes

$$\begin{aligned} P^*_{\chi^2} &= \sum_{i=1}^{n} \frac{(y_i - \widehat{\mu}_i)^2}{k\widehat{v}_i} \\ &= \frac{P_{\chi^2}}{k}. \end{aligned}$$

Since the χ^2_{n-p} distribution has expected value of $n - p$, it is natural to choose k so that $P^*_{\chi^2} = n - p$. That is, the value of k in (7.14) is taken to be

$$\widehat{k} = \frac{P_{\chi^2}}{n-p} \ .$$

The above variance adjustment can be interpreted as extending the Poisson or binomial model by using an estimated value, $\widehat{\phi} = \widehat{k}$, of the dispersion parameter (rather than $\phi = 1$) in their exponential family formulation $f(y; \psi, \phi)$ in (7.2). This follows from the property that exponential family models have variance of the form in (7.14), where v_i is a function of μ_i, and k is the dispersion parameter ϕ (see Equation (11.26) in Section 11.7.2). That is, the estimated dispersion is

$$\widehat{\phi} = \widehat{k} = \frac{1}{n-p} \sum_{i=1}^{n} \frac{(y_i - \widehat{\mu}_i)^2}{\widehat{v}_i} \ . \tag{7.15}$$

The square root of $\widehat{\phi}$ is called the estimated scale parameter. The dispersion parameter can also be estimated using the deviance in place of the Pearson chi-square statistic. However, the Pearson is more robust to sparsity in the data and is generally preferred.

Care must be taken with the interpretation of the exponential family formulation $f(y; \psi, \widehat{\phi})$ in (7.2), because it is no longer the case that $f(y; \psi, \widehat{\phi})$ necessarily corresponds to a density function. This is because it may no longer integrate to unity (when integrated with respect to y), and $f(y; \psi, \widehat{\phi})$ (as a function of ψ) is best regarded as a useful modification of the likelihood function. More general application of the notion of quasi-likelihood is covered in Chapter 8.

It is a trivial matter for GLM software to implement the above variance adjustment. For GLMs, the partial derivatives of the log-likelihood are given in

Equation (7.7), and are

$$\frac{\partial l(\boldsymbol{\beta};\mathbf{y})}{\partial \beta_j} = \sum_{i=1}^{n} \frac{\frac{\partial \mu_i}{\partial \beta_j}(y_i - \mu_i)}{v_i}, \quad j = 1, ..., p.$$

The variance adjustment in (7.14), with k estimated by $\widehat{\phi}$, is simply scaling these derivatives by a multiplicative constant,

$$\frac{\partial l(\boldsymbol{\beta};\mathbf{y})}{\partial \beta_j} = \frac{1}{\widehat{\phi}} \sum_{i}^{n} \frac{\frac{\partial \mu_i}{\partial \beta_j}(y_i - \mu_i)}{v_i}, \quad j = 1, ..., p. \tag{7.16}$$

Consequently, this does not alter the MLE, $\widehat{\boldsymbol{\beta}}$.

In effect, the multiplicative correction to the derivatives in (7.16) is saying that inference should be based on using the log-likelihood divided by $\widehat{\phi}$. Therefore, under this adjustment, likelihood ratio test statistics (and differences in deviance) must be divided by $\widehat{\phi}$ before being evaluated against the appropriate χ^2 distribution. The estimated standard errors of MLEs are inversely proportional to the magnitude of the second derivatives of the log-likelihood, and it therefore follows that the standard errors of the MLEs must be multiplied by the estimated scale parameter, $\widehat{\phi}^{\frac{1}{2}}$.

The quasi-likelihood correction to the AIC model-selection criterion gives the quasi-AIC

$$\text{QAIC} = -2\frac{l(\widehat{\boldsymbol{\theta}})}{\widehat{\phi}} + 2p. \tag{7.17}$$

Use of QAIC for model comparison makes sense only if a common value of $\widehat{\phi}$ is used (Burnham and Anderson 2002) within the collection of models being compared. This value will typically be obtained by fitting the full model, provided this retains sufficient degrees of freedom. Application of QAIC is demonstrated in Section 7.7, where the model comparison includes choosing between two different specifications of the link function.

7.6.2 Zero inflation and the negative binomial

When count (or proportion) data are not well modelled by the Poisson (or binomial) model, it may be due to extra variation arising because the data contain more zero observations than is plausible under that model. Then, one approach is to modify the model to explicitly include the excess of zeros. For example, a zero-inflated Poisson (ZIP) random variable is obtained as a probabilistic mixture of a zero and a Poisson random variable. Note that the value zero can be regarded as a degenerate random variable having density function (i.e, probability mass function) that takes the value zero with probability of one. This probability mass function can be denoted using the indicator function $1_{[y=0]}$. Then, using the mixture-model density function from

Equation (5.6), and with p denoting the probability that Y is sampled from the 'zero distribution', the density function of the ZIP(p, λ) distribution is a weighted sum of the $1_{[y=0]}$ and Pois(λ) density functions,

$$f(y; p, \lambda) = p1_{[y=0]} + (1-p)\frac{e^{-\lambda}\lambda^y}{y!}$$
$$= \begin{cases} p + (1-p)e^{-\lambda}\ , & y = 0, \\ (1-p)\frac{e^{-\lambda}\lambda^y}{y!}\ , & y = 1, 2, 3, ... \end{cases}$$

The ZIP model can be fitted using `PROC GENMOD` in SAS, by specifying the `MODEL` statement option `DIST=ZIP`. Using R, the ZIP and zero-inflated binomial models can be fitted using the `vglm` function within the `VGAM` package, using syntax analogous to that of the `glm` function. These software tools are able to model independent data $Y_i \sim \text{ZIP}(p_i, \lambda_i)$ using separate generalized linear model specifications for the zero-inflation probabilities p_i, and the Poisson means λ_i, $i = 1, ..., n$.

The negative binomial (NB) distribution is another alternative for modelling of over-dispersed count data, and can be very effective at accommodating excess zeros (Warton 2005). However, it also allows for extra-Poisson variation in the nonzero counts. For the purpose of modelling Y_i, $i = 1, ..., n$ there are alternative parameterizations of the NB distribution, but the most convenient choice is to assume that parameter m in the NB density function (see Equations (15.1) and (15.2)) is constant across all observations. With a constant value of m, var(Y_i) is a quadratic function of $\mu_i = E[Y_i]$,

$$\text{var}(Y_i) = \mu_i\left(1 + \frac{\mu_i}{m}\right)\ . \tag{7.18}$$

Moreover, for any fixed value of m, the NB is an exponential family distribution (Exercise 7.3). Hence it can be fitted efficiently via profile likelihood using a one-dimensional numerical optimization with respect to m (Section 5.4.1).

Negative binomial models that use the (m, μ_i) parameterization can be fitted using `PROC GENMOD`, by specifying the `MODEL` statement option `DIST=NEGBIN`. In R, one can use the `glm.nb` function in the `MASS` package, or the `vglm` function in the `VGAM` package. Function `vglm` includes additional flexibility, such as permitting non-constant m.

Example 7.7 continued. If the 15 fish counts are modelled as iid NB(m, μ) then $(\widehat{m}, \widehat{\mu}) = (2.766, 6.067)$, and the resulting deviance is 15.29 (Exercise 7.8). The degrees of freedom of the deviance remains at 14, since the saturated model uses the value of $\widehat{m} = 2.766$ from the fitted model. The Pearson chi-square is 14.71, on 13 d.o.f. (since two parameters were estimated). Neither of these statistics suggests lack of fit. □

When comparing negative binomial models, caution should be used when using the reported deviances of the fitted models, as given in (7.10). The deviance assumes that the likelihood of the saturated model is calculated using the same value of parameter m as the fitted model. Thus, the difference in deviance between two fitted models no longer corresponds to twice the difference in the log-likelihoods of those models (because the likelihood of the saturated models differ according to the value of m used).

Box 7.7

For negative binomial modelling of Y_i, $i = 1, \ldots, n$, a less-common parameterization is to use the form of (15.1), but with parameters (p, μ_i), where p is held constant. It is then the case that

$$\text{var}(Y_i) = \frac{\mu_i}{p},$$

and so this parameterization is useful if it is desired to model $\text{var}(Y_i)$ as proportional to μ_i. However, this parameterization corresponds to allowing the m parameter to vary across observations, and can no longer be fitted as a conventional GLM. It can be fitted using the `vglm` function in the `VGAM` package by specifying appropriate parameter constraints, else it may require the use of a general purpose optimizer.

While the negative binomial model is effective for dealing with an inflation in the number of zeros compared to the Poisson, it may not always be able to accommodate enough excess zeros to be able to adequately fit the data. In that case a zero-inflated negative binomial (ZINB) could be considered. The ZINB model can be fitted in R using the `vglm` function, and the iid ZINB model is demonstrated using the following code.

```
library(VGAM)
vglm(y~1,family=zinegbinomial)
```

`PROC GENMOD` does not currently (SAS 9.2) implement the zero-inflated negative binomial, but for those with access to the SAS/ETS module, it can be fitted using `PROC COUNTREG`. In the code below, the `ZEROMODEL` statement is used to specify a model for the zero-inflation probability, which of course is fixed under the iid ZINB model.

```
TITLE "Fit of iid ZINB model";
PROC COUNTREG;
  MODEL y = / DIST=ZINB;
  ZEROMODEL y ~;
RUN;
```

Example 7.8. The 40 counts of roots in the apple micro-propagation data of Example 4.8 are

0	0	0	0	0	0	0	0	0	0	0	0	0	0	0	0	0	0	0	1
1	2	2	3	3	3	3	4	4	4	5	6	6	6	6	7	7	7	9	9

It has previously been determined (from applying a G-test to a frequency table of these data in Example 3.4) that an iid Poisson model is not appropriate, due to many more zeros than plausible under the Poisson model. This is corroborated by fitting the Poisson model directly to these counts. The log-likelihoods of the saturated and Poisson models are -34.316 and -110.555, respectively, resulting in a deviance of $2(-34.316 - (-110.555)) = 152.477$, and the Pearson chi-square is 135.469. These are highly extreme compared to a χ^2_{39} distribution, and the Poisson model is again soundly rejected.

The ZIP, NB and ZINB models have log-likelihoods of -74.880, -81.000 and -74.598, respectively (Exercise 7.9). These correspond to AICs of 153.761, 166.000 and 155.196, respectively, and hence the ZIP model is preferred. The MLE is $(\widehat{p}, \widehat{\lambda}) = (0.4698, 4.621)$.

The mean and variance of $Y \sim \text{ZIP}(p, \lambda)$ are $\mu = (1 - p)\lambda$ and $\text{var}(Y) = (1 + p\lambda)\mu$ (see Exercise 13.11), from which it is possible to evaluate the Pearson chi-square statistic for the fitted ZIP model by using the MLE to obtain $\widehat{\mu}$ and $\widehat{\text{var}}(Y)$. This is $P_{\chi^2} = 42.725$ with 38 d.o.f. under the ZIP model. The p-value is approximately 0.28 and provides no evidence against lack of fit. □

7.7 Case study 2: Multiplicative vs additive models of over-dispersed counts in SAS

This case study presents an application of quasi-likelihood and negative binomial modelling to count data that were found to have extra-Poisson variation. ZIP and ZINB modelling could also have been investigated, and this is left as an exercise. SAS is used exclusively in this case study, but an example of quasi-likelihood analysis using R can be found in the analysis of the barley blotch data in Section 8.2.1.

This analysis includes a comparison of two link functions, the log and identity, because it was a primary research question to determine whether the effects of the covariates were best modelled as multiplicative or additive. That is, with η_i denoting the linear predictor defined in (7.4), it was of interest to know which choice, $E[Y_i] = \exp(\eta_i)$, or $E[Y_i] = \eta_i$, would provide the best model of these data.

7.7.1 Background

Willis and Millar (2005) counted snapper (*Pagrus auratus*) using a baited underwater video camera. Sampling was conducted in three marine reserves located on the

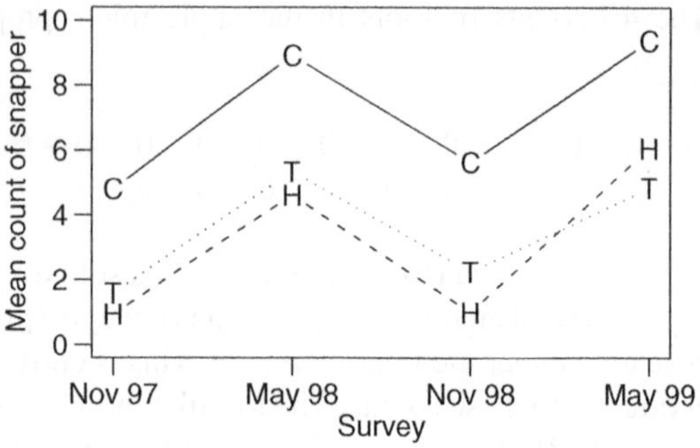

Figure 7.5 Mean counts of snapper from baited underwater video footage in three marine reserves at each of four sampling times. The reserves are labelled C (Cape Rodney-Okakari Point), H (Hahei) and T (Tawharanui).

NE coast of the North Island of New Zealand at four sampling times (Nov 1997, May 1998, Nov 1998 and May 1999). Between 12 and 24 replicate deployments of the camera were used at each combination of location and time. The total number of deployments was 204. The recorded count was the maximum number of snapper observed within any single video frame, over a period of 30 minutes of video footage.

The strong seasonal pattern in the counts (Figure 7.5) is due to the seasonal migration of snapper moving from deep offshore waters into the coastal reserves. This movement is initiated by an increase in coastal water temperatures that begins in the austral spring, typically in November or December, and peaks in late summer or early autumn.

In the context of GLM modelling of count data it is usual practice to fit a multiplicative model by specifying a log-link function. However, here, it was not appropriate to constrain the research question by this assumption. In particular, to gain greater understanding of the influence of marine reserves on the population dynamics of snapper, it was of interest to determine whether the seasonal migration had a multiplicative or additive effect on snapper density.

The data set `SnapperCounts.dat` contains the four variables, `reserve` (taking values 'CROP', 'Hahei' or 'Tawh'), `season` ('Autumn' or 'Spring' – austral seasons corresponding to the May and November measurements, respectively), `year` ('97-98' and '98-99'), and the snapper count, `y`. The analysis begins by fitting a Poisson GLM using the full model with all interaction terms.[4] The number of terms in the model was then reduced according to the sequence below, where R, S, and Y denote reserve, season and year effects, respectively, and $*$ denotes interaction.

[4] At the time of the study, these three reserves were the only no-take marine reserves on the NE coast of New Zealand, and it is therefore appropriate to treat them as fixed effects. Year is undoubtedly a random effect (see Chapter 10), but with only two replicates, it will be treated as fixed.

1. $R * S * Y$: The three-way interaction and all two-way interactions, i.e. the full model.
2. $R * Y + Y * S$: Season (i.e. migration) effect that is not dependent on reserve.
3. $R * Y + S$: Season effect that is not dependent on reserve or year.
4. $R + Y + S$: All main effects only.
5. $R + S$: Reserve and season main effects only.
6. S: Season effect only.
7. 1: No effects (intercept only), i.e. the null model.

Each of the above seven specifications of the model terms was fitted using both a log link (i.e. multiplicative model) and an identity link (i.e. additive model), resulting in a total of 14 model specifications for $E[Y_i]$. However, note that the null and full models are identical under both links, albeit with different parameterizations. In the case of the full model, this is because it fits an unrestricted specification of $E[Y_i]$ to each of the 12 reserve-season-year combinations, regardless of the link function employed.

7.7.2 Poisson and quasi-Poisson fits

The Poisson log-linear GLM fit of the full model with all interaction terms is obtained using the following SAS code.

```
***Poisson fit of full model***;
PROC GENMOD;
  CLASS Reserve Season Year;
  MODEL y=Reserve*Season*Year / DIST=POISSON;
```

In the SAS output (Figure 7.6), the log-likelihood of the fitted model is given in the row labelled `Full Log Likelihood`. (The `Log Likelihood` value should be ignored, since this is the log-likelihood calculated without the constant term in the Poisson density function.) The Pearson chi-square of 614.1 on 192 degrees of freedom clearly indicates lack of fit. The full model corresponds to modelling the replicate count data within each of the 12 reserve-year-season combinations as iid Poisson, and hence the lack of fit must be due to the inadequacy of the Poisson to describe the sampling variability in these replicate counts. Using the Pearson chi-square statistic, dispersion is estimated to be $\widehat{\phi} = 614.0835/192 = 3.1984$, from Equation (7.15).

To compare the seven model-term specifications listed above using quasi-AIC, it is necessary to use the common value of $\widehat{\phi} = 3.1984$ from the fit of the full model. The SAS code below demonstrates this for the quasi-Poisson fit of an additive model with $R + S$ terms. It uses the option `SCALE=1.7884` to specify $\widehat{\phi}^{\frac{1}{2}} = 1.7884$. The

Criteria for assessing goodness of fit			
Criterion	DF	Value	Value/DF
Deviance	192	702.3815	3.6582
Scaled Deviance	192	702.3815	3.6582
Pearson Chi-Square	192	614.0835	3.1984
Scaled Pearson X2	192	614.0835	3.1984
Log Likelihood		836.6711	
Full Log Likelihood		-627.9418	
AIC (smaller is better)		1279.8836	
AICC (smaller is better)		1281.5171	
BIC (smaller is better)		1319.7010	

Figure 7.6 The model fit table from `PROC GENMOD`*, for the full three-way Poisson GLM with all interaction terms.*

`LINK=IDENTITY` option specifies that the link function is the identity. That is, the Poisson means are linear (additive) functions of the regression coefficients.

```
***Quasi-Poisson fit of the R+S additive model with fixed scale***;
PROC GENMOD;
  CLASS Reserve Season Year;
  MODEL y=Reserve Season /  DIST=POISSON LINK=IDENTITY SCALE=1.7884;
```

The AIC value reported in the resulting model-fit table (not shown) is the desired QAIC, calculated according to (7.17). The QAIC values for the 14 models are shown in Table 7.2.

It is seen that an additive model with reserve and season main effects is preferred. The quasi-Poisson fit of this preferred model was then repeated, but with the `SCALE=1.7884` option replaced by `PSCALE`. This instructs `GENMOD` to use

Table 7.2 Model comparison results for the marine reserve data. The quasi-AIC values are computed using $\widehat{\phi} = 3.198$, and are not comparable to the AIC values from the negative binomial fit.

		Quasi-likelihood		Negative binomial	
		Mult.	Additive	Mult.	Additive
Terms	No. pars	QAIC	QAIC	AIC	AIC
1: Full	12	416.66	416.66	1057.31	1057.31
2: $R * Y + Y * S$	8	420.00	410.58	1061.76	1050.90
3: $R * Y + S$	7	418.19	408.60	1059.91	1048.97
4: $R + Y + S$	5	414.57	404.71	1056.06	1045.25
5: $R + S$	4	413.51	403.64	1055.07	1044.01
6: S	2	456.70	456.70	1091.39	1091.39
7: Null	1	502.76	502.76	1116.06	1116.06

Analysis of maximum likelihood parameter estimates								
Parameter		DF	Estimate	Standard Error	Wald 95 % Confidence Limits		Wald Chi-Square	Pr > ChiSq
Intercept		1	1.7422	0.4186	0.9218	2.5626	17.32	<.0001
Reserve	CROP	1	3.4773	0.6224	2.2575	4.6972	31.22	<.0001
Reserve	Hahei	1	-0.6953	0.5088	-1.6926	0.3020	1.87	0.1718
Reserve	Tawh	0	0.0000	0.0000	0.0000	0.0000	.	.
Season	Autumn	1	3.9067	0.5178	2.8918	4.9215	56.92	<.0001
Season	Spring	0	0.0000	0.0000	0.0000	0.0000	.	.
Scale		0	1.7660	0.0000	1.7660	1.7660		

Note: The scale parameter was estimated by the square root of Pearson's Chi-Square/DOF.

Figure 7.7 The parameter estimates table from the quasi-likelihood fit of the additive model with reserve and season main effects.

the $\widehat{\phi}$ value estimated from the specified model. This was found to be $\widehat{\phi} = 3.1187$, corresponding to a scale of $\widehat{\phi^{\frac{1}{2}}} = 1.7660$ (Figure 7.7).

The quasi-likelihood fit has identical parameter estimates to the Poisson fit, but the approximate standard errors have been inflated by a factor of 1.7660. The model is additive, so for example, the fitted expected count of snapper at the Tawharanui reserve in spring is approximately 1.74. This increases to approximately $1.74 + 3.91 = 5.65$ by autumn.

7.7.3 Negative binomial fits

The above quasi-Poisson model inflated the Poisson variance by the multiplicative factor $\widehat{\phi} = 3.1187$, and hence is implicitly assuming $\text{var}(Y_i) = 3.1187E[Y_i]$. However, a plot of variance versus mean (within each of the 12 reserve-year-season

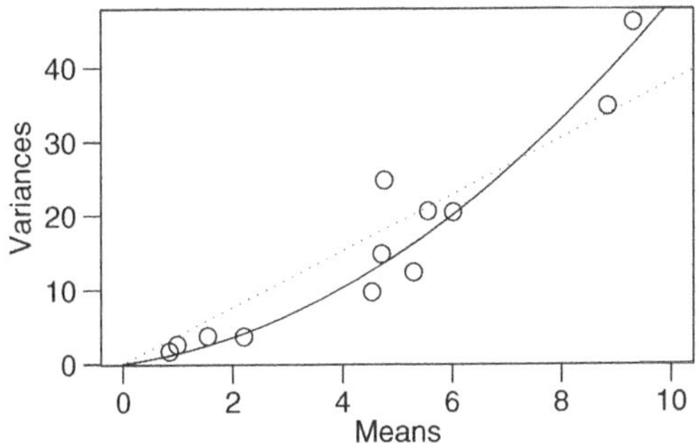

Figure 7.8 Plot of sample variances versus sample means of the snapper count data within each of the 12 combinations of reserve, year and season. The dashed line shows the least-squares fit of var(Y) $= b\mu$, and the solid line shows the least squares fit of var(Y) $= \mu + b\mu^2$.

combinations) reveals that this assumption is questionable (Figure 7.8). The plot suggests a quadratic relationship between the variances and the means, and hence it would be natural to explore the use of a negative binomial model for these data.

Negative binomial models were fitted under the seven specifications of model terms, and two choices of link function, and using AIC, the additive model with reserve and season main effects was again found to be the preferred model (Table 7.2). The following PROC GENMOD code demonstrates the fit of this model.

```
***Negative binomial additive fit of R+S model***;
PROC GENMOD;
  CLASS Reserve Season Year;
  MODEL y=Reserve Season /  DIST=NEGBIN LINK=IDENTITY;
```

The total number of parameters in the preferred model has increased from four to five, due to the addition of the negative binomial shape parameter, m. However, recall that m is regarded as fixed in calculation of the degrees of freedom (Figure 7.9) for the deviance (see Box 7.7), and so its d.o.f. is 200. PROC GENMOD uses the same d.o.f. for the Pearson chi-square, but it could be argued that it should be 199. The number of parameters is five for the purpose of calculating AIC.

The Pearson chi-square (Figure 7.9) does not show evidence against the fit, but the deviance does (p-value$\approx$ 0.023). It would be good practice to explore this further, and one possibility would be to determine if the zero-inflated negative binomial model provided a sufficient improvement in fit to warrant the extra complexity. Snippets of code for fitting the ZINB model were provided in Section 7.6.2, and their modification for use in this example is left as an exercise (Exercise 7.10).

Finally, the estimates from the fit of the additive negative binomial model with reserve and season main effects are shown in Figure 7.10. Note that there is fairly

Criteria for assessing goodness of fit			
Criterion	**DF**	**Value**	**Value/DF**
Deviance	200	241.6623	1.2083
Scaled Deviance	200	241.6623	1.2083
Pearson Chi-Square	200	165.1128	0.8256
Scaled Pearson X2	200	165.1128	0.8256
Log Likelihood		947.6084	
Full Log Likelihood		-517.0045	
AIC (smaller is better)		1044.0089	
AICC (smaller is better)		1044.3120	
BIC (smaller is better)		1060.5995	

Figure 7.9 The model fit table for the additive negative binomial model with reserve and season main effects.

Analysis of maximum likelihood parameter estimates								
Parameter		DF	Estimate	Standard Error	Wald 95% Confidence Limits		Wald Chi-Square	Pr > ChiSq
Intercept		1	1.8177	0.3672	1.0980	2.5374	24.50	<.0001
Reserve	CROP	1	3.3925	0.6845	2.0508	4.7342	24.56	<.0001
Reserve	Hahei	1	-0.8180	0.4269	-1.6547	0.0188	3.67	0.0554
Reserve	Tawh	0	0.0000	0.0000	0.0000	0.0000	.	.
Season	Autumn	1	3.9304	0.5894	2.7753	5.0855	44.47	<.0001
Season	Spring	0	0.0000	0.0000	0.0000	0.0000	.	.
Dispersion		1	0.5967	0.0942	0.4121	0.7813		

Note: The negative binomial dispersion parameter was estimated by maximum likelihood.

Figure 7.10 The parameter estimates table for the additive negative binomial model with reserve and season main effects.

good agreement of the MLEs and their standard errors between the quasi-Poisson and negative binomial fits.

7.8 Exercises

7.1 Show that the Pois(λ) distribution is an exponential family distribution.

7.2 Show that the Gamma(α, β) distribution is an exponential family distribution.

7.3 Show that the negative binomial distribution, NB(m, p), with known $m > 0$, and density function

$$f(y; p) = \frac{\Gamma(y+m)}{\Gamma(m)y!} p^m (1-p)^y, \quad y = 0, 1, ..., \quad 0 \leq p \leq 1 ,$$

is an exponential family distribution with $\psi = \log(1-p)$ and $\phi = 1$.

7.4 In Exercise 11.8 it is seen that for exponential family distributions, $\mu = b'(\psi)$ and $\text{var}(Y) = b''(\psi)a(\phi)$. Using these equivalences, and the results of Exercises 7.2 and 7.3, calculate the mean and variance of the following random variables

1. $Y \sim \text{Gamma}(\alpha, \beta)$
2. $Y \sim \text{NB}(m, p)$

7.5 Determine the form of the deviance when y is observed from a Bin(n, p) distribution.

7.6 In the continuation of Example 3.3 in Section 3.4, $y_1 = 10$ and $y_2 = 35$ were observed from Pois(λ_i) distributions, $i = 1, 2$, and the likelihood ratio test of $H_0 : \lambda_2 = 2\lambda_1$ required fitting the model subject to this restriction. Under

H_0, and using the log link, the linear predictors are

$$\eta_1 = \log \lambda_1 = \beta ,$$
$$\eta_2 = \log \lambda_2 = \beta + \log 2 .$$

This requires a fixed constant term of $\log 2$ to be included in the linear predictor η_2. This constant is called an offset, and can be specified as an option in GLM software. Hence, use R or SAS to fit the model under H_0, and verify that the fitted model has log-likelihood of -6.118.

7.7 Repeat the analysis of the haddock length frequency data in Table 7.1 by modelling these data as counts using a Poisson log-linear GLM, and confirm that $\widehat{\beta}_1, \widehat{\beta}_2$, and their estimated standard errors are unchanged. Note that under the Poisson model, the data are 74 counts and `forklen` and `gear` (codend or cover) are fitted as factors (see Examples 9.2 and 9.3). The effect of length on retention corresponds to the interaction between `gear` and `forklen`.

7.8 Use R or SAS to fit an iid negative binomial model to the fish count data in Example 7.7, and verify the deviance and Pearson chi-square statistics given in the continuation of that example in Section 7.6.2.

7.9 Use SAS or R to fit the iid ZIP, NB and ZINB models to the micro-propagation data in Example 7.8, and verify the stated AICs.

7.10 Extend the fit of the $R+S$ model in Section 7.7 by using the `vglm` function in the R package `VGAM`, or `PROC COUNTREG` in SAS/ETS, to fit a zero-inflated negative binomial model to the snapper count data. Fit the model with constant zero-inflation probability, and compare both log and identity links. (The additive ZINB fit of this model has AIC of 1028.4, and hence is preferred to the NB fit.)

8

Quasi-likelihood and generalized estimating equations

In theory there is no difference between theory and practice. In practice there is.
– Yogi Berra

8.1 Introduction

Section 7.6.1 introduced a form of quasi-likelihood for the situation where the variance of the observations clearly differed from that specified under the assumed generalized linear model. There, it was assumed that the true variance of each observation, var(Y_i), was proportional to the variance specified under the GLM. For example, if the observations are over-dispersed count data, and the underlying GLM assumes (erroneously) that they are Poisson distributed, then $v_i = \mu_i$ is the variance specified under the GLM. The quasi-likelihood approach assumes $\text{var}(Y_i) = \phi v_i$, and it can then be said that the data are being modelled as quasi-Poisson. However, it is important to note that the quasi-likelihood approach does not attempt to make any specification of the true sampling distribution of Y_i. That is, there is no specification of a 'quasi-Poisson' distribution. This is an example of performing likelihood-type inference in the absence of a formal statistical model for the data, and such inference is the focus of this chapter.

Maximum Likelihood Estimation and Inference: With Examples in R, SAS and ADMB, First Edition. Russell B. Millar.

Section 8.2 presents the more general form of quasi-likelihood that was popularized by Wedderburn (1974). This approach assumes that the observations are independent, however, it is no longer a requirement for $E[Y_i]$ and $\text{var}(Y_i)$ to be obtained from a dispersion-adjusted generalized linear model. Instead, it is enough that $E[Y_i]$ and $\text{var}(Y_i)$ are smooth functions of the parameter (vector) $\boldsymbol{\theta}$. When $E[Y_i]$ does follow the structure of a generalized linear model, but $\text{var}(Y_i)$ is otherwise arbitrary, then the quasi-likelihood estimator (QLE) can be obtained using the `glm` function in R or the SAS procedure GENMOD.

Liang and Zeger (1986) showed that Wedderburn's quasi-likelihoood approach could be extended to data that are correlated, and in particular, to the analysis of grouped data (Section 8.3). Such data could be obtained from repeated measurements on randomly chosen experimental subjects, say. This approach is commonly known as the method of generalized estimating equations (GEEs), and is demonstrated in an analysis of drug efficacy data in which observations are grouped within clinics (Section 8.3.1). These data are subsequently re-analyzed using a mixed-effects model in Section 10.6, and comparison with the GEE approach can be found there.

Quasi-likelihood and GEEs enable the modeler to make inference about $\boldsymbol{\theta}$ from specification of only the mean and variance (matrix) of the data. This is advantageous in situations where there may be no convenient statistical model from which to construct a likelihood, or when there is uncertainty or reluctance to specify a particular form of the density function $f(y; \boldsymbol{\theta})$. The analysis of leaf blotch data (Section 8.2.1) demonstrates a situation where it is unclear how to specify a likelihood for proportion data. A binomial model does not suit these data, because the proportion is measured by visual inspection, and there is no notion of the proportion having arisen from n iid Bernoulli trials.

Estimation using quasi-likelihood or GEEs can be considered an example of M-estimation (Huber 1964).[1] M-estimation can be regarded as a general methodology that emulates the salient features of maximum likelihood estimation, but without the strict need to use a likelihood function from a formal parametric model. It achieves this through the use of estimating functions in place of the score function (the derivative of the log-likelihood). So, instead of solving the likelihood equation (i.e. setting the score function to zero), a M-estimator is obtained by solving the estimating equation (i.e. setting the estimating function to zero). An estimating function is required to possess those properties of a score function that are sufficient to ensure the resulting M-estimator will be consistent, and have an approximate normal distribution under weak conditions. A brief look at the theory and properties of M-estimators, including the use of robust estimators of variance, is given in Part III of this text (Sections 12.2.5 and 12.3.2), and more detail can be found in Stefanski and Boos (2002). Indeed, for quasi-likelihood or GEE estimators to be

[1] Use of the terminology is not standardized, and some authors (e.g. Heyde 1997) use the term quasi-likelihood in place of M-estimation.

consistent, it is enough that $E[Y_i]$ be correctly specified (this follows from Equation (12.27) in Section 12.2.5).

There is a price to be paid for the flexibility and robustness offered by M-estimation. For example, without the notion of an underlying statistical model it is not possible to predict future observations, and there is limited ability to compare models. Moreover, its robustness comes at the price of efficiency, and the M-estimator has higher asymptotic variance than the MLE (Section 12.2.5). Of course, this is a price worth paying if the true form of the likelihood can not be well specified.

8.2 Wedderburn's quasi-likelihood

Let Y_i, $i = 1, \ldots, n$ be independent. It will be assumed that $\mu_i = E[Y_i]$ and $\text{var}(Y_i)$ are smooth (i.e. differentiable) functions of $\boldsymbol{\theta} \in \mathbb{R}^s$, but no further specification of the distribution of Y is required.

The quasi-likelihood estimate $\widetilde{\boldsymbol{\theta}}$ of the true unknown $\boldsymbol{\theta}_0$ is obtained as the solution of the s-dimensional set of estimating equations

$$\sum_{i=1}^{n} \frac{\frac{\partial \mu_i}{\partial \theta_j}(y_i - \mu_i)}{v_i} = 0 \,, \quad j = 1, \ldots, s, \tag{8.1}$$

where $v_i \propto \text{var}(Y_i)$. Note that v_i need only be specified up to a multiplicative constant, because this does not alter $\widetilde{\boldsymbol{\theta}}$.

Wedderburn (1974) showed that, for sufficiently large n, $\widetilde{\boldsymbol{\theta}}$ is approximately normally distributed with mean $\boldsymbol{\theta}_0$. More generally, this large-sample approximation holds even when v_i is misspecified (Section 12.3.2), but at the cost of an increase in the variance of $\widetilde{\boldsymbol{\theta}}$ (Exercise 12.5).

In the following analysis of proportion data, $p_i = \mu_i$ has the usual logistic model form, and so quasi-likelihood estimation can be implemented using the `glm` function in R. A residual plot from the fit of a quasi-binomial model indicates that the assumption $\text{var}(Y_i) \propto p_i(1 - p_i)$ is not appropriate, and a user-defined form of $\text{var}(Y_i)$ is then employed. Implementation in `PROC GENMOD` is analogous, with the specification of $\text{var}(Y_i)$ being implemented via the `VARIANCE` statement.

8.2.1 Quasi-likelihood analysis of barley blotch data in R

The data used herein are from the original quasi-likelihood example of Wedderburn (1974), and have been revisited many times (e.g. McCullagh and Nelder 1989). The observations are the proportion of leaf afflicted by blotch, measured on ten varieties of barley at each of nine sites. These data are available as the dataset `barley` in R package `gnm`, where the 90 observations are given in three columns labelled, `y`, `site`, and `variety`, respectively.

```
> data(barley,package="gnm")
> #Put the data in table format for presentation
> BarleyTable=matrix(barley$y,ncol=10,
+     dimnames=list(paste("site",1:9,sep=""),paste("vrty",1:10,sep="")))

> BarleyTable
       vrty1  vrty2  vrty3  vrty4 vrty5 vrty6 vrty7 vrty8 vrty9 vrty10
site1 0.0005 0.0150 0.0100 0.2000 0.250 0.080 0.050  0.05 0.050  0.250
site2 0.0000 0.0000 0.1270 0.3750 0.550 0.165 0.050  0.50 0.050  0.425
site3 0.0000 0.0005 0.0125 0.2625 0.050 0.295 0.100  0.10 0.250  0.500
site4 0.0010 0.0005 0.0125 0.0250 0.400 0.200 0.050  0.50 0.750  0.375
site5 0.0025 0.0030 0.0250 0.0050 0.055 0.435 0.500  0.25 0.500  0.950
site6 0.0005 0.0075 0.1660 0.0001 0.010 0.010 0.750  0.50 0.750  0.625
site7 0.0050 0.0030 0.0250 0.0300 0.060 0.050 0.050  0.75 0.750  0.950
site8 0.0130 0.0300 0.0250 0.0250 0.011 0.050 0.001  0.05 0.750  0.950
site9 0.0150 0.0750 0.0000 0.0001 0.025 0.050 0.050  0.10 0.175  0.950
```

The logistic model for the mean is

$$p_{ij} = \frac{\exp(\eta_{ij})}{1+\exp(\eta_{ij})}, \qquad i = 1, \ldots, 9,\ j = 1, \ldots, 10\,, \tag{8.2}$$

where $p_{ij} = \mu_{ij}$ denotes the expected proportion of blotch on variety j at site i. Here, η_{ij} is the linear predictor containing the main effects of site and variety.

Although the observations are proportions, there is no notion of the number of trials, n_{ij}, associated with the measurement of the proportion of blotch on a leaf, and hence it is unclear how to express the variance of these observations. The most straightforward approach would be to assume a quasi-binomial variance. That is, to use quasi-likelihood with $\text{var}(Y_{ij}) = \phi p_{ij}(1 - p_{ij})$ and to obtain the estimate $\widehat{\phi}$ using Equation (7.15). This approach is implemented in `glm` and simply requires specification of the `family=quasibinomial` option.

```
> quasibin.fit=glm(y~variety+site,family=quasibinomial,data=barley)
> #Use which=1 for a plot of deviance resids vs linear predictors
> plot(quasibin.fit,which=1)
> summary(quasibin.fit)

Coefficients:
            Estimate Std. Error t value Pr(>|t|)
(Intercept)  -8.0546     1.4219  -5.665 2.84e-07 ***
variety2      0.1501     0.7237   0.207 0.836289
variety3      0.6895     0.6724   1.025 0.308587
variety4      1.0482     0.6494   1.614 0.110910
variety5      1.6147     0.6257   2.581 0.011895 *
variety6      2.3712     0.6090   3.893 0.000219 ***
variety7      2.5705     0.6065   4.238 6.58e-05 ***
variety8      3.3420     0.6015   5.556 4.39e-07 ***
variety9      3.5000     0.6013   5.820 1.51e-07 ***
varietyX      4.2530     0.6042   7.039 9.38e-10 ***
siteB         1.6391     1.4433   1.136 0.259870
siteC         3.3265     1.3492   2.466 0.016066 *
```

```
siteD          3.5822     1.3444    2.664 0.009510 **
siteE          3.5831     1.3444    2.665 0.009493 **
siteF          3.8933     1.3402    2.905 0.004875 **
siteG          4.7300     1.3348    3.544 0.000697 ***
siteH          5.5227     1.3346    4.138 9.38e-05 ***
siteI          6.7946     1.3407    5.068 3.00e-06 ***
---
Signif. codes:  0 '***' 0.001 '**' 0.01 '*' 0.05 '.' 0.1 ' ' 1
(Dispersion parameter for quasibinomial family taken to be 0.088777)

    Null deviance: 40.803  on 89  degrees of freedom
Residual deviance:  6.126  on 72  degrees of freedom

> cat("Pearson chi-sq stat is",
+      sum(resid(quasibin.fit,type="pearson")^2),"\n")
Pearson chi-sq stat is 6.391999
```

The estimated coefficients from the above quasi-likelihood fit are identical to those that would be obtained from a conventional binomial logistic regression (i.e. using `family= binomial`). However, the dispersion parameter has been estimated from application of Equation (7.15), $\widehat{\phi} = P_{\chi^2}/72 = 6.392/72 = 0.0888$. That is, the quasi-binomial model is assuming $\text{var}(Y_{ij}) = 0.0888 p_{ij}(1 - p_{ij})$, and the above standard errors have been scaled by a factor of $\sqrt{0.0888}$ compared to those that would be obtained from a binomial logistic regression.

The residual plot (Figure 8.1) from this fit shows a clear increase in the magnitude of the residual with increasing value of the fitted linear predictor $\widehat{\eta}_{ij}$, followed by a possible decrease for values of $\widehat{\eta}_{ij}$ in excess of zero. This suggests that, compared to the quasi-binomial, there may be relatively more variability in the data for values of η near zero (corresponding to p near 0.5) than for extreme values of η (corresponding to p close to 0 or 1). Consequently, Wedderburn (1974) suggested using the variance function $\text{var}(Y_{ij}) \propto p_{ij}^2(1 - p_{ij})^2$.

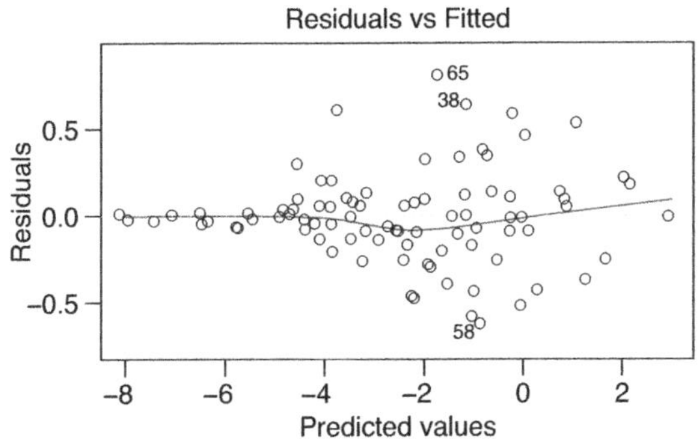

Figure 8.1 Plot of the deviance residuals (uncorrected for dispersion) versus the linear predictors from a quasi-binomial logistic regression fitted to the barley blotch data. This model assumes $\text{var}(Y_{ij}) \propto p_{ij}(1 - p_{ij})$.

The glm function permits the use of user-defined variance functions. This requires a list object to be written[2] with components that provide the necessary information required by glm. In the list object BarleyVar that is defined below, component varfun specifies the Wedderburn variance function. Another component of the list is a definition of function dev.resids for calculation of the squared deviance residual, that is, the contribution of each observation to the residual deviance. Here, since no explicit notion of a likelihood is utilized, it is more convenient to substitute this with the Pearson chi-square statistic. For more detail about the other components of BarleyVar, see the R documentation on family objects

```
> BarleyVar=list(
+   varfun=function(mu) (mu*(1-mu))^2,
+   validmu=function(mu) all(mu > 0) && all(mu < 1),
+   dev.resids=function(y, mu, wt) wt * ((y - mu)^2)/(mu*(1-mu))^2,
+   initialize=expression({
+           n <- rep.int(1, nobs)
+           mustart <- pmax(0.001, pmin(0.999, y)) }),
+   name="(mu(1-mu))^2" )
```

The quasi-likelihood fit with Wedderburn's form of variance is obtained by using the quasi function to specify a family object to be used in the call of glm. The arguments of the quasi function include specification of the link function and the list object created above.

```
> Wedderburn.fit=glm(y~site+variety,
+         family=quasi(link="logit",variance=BarleyVar),data=barley)

> plot(Wedderburn.fit,which=1) #Deviance resids vs linear predictors
> summary(Wedderburn.fit)

Coefficients:
            Estimate Std. Error t value Pr(>|t|)
(Intercept) -7.92238    0.44465 -17.817  < 2e-16 ***
siteB        1.38312    0.44465   3.111  0.00268 **
siteC        3.86006    0.44465   8.681 8.20e-13 ***
siteD        3.55700    0.44465   8.000 1.54e-11 ***
siteE        4.10786    0.44465   9.239 7.53e-14 ***
siteF        4.30536    0.44465   9.683 1.13e-14 ***
siteG        4.91810    0.44465  11.061  < 2e-16 ***
siteH        5.69489    0.44465  12.808  < 2e-16 ***
siteI        7.06763    0.44465  15.895  < 2e-16 ***
variety2    -0.46735    0.46870  -0.997  0.32204
variety3     0.07881    0.46870   0.168  0.86695
variety4     0.95408    0.46870   2.036  0.04547 *
variety5     1.35263    0.46870   2.886  0.00515 **
variety6     1.32854    0.46870   2.835  0.00595 **
variety7     2.34007    0.46870   4.993 4.01e-06 ***
```

[2] Some of these are provided in various R packages. For example, the gnm package includes a list object called wedderburn which can be used instead of the BarleyVar list object defined here.

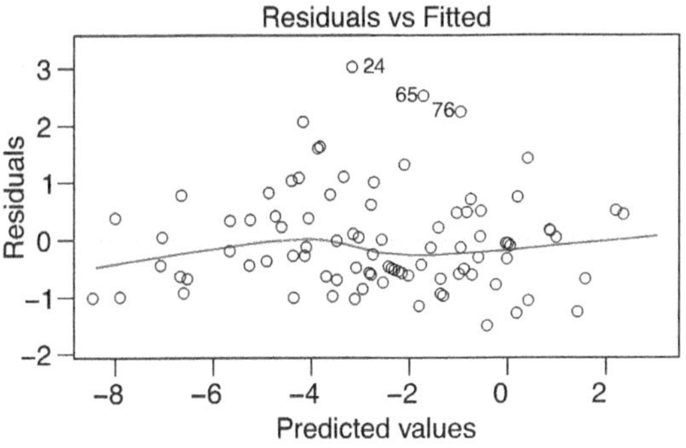

Figure 8.2 Plot of Pearson residuals versus linear predictors from a quasi-likelihood fit of a logistic model to the barley blotch data using Wedderburn's form of variance.

```
variety8      3.26258     0.46870     6.961 1.30e-09 ***
variety9      3.13549     0.46870     6.690 4.10e-09 ***
varietyX      3.88727     0.46870     8.294 4.34e-12 ***
---
Signif. codes:  0 '***' 0.001 '**' 0.01 '*' 0.05 '.' 0.1 ' ' 1
(Dispersion parameter for quasi family taken to be 0.9885464)

    Null deviance: 252.155  on 89  degrees of freedom
Residual deviance:  71.175  on 72  degrees of freedom
AIC: NA
```

The residual plot (Figure 8.2) from the Wedderburn quasi-likelihood fit shows a big improvement. The reported residual deviance of 71.175 is the Pearson chi-square statistic as defined in `BarleyVar`, which implicitly assumed $n_{ij} = 1$. It is pure coincidence that this provided a value very close to the 72 degrees of freedom. The standard errors in the above output assumed a dispersion parameter of $71.175/72 \approx 0.989$, that is, $\text{var}(Y_{ij}) = 0.989 p_{ij}^2(1 - p_{ij})^2$.

8.3 Generalized estimating equations

Suppose that m independent subjects are randomly sampled, and that n_i observations $\boldsymbol{y}_i = (y_{i1}, \ldots, y_{in_i})$ are recorded from subject i, $i = 1, \ldots, m$. Note that the total number of observations is $\sum_{i=1}^{m} n_i$. Here the terminology 'subject' is used generically to represent the independent experimental units from which the observations are taken. The subject could be an individual, but in the example in Section 8.3.1, the subjects are medical clinics from which observations on the success of a drug treatment are recorded.

There will typically be differences between the subjects, as might arise from uncontrolled causes that are specific to each subject. For example, a multi-clinic trial may utilize the in-house staff at each of the clinics, and these staff will differ in expertise. Any variability in subject i will affect all measurements made on that subject, and so it is to be anticipated that y_{ij}, $j = 1, \ldots, n_i$ will be positively correlated, in which case it is not appropriate to model all observations y_{ij}, $i = 1, \ldots, m$, $j = 1, \ldots, n_i$ as independent. The generalized estimating equations approach assumes that $\boldsymbol{Y}_i = (Y_{i1}, \ldots, Y_{in_i})$ is a multivariate random vector with expected values $(\mu_{i1}, \ldots, \mu_{in_i})$ depending on parameters $\boldsymbol{\theta}$, and variance matrix var($\boldsymbol{Y}_i$) that may depend on both $\boldsymbol{\theta}$ and additional association parameters, $\boldsymbol{\alpha}$, that specify the correlation structure.

GEEs make assumptions about the random vector $\boldsymbol{Y}_i$ at the marginal (i.e. population) level, through specification of the mean and variance structure. In contrast, hierarchical models specify the distribution of $\boldsymbol{Y}_i$ conditional on the random variability in subject i, that is, at the subject level. It may be that Y_{ij}, $j = 1, \ldots, n_i$ are independent conditional on the variability in subject i, yet unconditionally (i.e. marginally) they are correlated due to that shared variability. See Section 10.6.4 to see the distinction in practice.

Box 8.1

The correlation structure that is imposed on the observations within each subject should be chosen to match the manner in which the observations are made. For example, if observations y_{ij}, $j = 1, \ldots, n_i$ on subject i are collected sequentially (with increasing j) then it might be appropriate to assume an autoregressive correlation structure of the form $\text{cor}(Y_{ij}, Y_{ik}) = \alpha^{|j-k|}$, where $|\alpha| < 1$. In the absence of an ordering then it may be reasonable to assume exchangeability, corresponding to $\text{cor}(Y_{ij}, Y_{ik}) = \alpha$, for all $j \neq k$. An unstructured correlation matrix, $\text{cor}(Y_{ij}, Y_{ik}) = \alpha_{jk} = \alpha_{kj}$ is another possibility, but one that should be used cautiously due to the large number of correlation parameters required. It can also be useful to fit a GEE model with an assumed independence structure, $\text{cor}(Y_{ij}, Y_{ik}) = 0$, $j \neq k$, to take advantage of the robust variance estimator that is available using this approach. This is demonstrated in Section 8.3.1.

Given the assumed specification of the within-subject correlation structure, denoted $\text{cor}(\boldsymbol{Y}_i) = \mathbf{R}_i(\boldsymbol{\alpha})$, the assumed variance matrix for $\boldsymbol{Y}_i$ is

$$\mathbf{V}_i = \mathbf{D}(\boldsymbol{\theta})\mathbf{R}_i(\boldsymbol{\alpha})\mathbf{D}(\boldsymbol{\theta}),$$

where $\mathbf{D}(\boldsymbol{\theta})$ is a $n_i \times n_i$ diagonal matrix having the standard deviation of Y_{ij} in the jth diagonal position, $j = 1, \ldots, n_i$. These standard deviations are determined from the specified exponential family distribution. For example, if the Poisson is specified then $\text{sd}(Y_{ij}) = \sqrt{\mu_{ij}}$.

For a given value of the association parameters $\boldsymbol{\alpha}$, an estimate of $\boldsymbol{\theta}$ can be obtained by solving a multivariate version of the estimating equations given by (8.1) (see Example 12.9 for details). The residuals from this fit are then used to calculate a new value of $\boldsymbol{\alpha}$. Hardin and Hilbe (2003) give examples of explicit formulae for calculation of $\boldsymbol{\alpha}$ under several choices of the assumed correlation matrix. The iterative process of estimating $\boldsymbol{\alpha}$ and $\boldsymbol{\theta}$ is continued until convergence, giving the GEE estimate $\widetilde{\boldsymbol{\theta}}$.

The variance of $\widetilde{\boldsymbol{\theta}}$ is typically estimated using the sandwich estimator given in Equation (12.52). This is also called the empirical or robust estimator, because it is valid under misspecification of the variance structure (Section 12.3.2). In the case of an assumed independence structure, the correlation matrix $\mathbf{R}_i(\boldsymbol{\alpha})$ is then an identity matrix and $\mathbf{V}_i$ is just the variance under the specified exponential family distribution. It follows that the generalized estimating equation in (12.48) is equivalent to (7.7), and hence that $\widetilde{\boldsymbol{\theta}}$ is identical to the MLE $\widehat{\boldsymbol{\theta}}$ under that exponential family model. However, the GEE method has the advantage that it provides an estimate of variance that is robust to failure of the independence assumption.

GEEs are implemented in SAS using the `REPEATED` statement in the `GENMOD` procedure. Within R, the `geepack` package provides function `geeglm`. Hardin and Hilbe (2003) note that competing software packages may give differing fits due to nuances in the denominator degrees-of-freedom term used in calculation of the association parameters. This could be problematic when the number of subjects m is small, especially if a large number of association parameters are used to model the correlation structure. For the correlated binomial data investigated in Section 8.3.1, `GENMOD` would not converge using default options. However, it provided a $\widetilde{\boldsymbol{\theta}}$ value very close to that obtained from using R function `geeglm` when the `V6CORR` option was used in the `REPEATED` statement to specify a different method for calculation of the correlation parameter. In any case, the example below uses another variant that was proposed by Carey, Zeger and Diggle (1993) as being more appropriate for binomial data. This variant uses log odds-ratios rather than correlation to quantify the association between binomial responses, and additional description can be found in Hardin and Hilbe (2003, Section 3.2.5 therein).

8.3.1 GEE analysis of multi-centre data in SAS

The data are from Beitler and Landis (1985), and come from an experiment designed to determine the effectiveness of an active drug cream for treating infection. Trials were performed at eight clinics, and it is the clinics that are the experimental subjects here. At each clinic, one group of patients received the control cream and another group received the treatment cream. The observations recorded from each clinic were the proportion of favourable outcomes in each of the two groups. The treatment resulted in a higher proportion of favourable outcomes at each clinic (Figure 8.3), with the sole exception of the clinic having the smallest total sample size (Clinic 8). The positive correlation of the two observed proportions within each clinic is clearly evident.

The research objective (Beitler and Landis 1985, p. 992) required 'extended inferences from these data', meaning that it was not enough to model the clinics as fixed effects (say), because then inference would be confined to only the eight clinics that were used in the study. The GEE approach satisfies this

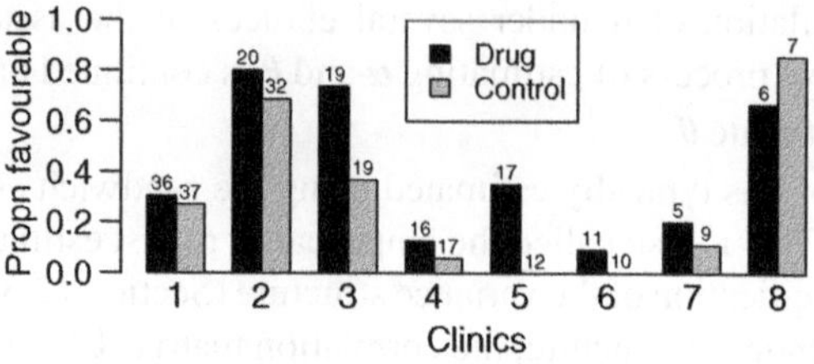

Figure 8.3 Proportion of favourable outcomes from a multi-centre trial investigating the effectiveness of two topical cream treatments for curing infection. Numbers above the bars give the sample size.

objective, by modelling the marginal (i.e. population) mean and variance of the outcomes (Box 8.1).

Prior to the GEE analysis, the first model to be fitted is a quasi-binomial logistic, using the code shown below. In the `DATA` step, this code uses the `@@` option on the `INPUT` statement so that multiple observations can be read from a single input line. Also, the `REF=FIRST` option in GENMOD's `CLASS` statement is used so that the lowest level of `trmt`, zero, will correspond to the intercept parameter. Since `trmt=0` corresponds to the control, the effect of the treatment cream will correspond to the `trmt` coefficient.

```
DATA infection;
INPUT clinic trmt y n @@;
LINES;
1 1 11 36   1 0 10 37   2 1 16 20   2 0 22 32
3 1 14 19   3 0 7 19    4 1 2 16    4 0 1 17
5 1 6 17    5 0 0 12    6 1 1 11    6 0 0 10
7 1 1 5     7 0 1 9     8 1 4 6     8 0 6 7
RUN;

***Quasi-binomial fit to the multi-center data***;
*Use REF=FIRST so that intercept coefficient is the control term;
PROC GENMOD DATA=infection;
      CLASS clinic trmt / PARAM=REF REF=FIRST;
      MODEL y/n = trmt / DIST=BINOMIAL PSCALE;
RUN;
```

The quasi-binomial logistic model estimates an intercept of −0.7142 (Figure 8.4). This is the linear predictor coefficient for the control cream, and hence the estimated probability of a favourable outcome is $e^{-0.7142}/(1+e^{-0.7142}) = 0.329$ for this cream. For the drug cream, the fitted value of the linear predictor is $-0.7142 + 0.4040 = -0.3102$ and the probability of a favourable outcome is $e^{-0.3102}/(1+e^{-0.3102}) = 0.423$. These are simply the observed proportions of favourable outcomes for each of the treatments when the data are combined over all clinics.

Analysis of maximum likelihood parameter estimates								
Parameter		DF	Estimate	Standard Error	Wald 95 % Confidence Limits		Wald Chi-Square	Pr > ChiSq
Intercept		1	-0.7142	0.4296	-1.5561	0.1277	2.76	0.0964
trmt	1	1	0.4040	0.6066	-0.7850	1.5931	0.44	0.5054
Scale		0	2.4130	0.0000	2.4130	2.4130		

Figure 8.4 The parameter estimates table from the quasi-binomial fit to the multi-centre trial data.

The estimated treatment effect of 0.4040 has a high p-value, suggesting little evidence of a genuine treatment effect. However, this p-value is erroneous, due to the between-clinic variability inflating the standard error of the estimated effect. Indeed, the Pearson χ^2 statistic from the logistic fit was 81.51, and dividing by the 14 degrees of freedom gives an estimated over-dispersion of $\widehat{\phi} = 5.822$. The quasi-binomial fit therefore inflates the estimated standard errors of $\widehat{\boldsymbol{\theta}}$ by a factor of $\widehat{\phi^{\frac{1}{2}}} = 2.413$ compared to a standard binomial logistic regression. A better analysis would use the grouping structure in the data to explain the over-dispersion. The following code fits a binomial GEE using the independence correlation structure, for the purpose of obtaining variance estimates that are robust to lack of independence.

```
***Independence GEE fit to Multi-center data***;
PROC GENMOD DATA=infection;
      CLASS clinic trmt / PARAM=REF REF=FIRST;
      MODEL y/n = trmt / DIST=BINOMIAL PSCALE;
  REPEATED SUBJECT=clinic / CORR=INDEP;
RUN;
```

The estimated $\widetilde{\boldsymbol{\theta}}$ is necessarily identical to the MLE $\widehat{\boldsymbol{\theta}}$ from the quasi-binomial fit (because of the independence assumption), but the reported standard errors now use the robust variance estimator (Figure 8.5). In particular, the standard error of the estimated treatment effect is much smaller, and the p-value is now starting to provide some mild evidence of a treatment effect.

The GEE with assumed independence is an improvement, because it provides better estimates of variances, but positive correlation of the binomial observations within each clinic is clearly evident (Figure 8.3), and should be modelled. With just two binomial observations per clinic, the only choice to model a non-independence correlation structure is

$$\mathbf{R}_i(\alpha) = \begin{pmatrix} 1 & \alpha \\ \alpha & 1 \end{pmatrix}, \quad i = 1, \ldots, 8. \tag{8.3}$$

Analysis of GEE parameter estimates							
Empirical Standard Error Estimates							
Parameter		Estimate	Standard Error	95 % Confidence Limits		Z	Pr > |Z|
Intercept		-0.7142	0.4505	-1.5971	0.1687	-1.59	0.1129
trmt	1	0.4040	0.2462	-0.0785	0.8866	1.64	0.1008

Figure 8.5 The parameter estimates table from the GEE fit to the multi-centre trial data, using an independence correlation structure.

Hardin and Hilbe (2003, Section 3.2.5 therein) note limitations in the ability to specify correlation structure on binomial proportions, and recommend instead using the approach of Carey *et al.* (1993) whereby the correlation is expressed using log odds-ratios. To use this approach, `PROC GENMOD` requires that the binomial data be in Bernoulli form with one row per Bernoulli outcome. That is, one row for each of the 273 patients involved in this study. The `LOGOR=EXCH` option can then be used to specify the correlation structure in (8.3).

```
***Create the binary response data from all 273 patients;
DATA binary;
 SET infection;
 KEEP clinic trmt outcome;
 DO i=1 TO y;
   outcome=1;
   OUTPUT;
 END;
 DO i=1 TO n-y;
   outcome=0;
   OUTPUT;
 END;
RUN;

***Exhangeable correlation GEE fit to Multi-center data***;
*Use DESCENDING so that trmt effect is for outcome=1;
PROC GENMOD DATA=binary DESCENDING;
      CLASS clinic trmt / PARAM=REF REF=FIRST;
      MODEL outcome = trmt / DIST=BINOMIAL PSCALE;
  REPEATED SUBJECT=clinic / LOGOR=EXCH;
RUN;
```

The correlation is estimated to be $\widetilde{\alpha} = 0.8965$ and the GEE estimate $\widetilde{\boldsymbol{\theta}}$ is no longer equal to the MLE $\widehat{\boldsymbol{\theta}}$ from a binomial logistic regression (Figure 8.6). However, the most important change is that, by appropriate modelling of the grouping structure, a more efficient estimator is obtained. This is seen here by the p-value for `trmt`, which now shows strong evidence of a treatment effect.

Note that compared to the quasi-binomial fit (Figure 8.4), the GEE fits shows a slight increase in the standard error of the intercept parameter, but a big decrease

Analysis of GEE parameter estimates							
Empirical standard error estimates							
Parameter		Estimate	Standard Error	95 % Confidence Limits		Z	Pr > |Z|
Intercept		-0.8740	0.4694	-1.7940	0.0460	-1.86	0.0626
trmt	1	0.5532	0.2318	0.0989	1.0074	2.39	0.0170
Alpha1		0.8965	0.4198	0.0738	1.7192	2.14	0.0327

Figure 8.6 The parameter estimates table from the GEE fit to the multi-centre data, with non-independence correlation structure assumed between the log odds-ratio of outcomes within a clinic.

in that of the treatment effect. This can be reasoned as follows – the lack of independence within each clinic is akin to a reduction in effective sample size, and hence there is a reduction in the total information about the overall probability of a favourable outcome. This results in the relatively high standard error on the intercept. However, the within-clinic correlation does not obfuscate the treatment differences within each clinic, and these differences are more clearly seen once this correlation structure is appropriately modelled.

8.4 Exercises

8.1 For the estrone data in Table 10.1, fit a GEE to the response variable $y = \log_{10}(\text{estrone})$. Assume an exchangeable correlation structure for the observations within the five subjects, and that $Y_{ij} \sim N(\mu, \sigma^2), i = 1, \ldots, 5,\ j = 1, \ldots, 16$. (Note that the only regression coefficient in the model is the mean, and the GEE approach is being used to make inference about this parameter in the presence of within-subject correlation.)

8.2 The method of moments uses an estimating equation that equates (non-central) sample moments to the population moments. For example, for Y iid from the zero-inflated Poisson distribution (see Exercise 2.11), it follows from Exercise 13.11 that $E[Y] = (1 - p)\lambda$ and $E[Y^2] = E[Y](1 + \lambda)$. An estimating equation for parameters λ and p is obtained as

$$\begin{pmatrix} \overline{y} - E[Y] \\ \overline{y^2} - E[Y^2] \end{pmatrix} = \begin{pmatrix} 0 \\ 0 \end{pmatrix}.$$

1. Solve this estimating equation to obtain the method-of-moments estimates, $\widetilde{\lambda}$ and $\widetilde{p}$, and apply to the ZIP data in Exercise 3.9.
2. Apply the bootstrap to the method-of-moments estimator, to obtain approximate 95 % confidence intervals for λ and p.

9

ML inference in the presence of incidental parameters

Divide each difficulty into as many parts as is feasible and necessary to resolve it.
– René Descartes

9.1 Introduction

Loosely speaking, this chapter presents methodology to protect the ML estimators of specified parameters from undesirable consequences that might arise from the estimation of other parameters. The iid normal model, $Y_i \sim N(\mu, \sigma^2)$, $i = 1, ..., n$, serves as a simple example. The ML estimator of the variance, $\widehat{\sigma}^2 = \sum(Y_i - \overline{Y})^2/n$, can be criticized for having n in the denominator rather than the $n - 1$ denominator of the unbiased sample variance estimator, S^2. It could be said that this is an undesirable consequence[1] of $\widehat{\sigma}^2$ failing to take into account the loss of one degree of freedom from estimation of μ by $\widehat{\mu} = \overline{Y}$. Using methods presented in this chapter, it is seen that S^2 can be obtained as the MLE from using a modified form of likelihood (Example 9.5).

The subject matter of this chapter is not of relevance if the data are independent and the sample size is sufficiently large, since then the statistical properties of ML inference will be adequately approximated using the theoretical results established in Chapter 12. This chapter is concerned with the pragmatics of ML inference when there is limited statistical information about some (or all) of the parameters. This

[1] Though, see Exercise 2.4 for an alternative view.

Maximum Likelihood Estimation and Inference: With Examples in R, SAS and ADMB, First Edition. Russell B. Millar.

could arise due to small sample size relative to the number of parameters, but can also occur when the sample size would otherwise appear to be sufficient. It is often the case that additional model complexity (beyond that relevant to the research question) is required to appropriately specify the sampling distribution of the data, and this can result in limited statistical information about the relevant parameters, even when the sample size is large. By way of example, in Section 10.4 a one-way random-effects model is fitted to 80 measurements of the hormone estrone. However, the data (Table 10.1) are in the form of 16 measurements from each of five randomly chosen subjects. Sampling variability in the MLE of mean estrone level is dependent on both within-subject and between-subject variability. However, measurements have been taken on only five subjects, and so the 80 measurements are not a large sample for the purpose of estimating the between-subject source of variability.

Complex models with many parameters are often required when the observations are not independent and the model is therefore required to capture correlation structure. The random-effects model for estrone (Section 10.4) is a form of repeated-measures model and falls into this category, as do models for longitudinal, temporal, or spatial data. This correlation structure may or may not be of direct interest, depending on the research question. The model parameters that are directly relevant to the underlying research questions and objectives will be called the parameters of interest, and the remaining parameters will be called incidental parameters. Here, the parameter vector will be denoted $\boldsymbol{\theta} = (\boldsymbol{\psi}, \boldsymbol{\lambda})$ where $\boldsymbol{\psi}$ contains the parameters of interest and $\boldsymbol{\lambda}$ contains the incidental parameters.

It should be noted that there are alternative modelling approaches, such as generalized estimating equations (Section 8.3), that can be used when observations are correlated. The GEE approach avoids the need to specify a potentially complex parametric model, but this is also its limitation if such a model is required to address the relevant questions of interest. Correlated binomial data were analyzed using GEEs in Section 8.3.1, and Section 10.6 re-analyses these data using a mixed-effects model. The mixed-effects analysis includes a demonstration of the integrated likelihood that is presented in Section 9.3, and a comparison between the GEE and mixed-effects analyses is provided in Section 10.6.4.

The approach taken in this chapter is to use a modified form of the likelihood function for inference about $\boldsymbol{\psi}$. It is desired to obtain a form of likelihood that is a function of $\boldsymbol{\psi}$ alone, and which encapsulates the information about $\boldsymbol{\psi}$ that is present in the (standard) likelihood $L(\boldsymbol{\theta})$. The focus here is on conditional likelihood, due to its theoretical underpinning (Bartlett 1936) and its wide use in mixed-effects modelling where it is more commonly known as restricted maximum likelihood (REML).

Section 9.1.1 presents the paired t-test in the context of modelling paired normally distributed data, and shows that it is an application of conditional inference. It is seen that equivalent inference can be obtained by fitting (using the standard normal theory) a two-way ANOVA model to the raw observations, but this is certainly not as natural as working with the within-pair differences. In that sense, the use of conditional inference is optional. However, this equivalence of inference between using the within-pair differences or the raw data does not extent to ML inference. It is seen that conditional likelihood inference (i.e. using the within-pair differences) must be employed, because ML inference using the raw observations results in inconsistent estimation of the variance parameter.

The profile log-likelihood (Section 3.6) is

$$l^*(\boldsymbol{\psi}) = l(\boldsymbol{\psi}, \widehat{\boldsymbol{\lambda}}_{\boldsymbol{\psi}}) = \max_{\boldsymbol{\lambda}} l(\boldsymbol{\psi}, \boldsymbol{\lambda}) ,$$

and is a form of 'modified' likelihood that is a function of $\boldsymbol{\psi}$ alone. However, profile likelihood is simply a partial maximization of $l(\boldsymbol{\theta}) = l(\boldsymbol{\psi}, \boldsymbol{\lambda})$, and is useful for obtaining likelihood-based confidence intervals and regions for $\boldsymbol{\psi}$. The limitation of profile likelihood is that it does not encapsulate the information loss due to the estimation of $\widehat{\boldsymbol{\lambda}}_{\boldsymbol{\psi}}$ (e.g. see Section 7.2.4 of McCullagh and Nelder 1989).

Box 9.1

9.1.1 Analysis of paired data: an intuitive use of conditional likelihood

Example 9.1 is based on Example 2 in Neyman and Scott (1948) and is commonly referred to as an instance of the Neyman-Scott problem. Here, it is presented under a more familiar guise. Specifically, the data measured on each randomly chosen subject are assumed to be a pair of normally distributed observations, and each observation in the pair receives one of the two treatments under investigation. If the research question is about the difference between the treatments, then this is the familiar scenario under which a paired t-test is appropriate.

It may not be readily apparent that inference based on the within-pair differences is a form of conditional inference. The details are left to the continuation of Example 9.1 in Section 9.2, and it will just be remarked here that inference based solely on the within-pair differences is obtained by conditioning on the within-pair averages.

Example 9.1. Analysis of paired data. Consider the model where, for $i = 1, ..., n$,

$$Y_{i1} \sim N\left(\mu_i - \frac{\delta}{2}, \tau^2\right) \tag{9.1}$$

$$Y_{i2} \sim N\left(\mu_i + \frac{\delta}{2}, \tau^2\right) , \tag{9.2}$$

are mutually independent. Suppose that it is of interest to make inference about δ. In particular, consider evaluation of the null hypothesis $H_0 : \delta = \delta_0$. Here, the $\mu_i, i = 1, ..., n$ are considered to be incidental parameters.

Analysis of the within-pair differences

This is a scenario for which using a statistical model of the within-pair differences $D_i = Y_{i2} - Y_{i1},\ i = 1, ..., n$ would appear to be a sensible alternative to modelling $Y_{ij},\ i = 1, ..., n,\ j = 1, 2$. Denoting $\sigma^2 = 2\tau^2$, these differences are iid $N(\delta, \sigma^2)$, which does not depend on the incidental parameters, $\mu_i, i = 1, ..., n$.

Exact normal-based theory leads to the paired t-test of H_0. That is, a (one-sample) t-test applied to the differences D_i. Maximum likelihood applied to the iid $N(\delta, \sigma^2)$ model for D_i would lead to similar conclusions (for moderate n). In particular, due to the n denominator in the ML estimator $\widehat{\sigma}^2$, the square root of the Wald test statistic is equal to the t-statistic multiplied by $\sqrt{n/(n-1)}$.

Analysis of the raw data

Now, suppose that it was not recognized that H_0 could be tested by working with just the differences D_i, and instead inference was based on the model for $Y_{ij}, i = 1, ..., n, j = 1, 2$, as given in (9.1) and (9.2). This model can be fitted using a two-way ANOVA with explanatory variable `subject` being a factor variable having n distinct levels (corresponding to $i = 1, ..., n$), and `trmt` taking the value -0.5 when $j = 1$, and 0.5 when $j = 2$. The two-way ANOVA uses exact normal theory, and it correctly determines that there are $n - 1$ degrees of freedom since there are $2n$ observations and $n + 1$ regression coefficients. Moreover, it can be shown (Exercise 9.1) that the test of $H_0 : \delta = \delta_0$ is equivalent to the paired t-test.

However, things go horribly wrong if maximum likelihood inference is applied to the model specified by (9.1) and (9.2), because the loss of $n + 1$ degrees of freedom is not taken into account. In particular, the ML estimator of τ^2 (see Box 2.4) is

$$\widehat{\tau}^2 = \frac{1}{2n}\sum_{i=1}^{n}\sum_{j=1}^{2}(Y_{ij} - \widehat{\mu}_{ij})^2 , \tag{9.3}$$

whereas the unbiased estimator of τ^2 has a divisor of $n - 1$. Thus $E[\widehat{\tau}^2] < \tau^2/2$, and $\widehat{\tau}^2$ is an inconsistent estimator of τ^2. Moreover, the square root of the Wald test statistic is equal to the t-statistic multiplied by $\sqrt{2n/(n-1)}$, and hence has an approximate $N(0, 2)$ distribution when H_0 is true, rather than a distribution that is approximately $N(0, 1)$. Similarly, it can be shown that the likelihood ratio test statistic of H_0 has an approximate $2\chi_1^2$ distribution rather than the approximate χ_1^2 distribution that would be obtained under standard conditions. □

In Example 9.1, equivalent inference was obtained from modelling the raw values Y_{ij}, or the differences $D_i = Y_{i2} - Y_{i1}$, when the exact sampling theory was used. However, maximum likelihood failed when applied to the raw values. The problem facing ML is that there are only two observations for each of the incidental parameters μ_i, and hence properties of ML inference that depend on sufficient sample size are not valid. Indeed, the asymptotic theory of MLEs (Chapter 12) is not applicable because it assumes that the specification of the parametric model is fixed. This is not the case here, since the dimension of $\boldsymbol{\theta} = (\delta, \tau, \mu_1, ..., \mu_n)$ increases with increasing sample size.

The Rasch model is analogous to Example 9.1, but it is a model for paired (or grouped) Bernoulli data, $Y_{ij} \sim \text{Bernoulli}(p_{ij})$, and does not have the luxury of a tractable exact theory. Conditional likelihood provides an essential methodology to avoid the inconsistencies (Ghosh 1995) that would result from naive application of ML inference to this model.

Section 9.2 presents conditional likelihood and several examples of its use, including REML (Section 9.2.1). Conditional likelihood is not a generally applicable methodology because it requires an appropriate tractable factorization of $L(\boldsymbol{\theta})$. Integrated likelihood (Section 9.3) is presented as a more generally applicable methodology that, with due care, can be considered an approximation to conditional likelihood. Indeed, integrated likelihood can be used to generalize REML beyond its usual application to linear normal mixed models. Generalized REML (GREML) is demonstrated in Section 10.6.

9.2 Conditional likelihood

The use of conditional likelihood in the context of eliminating incidental parameters was first formalized by Bartlett (1936). The notation used here assumes that $(\boldsymbol{u}, \boldsymbol{v}) = h(\boldsymbol{y})$ is a suitable invertible transformation of the data $\boldsymbol{y}$. For example, in Example 9.5, the data $\boldsymbol{y} = (y_1, ..., y_n)$ are transformed to $(\boldsymbol{u}, \boldsymbol{v})$ where $\boldsymbol{u} = \overline{y}$ and $\boldsymbol{v} = (y_1 - \overline{y}, ..., y_{n-1} - \overline{y})$. Note that the density function for $(\boldsymbol{u}, \boldsymbol{v})$ is the same as that of $\boldsymbol{y}$ to within a multiplicative constant (see Box 13.1). Hence, using the likelihood principle (Section 14.4), and the fact that it is the same experimental situation that leads to observation of $(\boldsymbol{u}, \boldsymbol{v})$ and $\boldsymbol{y}$, it can be argued that $(\boldsymbol{u}, \boldsymbol{v})$ and $\boldsymbol{y}$ are equivalent for the purpose of making inference about $\boldsymbol{\theta} = (\boldsymbol{\psi}, \boldsymbol{\lambda})$.

> Conditional likelihood is conceptually analogous to the conditionality principle discussed in Section 14.3. The general idea of the conditionality principle is to condition on (that is, take as fixed) those random aspects of the experiment that contain no information about the parameter. In the present context, the idea is to condition upon statistics that contain all of the information about the incidental parameter $\boldsymbol{\lambda}$, but contain no (or little) information about the parameter of interest, $\boldsymbol{\psi}$.

Box 9.2

The density function $f(\boldsymbol{u}, \boldsymbol{v}; \boldsymbol{\theta})$ can be written as the product of the marginal density of $\boldsymbol{u}$ and the conditional density of $\boldsymbol{v}$ given $\boldsymbol{u}$. That is,

$$f(\boldsymbol{u}, \boldsymbol{v}; \boldsymbol{\theta}) = f(\boldsymbol{u}; \boldsymbol{\theta}) f(\boldsymbol{v}|\boldsymbol{u}; \boldsymbol{\theta}) \equiv f(\boldsymbol{u}; \boldsymbol{\psi}, \boldsymbol{\lambda}) f(\boldsymbol{v}|\boldsymbol{u}; \boldsymbol{\psi}, \boldsymbol{\lambda}), \tag{9.4}$$

where the notation $f()$ is used generically.[2]

The most convenient case arises when the marginal density of $\boldsymbol{u}$ depends only on $\boldsymbol{\lambda}$, and the conditional density of $\boldsymbol{v}$ (given $\boldsymbol{u}$) depends only on $\boldsymbol{\psi}$ (see Example 9.2). Then (9.4) can be written

$$f(\boldsymbol{u}, \boldsymbol{v}; \boldsymbol{\theta}) = f(\boldsymbol{u}; \boldsymbol{\lambda}) f(\boldsymbol{v}|\boldsymbol{u}; \boldsymbol{\psi}). \tag{9.5}$$

It can be said that $\boldsymbol{u}$ is partially sufficient for $\boldsymbol{\lambda}$, because the conditional distribution of $\boldsymbol{v}$ (given $\boldsymbol{u}$) does not depend on $\boldsymbol{\lambda}$. It follows that all of the information about the parameters of interest, $\boldsymbol{\psi}$, that is contained in the likelihood $L(\boldsymbol{\theta}) = f(\boldsymbol{u}, \boldsymbol{v}; \boldsymbol{\theta})$ is contained in the so-called conditional likelihood

$$L_c(\boldsymbol{\psi}) = f(\boldsymbol{v}|\boldsymbol{u}; \boldsymbol{\psi}).$$

The most common usage of conditional likelihood is the situation where $\boldsymbol{u}$ is partially sufficient for $\boldsymbol{\lambda}$, but the marginal density of $\boldsymbol{u}$ depends on both $\boldsymbol{\psi}$ and $\boldsymbol{\lambda}$. That is,

$$f(\boldsymbol{u}, \boldsymbol{v}; \boldsymbol{\theta}) = f(\boldsymbol{u}; \boldsymbol{\psi}, \boldsymbol{\lambda}) f(\boldsymbol{v}|\boldsymbol{u}; \boldsymbol{\psi}). \tag{9.6}$$

In this case, if inference about $\boldsymbol{\psi}$ is based solely on the conditional likelihood $L_c(\boldsymbol{\psi})$, then any information about $\boldsymbol{\psi}$ that is provided by the marginal density of $\boldsymbol{u}$, $f(\boldsymbol{u}; \boldsymbol{\psi}, \boldsymbol{\lambda})$, is lost. Conditional likelihood is therefore only appropriate when it can be argued that there is little, or no, loss of information about $\boldsymbol{\psi}$ from ignoring $f(\boldsymbol{u}; \boldsymbol{\psi}, \boldsymbol{\lambda})$. A formal basis for this argument is elusive (but see Bartlett 1936, Sprott 1975, Jørgensen 1993, Pace and Salvan 1997) and lies within the realm of fundamental statistical philosophy (see Chapter 14 for a taste of this). This issue is considered individually in each of the examples below.

Marginal likelihood. A further variation of (9.4) arises when the density function of $\boldsymbol{u}$ depends on $\boldsymbol{\psi}$ alone. That is,

$$f(\boldsymbol{u}, \boldsymbol{v}; \boldsymbol{\theta}) = f(\boldsymbol{u}; \boldsymbol{\psi}) f(\boldsymbol{v}|\boldsymbol{u}; \boldsymbol{\psi}, \boldsymbol{\lambda}). \tag{9.7}$$

If it can be argued that the conditional density $f(\boldsymbol{v}|\boldsymbol{u}; \boldsymbol{\psi}, \boldsymbol{\lambda})$ contains no information about $\boldsymbol{\psi}$, then inference should be performed using only the marginal likelihood, $L_m(\boldsymbol{\psi}) = f(\boldsymbol{u}; \boldsymbol{\psi})$.

Box 9.3

[2] Throughout this chapter the density function notation $f()$ is used generically. That is, the three density functions denoted by f in (9.4) are different. The distinction between density functions is explicit from their arguments.

Example 9.2. Equivalence of multinomial and Poisson models. For ease of notation, a form of equivalence between the Poisson and binomial models is established here. The equivalence between Poisson and multinomial models follows naturally (e.g. see Section 6.4.2 of McCullagh and Nelder 1989).

Let Y_1 and Y_2 be Poisson distributed with means μ_1 and μ_2, and suppose that interest lies in inference about $p = \mu_1/(\mu_1 + \mu_2)$. The density function of $\mathbf{y} = (y_1, y_2)$ is

$$f(\mathbf{y}; \mu_1, \mu_2) = \frac{\mu_1^{y_1} e^{-\mu_1}}{y_1!} \frac{\mu_2^{y_2} e^{-\mu_2}}{y_2!}$$

$$= \frac{(\mu_1 + \mu_2)^{y_1+y_2} e^{-(\mu_1+\mu_2)}}{(y_1 + y_2)!} \frac{(y_1 + y_2)!}{y_1!y_2!} \left(\frac{\mu_1}{\mu_1 + \mu_2}\right)^{y_1} \left(\frac{\mu_2}{\mu_1 + \mu_2}\right)^{y_2}.$$

Using the re-parameterization $p = \mu_1/(\mu_1 + \mu_2)$ and $\lambda = \mu_1 + \mu_2$, the above density can be written in the more compact form

$$f(\mathbf{y}; p, \lambda) = \frac{\lambda^n e^{-\lambda}}{n!} \times \frac{n!}{y_1!y_2!} p^{y_1}(1 - p)^{y_2}, \qquad (9.8)$$

where $n = y_1 + y_2$.

The second term on the right-hand side of (9.8) can be recognized as the density function of a Bin(n, p) distribution, and the first term as that of a Pois(λ) distribution. Since y_1 and y_2 take discrete values, the density function for $\mathbf{y} = (y_1, y_2)$ is the probability of observing (y_1, y_2), which is identical to the probability (density function) for (n, y_1). Thus, (9.8) can be written

$$\begin{aligned} f(\mathbf{y}; p, \lambda) &= f(n, y_1; p, \lambda) \\ &= f(n; \lambda) f(y_1|n; p) . \qquad (9.9) \end{aligned}$$

This is the ideal case where the density function can be expressed in the form of (9.5) and it is explicit that the information about p resides solely in the conditional distribution of y_1 given n. Thus, for inference about p, the Poisson modelling of Y_1 and Y_2 is equivalent to the Bin(n, p) model for Y_1. □

The above example also appears in Chapter 14 (see Exercise 14.7 and Equation (15.12)) where the assertion that there is no loss of information about p from using only the conditional likelihood $L_c(p) = f(y_1|n; p)$ is formally justified using the conditionality principle.

Example 9.3. Revisiting the logistic regression case study. The equivalence of binomial and Poisson models was put into practice in Exercise 7.7, where it was required to refit the binomial logistic model used in the case study of Section 7.5, by using its alternative representation as a log-linear Poisson model. Let y_{i1} and y_{i2} denote the observed counts of length-class i fish in the codend and cover, respectively, $i = 1, ..., 37$. Under the Poisson model specified in Exercise 7.7, the expected counts are

$$\mu_{i1} = \exp(\beta_1 + \beta_2 l_i + \alpha_i) \tag{9.10}$$

$$\mu_{i2} = \exp(\alpha_i) , \tag{9.11}$$

where l_i is the fish length corresponding to length-class i, and α_i, $i = 1, ..., 37$, are incidental parameters that are not of interest,

Using the notation

$$p_i = \frac{\mu_{i1}}{\mu_{i1} + \mu_{i2}} ,$$

note that

$$p_i = \frac{\exp(\beta_1 + \beta_2 l_i)}{1 + \exp(\beta_1 + \beta_2 l_i)} . \tag{9.12}$$

From Example 9.2, it follows that the information about p_i from using the Poisson model is equivalent to assuming that y_{i1} are observations under a $\text{Bin}(n_i, p_i)$ model, where $n_i = y_{i1} + y_{i2}$. This is precisely the logistic regression model used in Section 7.5. □

Now, the above example is not entirely complete, because it has glossed over the fact that inference is not about p_i, but is to be made about parameters β_1 and β_2. The crucial point is that n_i, $i = 1, ..., 37$, are observations from a distribution that depends on these parameters. Specifically, since Y_{i1} and Y_{i2} are independent Poisson random variables, it follows that n_i is an observation from a Poisson distribution, with mean λ_i equal to

$$\begin{aligned} \lambda_i &= \mu_{i1} + \mu_{i2} \\ &= \exp(\alpha_i)(1 + \exp(\beta_1 + \beta_2 l_i)) . \end{aligned} \tag{9.13}$$

This is an example of the density partitioning as in (9.6), where the marginal density of $\boldsymbol{u}$ (in this case $\boldsymbol{u} = (n_1, ..., n_{37})$) depends on both the parameters of interest and the incidental parameters. Thus, it needs to be explicitly argued that inference can ignore the density function of n_i, and use only that of y_{i1} (conditional on n_i being fixed). To argue this point, note that (9.13) is a non-identifiable parameterization since there are 37 values of λ_i and 39 parameters in total, $(\beta_1, \beta_2, \alpha_1, ..., \alpha_{37})$. Moreover, for any fixed value of $\boldsymbol{\beta} = (\beta_1, \beta_2)$, the collection of distributions indexed by the Poisson model for n_i is the same. It can be concluded that the model in (9.13) contains no information about $\boldsymbol{\beta}$, and hence that inference about $\boldsymbol{\beta}$ using the binomial conditional likelihood (i.e. logistic regression) is equivalent to that using the full Poisson likelihood.

An alternative to the above argument is provided by noting that the Poisson model for Y_{i1} and Y_{i2} could be re-parameterized using $(\beta_1, \beta_2, \lambda_1, ..., \lambda_{37})$, and under this parameterization the likelihood factorizes under the ideal case given in (9.5).

The next example shows that the paired analysis in Example 9.1 arises from the form of conditioning in (9.6), and again it is necessary to make the argument that there is no loss of information from using the conditional likelihood.

Example 9.1 continued. Paired analysis from conditioning. Here, the observed data are y_{ij}, $i = 1, ..., n$, $j = 1, 2$, and it was seen earlier that it is the incidental parameters $\boldsymbol{\mu} = (\mu_1, ..., \mu_n)$ that are problematic in this model. It is desired to find a conditional likelihood formulation that eliminates these incidental parameters, and is a function of only $\boldsymbol{\psi} = (\delta, \tau)$.

The observed data are equivalent to $(\boldsymbol{u}, \boldsymbol{v})$ where $\boldsymbol{u} = \overline{\boldsymbol{y}} = (\overline{y}_1, ..., \overline{y}_n)$ are the within-pair means, and $\boldsymbol{v} = \boldsymbol{d} = (d_1, ..., d_n)$ are the within-pair differences. Since the random variables D_i and $\overline{Y}_i$ are independent, the conditional likelihood $f(\boldsymbol{d}|\overline{\boldsymbol{y}}; \delta, \tau)$ is simply the marginal density of $\boldsymbol{d}$ from the model $D_i \sim N(\delta, \sigma^2)$, where $\sigma^2 = 2\tau^2$. That is, the conditional likelihood is the likelihood obtained from fitting an iid normal model to the within-pair differences (and see Section 9.2.1 to apply conditional ML to obtain an unbiased estimator of σ^2).

The term that is ignored, $f(\overline{\boldsymbol{y}}; \boldsymbol{\mu}, \tau)$ is the marginal density of the within-pair sample means. These sample means are iid $N(\mu_i, \tau^2/2)$, and so have distribution that depends on τ. However, note that this is a saturated model with a parameter μ_i for each observed $\overline{y}_i$, and in the absence of knowledge or structure imposed upon $\boldsymbol{\mu}$, it is reasonable to assert that no information about τ is lost by ignoring this term. Throwing these n marginal terms away, in effect, reduces the sample size to n, and avoids the erroneous $2n$ that appears in the divisor in the inconsistent ML estimator of τ^2 in (9.3). □

Example 9.4. Cox's partial likelihood. Cox's proportional-hazards model (Section 6.3.4) uses partial likelihood to eliminate the need to specify the baseline hazard function. Loosely speaking, the argument proceeds by conditioning on the actual times at which failures occurred. These times act as partially sufficient statistics for the baseline hazard $\lambda(t)$, with the result that the conditional likelihood does not depend on $\lambda(t)$. The rationale for the conditioning was that, in the absence of knowledge about $\lambda(t)$, the times of failure contain no information about the parameters of interest $\boldsymbol{\beta}$.

In his original paper on this approach, Cox (1972) referred to the likelihood obtained from the above argument as conditional likelihood. Critics of Cox's paper noted that this did not fit the standard definition of conditional likelihood due to the fact that the temporal nature of the data required sequential application of conditioning. Thus, Cox (1975) introduced the notion of partial likelihood, to generalize the application of conditional likelihood to sequential models of this type. □

9.2.1 Restricted maximum likelihood

The application of conditional likelihood in the continuation of the paired-data example (Example 9.1 continued, in Section 9.2) can be extended for general application to linear normal regression models, including linear mixed-effects models, and in these applications it is known as restricted (or residual) maximum likelihood (REML). Only a brief statement of REML is provided below, and a random-effects application of REML is demonstrated in Section 10.4.

Let $\boldsymbol{Y}$ be a n-dimensional multivariate normal random vector with mean vector $E[\boldsymbol{Y}] = \mathbf{X}\boldsymbol{\beta}$, where $\mathbf{X}$ is $n \times p$ of rank p, and with variance matrix $\boldsymbol{\Sigma}(\boldsymbol{\psi})$ parameterized by $\boldsymbol{\psi}$. The motivation for REML is that the regression coefficients $\boldsymbol{\beta}$ can be considered incidental to estimation of the variance parameters $\boldsymbol{\psi}$. It is convenient to consider $(\boldsymbol{u}, \boldsymbol{v})$ defined by

$$\boldsymbol{u} = \left(\mathbf{X}^T\boldsymbol{\Sigma}^{-1}\mathbf{X}\right)^{-1}\mathbf{X}^T\boldsymbol{\Sigma}^{-1}\boldsymbol{y} = \widehat{\boldsymbol{\beta}} ,$$

and

$$\boldsymbol{v} = \left(\mathbf{I}_n - \mathbf{X}\left(\mathbf{X}^T\mathbf{X}\right)^{-1}\mathbf{X}^T\right)\boldsymbol{y} ,$$

where $\mathbf{I}_n$ is the $n \times n$ identity matrix. Vector $\boldsymbol{u} \in \mathbb{R}^p$ is a function of $\boldsymbol{\Sigma}$, and is the MLE (and generalized least-squares estimate) of $\widehat{\boldsymbol{\beta}}$ for known $\boldsymbol{\Sigma}$ (Seber and Lee 2003). Although vector $\boldsymbol{v}$ is of length n, and hence $(\boldsymbol{u}, \boldsymbol{v})$ is of length $n + p$, it can be shown that there is a one-to-one mapping between $\boldsymbol{y}$ and $(\boldsymbol{u}, \boldsymbol{v})$ (see Smyth and Verbyla 1996). This arises because $\boldsymbol{v}$ is a reduced-rank multivariate normal with rank $n - p$. Indeed, $\boldsymbol{v}$ is the vector of residuals from an ordinary least squares fit, hence the alternative appellation 'residual' maximum likelihood for REML.

Since $\boldsymbol{u}$ and $\boldsymbol{v}$ are linear functions of $\boldsymbol{y}$, it is straightforward to show that $\boldsymbol{u}$ and $\boldsymbol{v}$ are independent. Thus the conditional distribution of $\boldsymbol{v}$ given $\boldsymbol{u}$ is also the marginal distribution of $\boldsymbol{v}$. It is for this reason that REML has its earliest origins as an application of marginal likelihood (Patterson and Thompson 1971, Harville 1977). The distribution of $\boldsymbol{v}$ depends only on $\boldsymbol{\Sigma}(\boldsymbol{\psi})$, and hence only on $\boldsymbol{\psi}$. Moreover, the arguments of Bartlett (1936) and Sprott (1975) establish that no information about $\boldsymbol{\psi}$ is contained in $f(\boldsymbol{u}; \boldsymbol{\beta}, \boldsymbol{\psi})$.

Since $\boldsymbol{v}$ is less than full rank, some technical algebraic manipulations are required to obtain its density function (see Harville 1977, Smyth and Verbyla 1996, LaMotte 2007, for details). The log-density of $\boldsymbol{v}$ is the conditional log-likelihood for $\boldsymbol{\psi}$, and for an ordinary linear regression where $\boldsymbol{\Sigma} = \sigma^2\mathbf{I}_n$, it can be shown that it reduces to the form[3]

$$\begin{aligned} l_c(\sigma) &= \log f(\boldsymbol{v}|\boldsymbol{u}; \sigma) = \log f(\boldsymbol{v}; \sigma) \\ &= -(n-p)\log\sigma - \frac{\sum_{i=1}^n (y_i - \widehat{y}_i)^2}{2\sigma^2} , \end{aligned}$$

[3] More precisely, this is the log-likelihood from any $n - p$ dimensional subset of $\boldsymbol{v}$.

to within an additive constant. Here, $\widehat{\boldsymbol{y}} = (\widehat{y}_1, ..., \widehat{y}_n) = \mathbf{X}\widehat{\boldsymbol{\beta}}$ are the fitted values and $\widehat{\boldsymbol{\beta}}$ is the ordinary least-squares estimate (and MLE, Section 11.7.3) of $\boldsymbol{\beta}$. Maximization of $l_c(\sigma)$ gives the REML estimate of σ^2,

$$\widehat{\sigma}_c^2 = \frac{\sum_{i=1}^{n}(y_i - \widehat{y}_i)^2}{n - p} \, .$$

That is, the REML estimator of σ^2 is the usual unbiased estimator from least-squares theory. In contrast, the usual MLE of σ^2 has a n in the denominator and hence is negatively biased (see Box 2.4).

Example 9.5. As a special case, if Y_i are iid $N(\mu, \sigma^2)$ then the REML estimator of σ^2 is the usual unbiased sample variance estimator $S^2 = \sum(Y_i - \overline{Y})^2/(n - 1)$. □

When using REML, it should be remembered that the likelihood L_c is the likelihood of the residuals $\boldsymbol{v}$, and that these residuals depend on the design matrix $\mathbf{X}$. Thus, the REML likelihood can not be used to compare models (using AIC, say) that have different sets of fixed effects, or to perform likelihood ratio tests of nested models with differing fixed effects, since these REML likelihoods are effectively being constructed from different 'data'. Such comparisons or tests should use the standard likelihood.

Use of conditional likelihood not only requires the existence of an appropriate statistic $\boldsymbol{u}$ upon which to condition, but also that it is tractable to calculate $L_c(\boldsymbol{\psi}) = f(\boldsymbol{v}|\boldsymbol{u}; \boldsymbol{\psi})$. Unfortunately, this precludes its general use in practice. The next section provides a more generally applicable methodology, and the motivation for this methodology is that it can be considered an approximation to conditional likelihood.

9.3 Integrated likelihood

It can be shown that approximate conditional likelihood can be obtained using an integrated likelihood of the form

$$L_{\partial\lambda}(\boldsymbol{\psi}) = \int L(\boldsymbol{\psi}, \boldsymbol{\lambda})d\boldsymbol{\lambda} \, . \tag{9.14}$$

A justification of this approximation is provided below, and more complete detail can be found in Pawitan (2001, Section 10.6 therein) or Lee *et al.* (2006, Section 1.9 therein). This argument is rather technical, and is not required at a first reading.

9.3.1 Justification

For simplicity, it will be assumed that the MLE $\widehat{\boldsymbol{\lambda}}$ is partially sufficient for $\boldsymbol{\lambda}$, so that $\widehat{\boldsymbol{\lambda}}$ takes the role of $\boldsymbol{u}$ in Equation (9.5) (see Pawitan 2001, for extension to the general case). Pawitan (2001) shows that the conditional log-likelihood can be approximated as

$$l_c(\boldsymbol{\psi}) \approx l(\boldsymbol{\psi}, \widehat{\boldsymbol{\lambda}}_{\psi}) - \frac{1}{2}\log\det\left(-\mathbf{H}(\widehat{\boldsymbol{\lambda}}_{\psi})\right) + c(\boldsymbol{\psi}, \widehat{\boldsymbol{\lambda}}, \widehat{\boldsymbol{\lambda}}_{\psi}), \tag{9.15}$$

where $l(\boldsymbol{\psi}, \widehat{\boldsymbol{\lambda}}_{\psi}) = l^*(\boldsymbol{\psi})$ is the profile log-likelihood (Section 3.6), and

$$\mathbf{H}(\widehat{\boldsymbol{\lambda}}_{\psi}) = \left.\frac{\partial^2 l(\boldsymbol{\psi}, \boldsymbol{\lambda})}{\partial \boldsymbol{\lambda}^2}\right|_{\boldsymbol{\lambda}=\widehat{\boldsymbol{\lambda}}_{\psi}} \tag{9.16}$$

is the negative of the observed Fisher information matrix for $\boldsymbol{\lambda}$ (evaluated at $\widehat{\boldsymbol{\lambda}}_{\psi}$) when $\boldsymbol{\psi}$ is known. More will be said about the $c(\boldsymbol{\psi}, \widehat{\boldsymbol{\lambda}}, \widehat{\boldsymbol{\lambda}}_{\psi})$ term below. The right-hand side of (9.15) is commonly called the modified profile log-likelihood because the additional terms modify the standard profile log-likelihood for the loss of information about $\boldsymbol{\psi}$ from estimation of $\widehat{\boldsymbol{\lambda}}$.

If the $c(\boldsymbol{\psi}, \widehat{\boldsymbol{\lambda}}, \widehat{\boldsymbol{\lambda}}_{\psi})$ term in (9.15) is negligible, then

$$l_c(\boldsymbol{\psi}) \approx l(\boldsymbol{\psi}, \widehat{\boldsymbol{\lambda}}_{\psi}) - \frac{1}{2}\log\det\left(-\mathbf{H}(\widehat{\boldsymbol{\lambda}}_{\psi})\right). \tag{9.17}$$

The term on the right-hand of (9.17) is often called the adjusted profile log-likelihood. To within an additive constant, the adjusted profile log-likelihood corresponds to the log of the Laplace approximation to $L_{\partial\boldsymbol{\lambda}}(\boldsymbol{\psi})$. This approximation is given in Equation (10.12), and here integration of $L(\boldsymbol{\theta}) \equiv L(\boldsymbol{\psi}, \boldsymbol{\lambda})$ is with respect to $\boldsymbol{\lambda}$. That is, the Laplace approximation is

$$L_{\partial\boldsymbol{\lambda}}(\boldsymbol{\psi}) = \int L(\boldsymbol{\psi}, \boldsymbol{\lambda})d\boldsymbol{\lambda} \approx L(\boldsymbol{\psi}, \widehat{\boldsymbol{\lambda}}_{\psi})(2\pi)^{\frac{q}{2}}\det\left(-\mathbf{H}(\widehat{\boldsymbol{\lambda}}_{\psi})\right)^{-\frac{1}{2}}, \tag{9.18}$$

where q is the dimension of $\boldsymbol{\lambda}$. Thus, if the Laplace approximation is good, and ignoring the constant $(2\pi)^{\frac{q}{2}}$ term in (9.18),

$$l_c(\boldsymbol{\psi}) \approx \log L_{\partial\boldsymbol{\lambda}}(\boldsymbol{\psi}). \tag{9.19}$$

It is relevant to note that the conditional log-likelihood $l_c(\boldsymbol{\psi})$ is invariant to re-parametrization of $\boldsymbol{\lambda}$ (since it *is* a log-likelihood, albeit one based on a conditional distribution), but the adjusted profile log-likelihood in (9.17), and integrated log-likelihood, $\log L_{\partial\boldsymbol{\lambda}}(\boldsymbol{\psi})$, are not. For example, to see the latter, let $\boldsymbol{\zeta} = g(\boldsymbol{\lambda})$ be a differentiable and invertible transformation, and denote the likelihood parameterized using $(\boldsymbol{\psi}, \boldsymbol{\zeta})$ by $L^{(g)}(\boldsymbol{\psi}, \boldsymbol{\zeta}) = L(\boldsymbol{\psi}, \boldsymbol{\lambda})$. Then, under the $(\boldsymbol{\psi}, \boldsymbol{\zeta})$ parameterization, the integrated likelihood becomes

$$L^{(g)}_{\partial\zeta}(\boldsymbol{\psi}) = \int L^{(g)}(\boldsymbol{\psi}, \boldsymbol{\zeta})d\boldsymbol{\zeta}$$
$$= \int L(\boldsymbol{\psi}, \boldsymbol{\lambda})\left|\det\left(g'(\boldsymbol{\lambda})\right)\right|d\boldsymbol{\lambda}\,,$$

which is not equal to $L_{\partial\lambda}(\boldsymbol{\psi})$ in general. The preservation of parameter invariance on the right-hand side of (9.15) is the role of the $c(\boldsymbol{\psi}, \widehat{\boldsymbol{\lambda}}, \widehat{\boldsymbol{\lambda}}_{\boldsymbol{\psi}})$ term.

The $c(\boldsymbol{\psi}, \widehat{\boldsymbol{\lambda}}, \widehat{\boldsymbol{\lambda}}_{\boldsymbol{\psi}})$ term is generally intractable, but it can be shown that it is negligible if $\boldsymbol{\psi}$ and $\boldsymbol{\lambda}$ are parameter orthogonal (Section 11.6.1). See Pawitan (2001, p. 287) or Lee *et al.* (2006, p. 119) for more details. Thus, if adjusted profile likelihood or integrated likelihood is to be used as an approximation to conditional likelihood, a parameterization should be used such that $\boldsymbol{\psi}$ and $\boldsymbol{\lambda}$ are at least approximately orthogonal. This is discussed at length in Lee *et al.* (2006). See also Cox and Reid (1987).

9.3.2 Uses of integrated likelihood

Integrated likelihood has been advocated on fundamental philosophical grounds, as a coherent hybrid approach that combines elements of both the Bayesian and frequentist philosophies to inference (Berger, Liseo and Wolpert 1999, Severini 2007). This is the view that is taken here, with the added reassurance that the integrated likelihood will be a good approximation to the conditional likelihood for sensible parameterizations. The adjusted profile likelihood (i.e. Laplace approximation to the integrated likelihood) can be regarded as a pragmatic method for numerical approximation of the integrated likelihood, but more accurate approximations are viable in some situations (Section 10.3).

For normal linear models (including mixed-effects models), parameters $\boldsymbol{\psi}$ (the variance parameters) and $\boldsymbol{\lambda} \equiv \boldsymbol{\beta}$ (the regression coefficients) are orthogonal, and the Laplace approximation in (9.18) is exact since the log-likelihood is quadratic in $\boldsymbol{\lambda}$. It follows that REML is equivalent to integrated likelihood in these models, This equivalence was first reported by Harville (1974), and is utilized in Section 10.4 where ADMB is used to verify the fitted REML model obtained using R and SAS.

Conditional likelihood is not tractable in the context of generalized linear mixed models (GLMMs), and an extension of REML to these models is not straightforward. However, approximate REML can be implemented via integrated likelihood, and herein this will be called generalized restricted maximum likelihood (GREML). An example of GREML is provided in Section 10.6. The reader is referred to Lee *et al.* (2006) for more detailed coverage about the use of adjusted and integrated likelihood in GLMMs.

Noh and Lee (2007) used a crossed mixed-effects model for binary data on salamander matings, and used simulations to compare the performance of several likelihood variants. The adjusted profile likelihood was found to perform well and to be computationally viable. Code is available to reproduce this analysis using ADMB, at `http://www.stat.auckland.ac.nz/~millar`, notwithstanding that there are some small differences in parameter estimates due to the computational approximations employed by Noh and Lee (2007).

9.4 Exercises

9.1 In Example 9.1, the ML and two-way ANOVA analyses of the observed data both estimate the regression coefficients by minimizing the residual sum-of-squares

$$\sum_{i=1}^{n}\sum_{j=1}^{2}(y_{ij}-\mu_{ij})^2=\sum_{i=1}^{n}\left\{(y_{i1}-[\mu_i-\delta/2])^2+(y_{i2}-[\mu_i+\delta/2])^2\right\}.$$

Denoting $d_i = y_{i2} - y_{i1}$, show that

$$\widehat{\delta}=\frac{1}{n}\sum d_i=\overline{d}$$

$$\widehat{\mu}_i=\frac{y_{i1}+y_{i2}}{2}=\overline{y}_i\,,$$

and hence that the minimized value of the residual sum-of-squares is

$$\text{RSS}_Y=\sum_{i=1}^{n}\sum_{j=1}^{2}(y_{ij}-\widehat{\mu}_{ij})^2=\frac{1}{2}\sum_{i=1}^{n}(d_i-\overline{d})^2\,. \tag{9.20}$$

Note that $S_Y^2=\sum_{i=1}^{n}\sum_{j=1}^{2}(Y_{ij}-\widehat{\mu}_{ij})^2/(n-1)$ is an unbiased estimator of $\tau^2=\text{var}(Y)$, and $S_D^2=\sum(D_i-\overline{D})^2/(n-1)=2S_Y^2$ is an unbiased estimator of $\sigma^2=\text{var}(D)$.

9.2 In Example 9.1, use (9.7) to show that inference based on the within-pair differences can be justified as a form of marginal inference.

10

Latent variable models

The intangible is the seed of the tangible. – Bruce Lee

10.1 Introduction

Latent variable models provide a rich and widely-used collection of models that are especially suited for inference when the data are observed from experiments where more than one source of variability is present. In many cases this will be because the data are grouped. For example, while it may be desired to test a drug treatment on 1000 patients, these patients might be selected from a much smaller number of hospitals. If there are any differences in the efficacy of the drug between the hospitals (perhaps due to staff skills, socio-economic status of patients, selection criteria for participation in the study, etc.) then the 1000 patients can not be considered an iid sample from a single population. Similarly, biological experiments face constraints over the availability of equipment, so an experiment measuring the growth of a large number of fish may only be able to utilize a handful of tanks in which to rear them. While all reasonable attempts can be made to keep the tanks equal, this can rarely be achieved – one dead fish in a tank could affect the water quality for all other fish in that tank. In ecological field experiments, a large number of samples may necessarily have to be taken from a relatively small number of locations. These locations can be chosen to be similar, but they can never be identical.

Failure to take into account any grouping that is inherent in the data can lead to erroneous inference. Indeed, Hurlbert (1984) caused quite a clamour by revealing that about one half of all quantitative results from manipulative ecological field experiments (published between 1960 and 1981) contained questionable statistical

Maximum Likelihood Estimation and Inference: With Examples in R, SAS and ADMB, First Edition. Russell B. Millar.

inference due to failure to accommodate grouping structure in the data. Hurlbert (1984) referred to this as pseudo-replication.

More generally, the class of latent variable models subsumes many other types of model, including classes of models that are variously called mixed-effects models, random-effects models, hierarchical models, state-space models and volatility models. Latent variable models also include semi-parametric regression models, by virtue of the mixed-effects model formulation of penalized smoothing splines (Wand 2003). They also include mixture models, and the next section uses the binormal mixture model introduced in Section 2.4.5 to develop the notation for likelihood-based inference from latent variable models.

10.2 Developing the likelihood

Throughout this text, the notation $f(\boldsymbol{y};\boldsymbol{\theta})$ represents the density function of $\boldsymbol{Y}$ under repetition of the experiment from which the observed data, $\boldsymbol{y}$, were generated. Regarded as a function of $\boldsymbol{\theta}$, $f(\boldsymbol{y};\boldsymbol{\theta})$ is the all-important likelihood. Except in special cases, the density functions arising from latent variable models can not be expressed in closed form, because the distribution of $\boldsymbol{Y}$ may depend on two or more sources of randomness in such a way that $f(\boldsymbol{y};\boldsymbol{\theta})$ can not be expressed using simple algebraic functions.

The binormal model for the waiting times of the Old Faithful geyser (Example 2.9) was presented in the form of a latent variable model. In that example, the binormal distribution was expressed as a mixture of two normal distributions where, conditional on an unobserved B_i from a Bernoulli(p) experiment, Y_i was observed from a $N(\mu, \sigma^2)$ distribution when $B_i = 1$, else from a $N(\nu, \tau^2)$ distribution. Here, the B_i are latent random variables, in the sense that they remain hidden, yet are relevant to determination of the likelihood function. In this particular example, it was straightforward to obtain the likelihood for parameters $\boldsymbol{\theta} = (p, \mu, \sigma, \nu, \tau)$ because the latent variables are dichotomous, and hence the (marginal) density of Y_i could be obtained as a weighted sum of the (conditional) densities of Y_i given B_i. Specifically, for each observed y_i,

$$
\begin{aligned}
f(y_i;\boldsymbol{\theta}) &= \sum_{b_i\in\{0,1\}} f(y_i, b_i;\boldsymbol{\theta}) && (10.1)\\
&= \sum_{b_i\in\{0,1\}} f(y_i|b_i;\boldsymbol{\theta})P(B_i = b_i;\boldsymbol{\theta}) && (10.2)\\
&= \sum_{b_i\in\{0,1\}} f(y_i|b_i;\mu, \sigma, \nu, \tau)P(B_i = b_i; p) &&\\
&= p\phi(y_i;\mu, \sigma^2) + (1-p)\phi(y_i;\nu, \tau^2) && (10.3)
\end{aligned}
$$

where $\phi(y;\mu, \sigma^2)$ denotes the $N(\mu, \sigma^2)$ density function evaluated at y.

The latent variable models considered in this chapter have a likelihood of the same form as (10.2), except that the marginalization over the latent variable involves an integral rather than a summation. In the binormal model the latent variables were the Bernoulli random variables B_i, $i = 1, \ldots, n$, but the notation used below will use $\boldsymbol{U}$ to generically denote all unobserved random variables in the model, and their (marginal) density function will be denoted $f(\boldsymbol{u};\boldsymbol{\theta})$.[1]

Letting $f(\boldsymbol{y}|\boldsymbol{u};\boldsymbol{\theta})$ denote the conditional distribution of $\boldsymbol{Y}$ given $\boldsymbol{U}$, the joint density of $\boldsymbol{Y}$ and $\boldsymbol{U}$ is

$$f(\boldsymbol{y}, \boldsymbol{u};\boldsymbol{\theta}) = f(\boldsymbol{y}|\boldsymbol{u};\boldsymbol{\theta})f(\boldsymbol{u};\boldsymbol{\theta}) . \tag{10.4}$$

The likelihood function of $\boldsymbol{\theta}$ arising from observation of $\boldsymbol{y}$ is given by marginalization of the joint density with respect to $\boldsymbol{u}$,

$$L(\boldsymbol{\theta}; \boldsymbol{y}) = f(\boldsymbol{y};\boldsymbol{\theta}) = \int f(\boldsymbol{y}, \boldsymbol{u};\boldsymbol{\theta})d\boldsymbol{u} . \tag{10.5}$$

In general, the integral in (10.5) can not be expressed in closed form. A notable exception is the linear mixed model (LMM) in which the latent variables $\boldsymbol{U}$, and $\boldsymbol{Y}|\boldsymbol{U}$ have a multivariate normal distribution, and $E[\boldsymbol{Y}]$ is a linear function of $\boldsymbol{U}$ and the regression coefficients. In that case, the marginal distribution of $\boldsymbol{Y}$ is also multivariate normal (Harville 1977, Laird and Ware 1982, McCulloch, Searle and Neuhaus 2008). The acronyms GLMM and NLMM are used for generalized linear mixed model and nonlinear mixed model, whereby $\boldsymbol{Y}|\boldsymbol{U}$ follows a generalized linear, or nonlinear model, respectively. State-space models for time-series data are another form of latent variable model, and in such models it is often the case that the distribution of $\boldsymbol{Y}|\boldsymbol{U}$ takes the form of an autoregressive model. The Kalman filter provides a closed-form iterative algorithm for determining the likelihood for linear normal state-space models (Meinhold and Singpurwalla 1983, Wei 2006).

This chapter focuses on demonstration of the current capabilities of R, SAS and ADMB for maximum likelihood inference from a likelihood of the form in Equation (10.5). In the examples that follow (Sections 10.4–10.7), the latent variables $\boldsymbol{U}$ are assumed to be normally distributed. The use of alternative distributions for the latent variables can be accommodated by ADMB because this software requires the user to code the equation for $f(\boldsymbol{y}, \boldsymbol{u};\boldsymbol{\theta})$ within the template file, and places no explicit restrictions on its form.

10.3 Software

10.3.1 Background

When the likelihood in (10.5) possesses a closed form then standard methods of optimization can be used. For example, to fit linear mixed models, `PROC MIXED` in SAS, and the `lmer` function in R package `lme4`, use an iterative estimating

[1] For ease of notation, $f()$ is used generically to denote density functions. The distinction between different density functions is implicit from their arguments.

equations approach (Harville 1977) that can be considered a variant of the Newton-Raphson algorithm.

When the likelihood can not be obtained in closed form then the Expectation-Maximization (EM) algorithm could be considered. This algorithm assumes that the data are, in some sense, 'incomplete', and this is precisely the scenario of a latent variable model, whereby the data would be 'complete' if $\boldsymbol{U}$ were observed. Indeed, the binormal mixture model was used as a simple example of the EM algorithm in Section 5.3. Of course, the binormal likelihood has closed form and was maximized using standard optimizers in Section 3.3.4. However, the EM algorithm would become more useful for mixtures of many component distributions, where its robustness would guarantee convergence to a maximum of the likelihood, but other optimizers might be challenged by the high-dimensional maximization. The EM algorithm has been used to fit linear mixed models (e.g. Laird and Ware 1982), but it may not be as convenient to implement in more general latent variable models if the E and M steps are not available in closed form.

Likelihoods for GLMMs and NLMMs can not generally be obtained in closed form, and likelihood-based methods for fitting these models have historically used an assortment of modified likelihoods obtained through various linear approximations to the model (e.g. Stiratelli, Laird and Ware 1984, Lindstrom and Bates 1990, Schall 1991, Breslow and Clayton 1993, Wolfinger and O'Connell 1993, Littell, Milliken, Stroup and Wolfinger 1996). A common feature of these methods is a step in the fitting algorithm in which a linearized approximation to the model is fitted using linear mixed models. For example, the pseudo-likelihood approach of Wolfinger and O'Connell (1993) fits a GLMM using alternating fits of a generalized linear model (using `PROC GENMOD`) and a linear mixed model (using `PROC MIXED`). This algorithm was implemented in the SAS macro `GLIMMIX`, and a version for nonlinear mixed models was implemented in the SAS macro `NLINMIX` (Littell *et al.* 1996). Within R, the leading package for mixed-effects modelling is `lme4`, and earlier versions of this package used the penalized quasi-likelihood method of Breslow and Clayton (1993).

While use of these modified likelihoods were popular at the time, it was known that they did not enjoy the full advantages of working with the true likelihood given by (10.5). For example, they can produce inconsistent estimators in situations where true ML estimators would be consistent (Breslow and Lin 1995, Lin and Breslow 1996, Millar and Willis 1999), and they can be of dubious use for the purposes of model comparison and selection. McCulloch *et al.* (2008) called for an end to the use of modified likelihoods, and their use has indeed waned. The `NLINMIX` and `GLIMMIX` macros of Littell *et al.* (1996) have now been replaced by the `NLMIXED` and `GLIMMIX` procedures (in SAS version 9.2). Both of these procedures perform numerical evaluation of the likelihood in (10.5), although pseudo-likelihood remains an option in `GLIMMIX`. The `NLMIXED` procedure and the R package `lme4` now use only methods based on numerical evaluation of (10.5).

The next section presents the Laplace approximation for numerical evaluation of the likelihood $L(\boldsymbol{\theta}; \boldsymbol{y})$ given by the integral in (10.5). It has been established that the Laplace approximation works extremely well in a wide variety of latent variable models (e.g. see Skaug and Fournier 2006, Rue, Martino and Chopin 2009). Moreover, the accuracy of the approximation can be improved by the use of importance sampling (Section 10.3.3). In addition, for low-dimensional integrals, Gauss-Hermite quadrature can be used to provide a higher order of approximation.

It has been shown that penalized quasi-likelihood (Breslow and Clayton 1993) and pseudo-likelihood (Wolfinger and O'Connell 1993) have much in common with the Laplace approximation (Wolfinger 1993, Vonesh 1996, McCulloch *et al.* 2008). However, the weakness in these ad hoc likelihood methods is that they use additional approximations to avoid the optimization and second derivative calculations required by the Laplace approximation (Section 10.3.2).

Box 10.1

10.3.2 The Laplace approximation and Gauss-Hermite quadrature

For ease of presentation, the Laplace approximation will be derived for the scalar case $u \in \mathbb{R}$, and it will be assumed that the domain of integration is the entire real line.

For any fixed $\boldsymbol{y}$ and $\boldsymbol{\theta}$, and with the simplifying notation $h(u) = f(\boldsymbol{y}, u; \boldsymbol{\theta})$, the integral in (10.5) can be written

$$L(\boldsymbol{\theta}; \boldsymbol{y}) = \int_{\mathbb{R}} h(u)du = \int_{\mathbb{R}} e^{\log h(u)}du \,. \tag{10.6}$$

Let $\widehat{u}$ denote the value that maximizes $h(u)$, and hence also $\log h(u)$. The first derivative of $\log h(u)$ is zero at $\widehat{u}$, and so a second-order Taylor series expansion of $\log h(u)$ around $\widehat{u}$ yields

$$\log h(u) \approx \log h(\widehat{u}) - \frac{c(u-\widehat{u})^2}{2} \,,$$

where $c > 0$ is given by

$$c = -\left.\frac{\partial^2 \log h(u)}{\partial u^2}\right|_{u=\widehat{u}} .$$

Substituting this expansion into (10.6) gives

$$L(\boldsymbol{\theta}; \boldsymbol{y}) \approx h(\widehat{u}) \int_{\mathbb{R}} \exp\left(-\frac{c(u-\widehat{u})^2}{2}\right) du \,. \tag{10.7}$$

As a function of u, the integrand in (10.7) is, to within a constant, the density function of a normal random variable with mean $\widehat{u}$ and variance c^{-1}. Hence, the integral is obtained from the normalizing constant of this density function, that is

$$\int_{\mathbb{R}} \exp\left(-\frac{(u-\widehat{u})^2}{2/c}\right) = \sqrt{\frac{2\pi}{c}}\,.$$

This leads immediately to the Laplace approximation,

$$L(\boldsymbol{\theta}; \boldsymbol{y}) \approx h(\widehat{u})\sqrt{\frac{2\pi}{c}} \tag{10.8}$$

$$\equiv f(\boldsymbol{y}, \widehat{u}_{\boldsymbol{\theta}}; \boldsymbol{\theta})\sqrt{\frac{2\pi}{c_{\boldsymbol{\theta}}}}\,, \tag{10.9}$$

where a $\boldsymbol{\theta}$ subscript has been added to the $\widehat{u}$ and c notation to convey the dependence of these quantities on $\boldsymbol{\theta}$.

In the multi-dimensional case where $\boldsymbol{u} \in \mathbb{R}^q$, Equation (10.8) extends in the natural way. The Laplace approximation is then

$$\int_{\mathbb{R}^q} h(\boldsymbol{u})d\boldsymbol{u} \approx h(\widehat{\boldsymbol{u}})(2\pi)^{\frac{q}{2}} \det(-\mathbf{H}(\widehat{\boldsymbol{u}}))^{-\frac{1}{2}} \tag{10.10}$$

where $\det(-\mathbf{H}(\widehat{\boldsymbol{u}}))$ is the determinant of the negative of the $q \times q$ Hessian matrix of second derivatives of $h(\boldsymbol{u})$, evaluated at $\widehat{\boldsymbol{u}}$, given by

$$\mathbf{H}(\widehat{\boldsymbol{u}}) = \left.\frac{\partial^2 \log h(\boldsymbol{u})}{\partial \boldsymbol{u}^2}\right|_{u=\widehat{u}}. \tag{10.11}$$

In the present context $h(\boldsymbol{u}) = f(\boldsymbol{y}, \boldsymbol{u}; \boldsymbol{\theta})$, and the Laplace approximation is then

$$L(\boldsymbol{\theta}; \boldsymbol{y}) \approx f(\boldsymbol{y}, \widehat{\boldsymbol{u}}_{\boldsymbol{\theta}}; \boldsymbol{\theta})(2\pi)^{\frac{q}{2}} \det(\mathbf{O}(\widehat{\boldsymbol{u}}_{\boldsymbol{\theta}}))^{-\frac{1}{2}}\,, \tag{10.12}$$

where $\mathbf{O}(\widehat{\boldsymbol{u}}_{\boldsymbol{\theta}})$ is the matrix

$$\mathbf{O}(\widehat{\boldsymbol{u}}_{\boldsymbol{\theta}}) = -\left.\frac{\partial^2 \log f(\boldsymbol{y}, \boldsymbol{u}; \boldsymbol{\theta})}{\partial \boldsymbol{u}^2}\right|_{u=\widehat{u}_{\theta}},$$

and can be regarded as the observed Fisher information (Section 12.3.1) about $\boldsymbol{u}$ when $\boldsymbol{\theta}$ is known.

The Laplace approximation is exact when the log-integrand, $\log h(\boldsymbol{u})$, is of quadratic form. Gauss-Hermite quadrature approximation provides an improvement to the Laplace approximation that is exact for a wider class of integrands. In the one-dimensional case, Gauss-Hermite quadrature is able to evaluate the integral exactly if the integrand can be expressed in the form

$$h(u) = p(u)e^{-u^2}\,, \tag{10.13}$$

where $p(u)$ is a polynomial in u (McCulloch *et al.* 2008). The approximation takes the form of a weighted sum of values of $p(u)$ evaluated at pre-determined quadrature points. If d quadrature points are used then the integral of (10.13) is exact for polynomials $p(u)$ up to order $2d - 1$. The Laplace approximation is equivalent to Gauss-Hermite quadrature with a single quadrature point.

Gauss-Hermite quadrature is only computationally efficient for integrals of low dimension. For this reason, in the examples of Section 10.4–10.7, Gauss-Hermite quadrature was only available in SAS, R and ADMB for the models in which the likelihood in (10.5) could be expressed as a product of low-dimensional integrals (see Section 10.3.4).

10.3.3 Importance sampling

The Laplace approximation assumes that, to within a multiplicative constant, the integrand $h(\boldsymbol{u}) = f(\boldsymbol{y}, \boldsymbol{u}; \boldsymbol{\theta})$ is well approximated by the density function of a q-dimensional multivariate normal centred at the value $\widehat{\boldsymbol{u}}_{\boldsymbol{\theta}}$. Importance sampling provides a technique for polishing this approximation.

Importance sampling is derived from the simple identity

$$\begin{aligned}\int_{\mathbb{R}^q} h(\boldsymbol{u})d\boldsymbol{u} &= \int_{\mathbb{R}^q} \frac{h(\boldsymbol{u})}{g(\boldsymbol{u})} g(\boldsymbol{u})d\boldsymbol{u} \\ &= \int_{\mathbb{R}^q} r(\boldsymbol{u})g(\boldsymbol{u})d\boldsymbol{u}\ ,\end{aligned} \tag{10.14}$$

where $r(\boldsymbol{u}) = h(\boldsymbol{u})/g(\boldsymbol{u})$. The choice of $g(\boldsymbol{u})$ is arbitrary, other than being non-negative and having support that contains the support of $h(\boldsymbol{u})$ (i.e. if $h(\boldsymbol{u}) > 0$ then $g(\boldsymbol{u}) > 0$) so that $r(\boldsymbol{u})$ is well defined. If $g(\boldsymbol{u})$ is the density function of a q-dimensional random variable, $\boldsymbol{U}$, then (10.14) is the expected value of $r(\boldsymbol{U})$, $E[r(\boldsymbol{U})]$. Thus, if $r(\boldsymbol{U})$ has finite variance σ_r^2, the sample mean of m iid realizations of $r(\boldsymbol{U})$ (with $\boldsymbol{U}$ generated from the distribution with density $g(\boldsymbol{u})$) is a consistent estimator of (10.14) and has variance equal to σ_r^2/m. This estimate of (10.14) is simply

$$\bar{r} = \frac{1}{m}\sum_{i=1}^{m} r(\boldsymbol{u}_{(i)})\ ,$$

where $\boldsymbol{u}_{(i)}$ denotes the ith realization of $\boldsymbol{U}$.

Importance sampling works best when $g(\boldsymbol{u})$ is chosen so that it is close to proportional to $h(\boldsymbol{u})$, because then $r(\boldsymbol{U})$ will have little variability and hence σ_r^2 will not be unduly large. The q-dimensional multivariate normal used in the Laplace approximation makes a convenient initial choice for $g(\boldsymbol{u})$. With this choice, importance sampling can be regarded as an improvement to the Laplace approximation that is achieved by evaluating the difference between the integrand $h(\boldsymbol{u})$ and the approximating normal density function at a large number of randomly chosen values of $\boldsymbol{u} \in \mathbb{R}^q$.

However, choosing $g(\boldsymbol{u})$ to be the multivariate normal approximation to $h(\boldsymbol{u})$ does sometimes result in σ_r^2 being extremely large. This can occur if the tails of $h(\boldsymbol{u})$ are fatter than those of $g(\boldsymbol{u})$, because values of $\boldsymbol{u}$ in the tails of density $g(\boldsymbol{u})$ will have very small values of $g(\boldsymbol{u})$ and hence will correspond to extremely high values of $r(\boldsymbol{u})$. This problem can usually be cured by replacing the approximating multivariate normal density with a density having fatter tails. A multivariate t-density, or a mixture of multivariate normals with covariance matrices of differing scale, are suitable for this purpose.

At present, ADMB provides the option of importance sampling, and this is requested by addition of the `-isb` command line option. `PROC NLMIXED` can also be instructed to use importance sampling, with the procedure option `METHOD=ISAMP`.

10.3.4 Separability

In many classes of latent variable models, the computational burden of the q-dimensional marginalization in (10.5) can be reduced by rewriting the integral as a product of lower-dimensional integrals. For example, suppose that $\boldsymbol{u} \in \mathbb{R}^q$ and that $\boldsymbol{Y}$ can be partitioned into q subsets $\boldsymbol{Y}_{(1)}, ...\boldsymbol{Y}_{(q)}$ such that $\boldsymbol{Y}_{(i)}|\boldsymbol{u}, i = 1, ..., q$ depends only on u_i, and $\boldsymbol{Y}_{(i)}|u_i, i = 1, ..., q$ are independent. Furthermore, if $u_i, i = 1, ..., q$ are independent, then the joint density function $f(\boldsymbol{y}, \boldsymbol{u}; \boldsymbol{\theta})$ can be expressed

$$\begin{aligned} f(\boldsymbol{y}, \boldsymbol{u}; \boldsymbol{\theta}) &= f(\boldsymbol{y}|\boldsymbol{u}; \boldsymbol{\theta}) f(\boldsymbol{u}; \boldsymbol{\theta}) \\ &= \prod_{i=1}^{q} f(\boldsymbol{y}_{(i)}|\boldsymbol{u}; \boldsymbol{\theta}) \prod_{i=1}^{q} f(u_i; \boldsymbol{\theta}) \\ &= \prod_{i=1}^{q} \left(f(\boldsymbol{y}_{(i)}|u_i; \boldsymbol{\theta}) f(u_i; \boldsymbol{\theta}) \right) \\ &= \prod_{i=1}^{q} f(\boldsymbol{y}_{(i)}, u_i; \boldsymbol{\theta}) . \end{aligned}$$

This joint density has the form of a separable integrand, in the sense that its integral with respect to $\boldsymbol{u}$ can be separated into the product of q one-dimensional integrals. That is,

$$L(\boldsymbol{\theta}; \boldsymbol{y}) = \int_{\mathbb{R}^q} f(\boldsymbol{y}, \boldsymbol{u}; \boldsymbol{\theta}) d\boldsymbol{u} = \prod_{i}^{q} \int_{\mathbb{R}} f(\boldsymbol{y}_{(i)}, u_i; \boldsymbol{\theta}) du_i . \qquad (10.15)$$

Separating the integral into a product of low-dimensional integrals can enable the efficient use of Gauss-Hermite quadrature approximation. The `NLMIXED` procedure requires separability, and Gauss-Hermite quadrature is its default method of integral approximation. However, this procedure is restricted to only one level of grouping structure. Similarly, `PROC GLIMMIX`, the `lme4` package and ADMB[2] all provide quadrature approximation for separable models, but otherwise revert to the Laplace approximation.

It is also the case that integrals of low dimension can often be accurately evaluated using numerical integration functions provided within R. The `integrate` function evaluates integrals of one dimension, and the `adapt` function (within the package of the same name) is able to evaluate integrals of low dimension. Thus, for latent variable models with joint density functions that can be expressed in separable form, it will often be possible to write an R function that returns the evaluated value of $L(\boldsymbol{\theta}; \boldsymbol{y})$. This can then be maximized using the optimizing function `optim`.

10.3.5 Overview of examples

Sections 10.4–10.7 present examples of linear mixed models, nonlinear mixed models, generalized linear mixed models, and Poisson state-space models, respectively. The LMM example uses data that were measured according to a multilevel hierarchical design. However, for ease of demonstration, the model fitted in Section 10.4 uses only the top level of grouping, and the intercept term is the sole regression parameter. As there are no fixed effects, this model is called a random-effects model, rather than a mixed-effects model. The addition of a second level of grouping is left as an exercise (see Exercises 10.1 and 10.2).

The second example (Section 10.5) finds fault in the fit of a NLMM to orange tree circumference data. The faulty model includes only a single grouping variable. Instead, a crossed-effects NLMM is postulated and fitted using ADMB. In the GLMM example (Section 10.6), dichotomous treatment outcomes from a clinical trial are modelled using a binomial model with nested random effects, and it is found that the nested random effects are not actually required. These data were previously analyzed using generalized estimating equations (Section 8.3.1) and comparison between the GLMM and GEE approaches is made in Section 10.6.4. The final example uses ADMB to fit a Poisson state-space model to a time series of disease counts (Section 10.7).

10.4 One-way linear random-effects model

The data presented in Table 10.1 are from an experiment to study the reproducibility of measurements of estrone. Sixteen vials of serum were taken from each of five subjects. There were additional levels of grouping in the design of this experiment (Gail *et al.* 1996), but only the grouping by subject will be considered for the purposes of this example, and hence the vials will be regarded as replicate observations on each subject.

[2] ADMB requires the user to explicitly code the joint density in separable form – see ADMB-project (2008b, or later version).

Table 10.1 Sixteen estrone measurements (pg/mL), from each of five postmenopausal women. The data are from Fears *et al.* (1996).

	Subject				
Replicate	1	2	3	4	5
1	23	25	38	14	46
2	23	33	38	16	36
3	22	27	41	15	30
4	20	27	38	19	29
5	25	30	38	20	36
6	22	28	32	22	31
7	27	24	38	16	30
8	25	22	42	19	32
9	22	26	35	17	32
10	22	30	40	18	31
11	23	30	41	20	30
12	23	29	37	18	32
13	27	29	28	12	25
14	19	37	36	17	29
15	23	24	30	15	31
16	18	28	37	13	32

Gail *et al.* (1996) assumed that the data were normally distributed on the log scale. They used $\log_{10}$ (log base 10) rather than natural log, and the same is done here. The five subjects were considered to be randomly chosen from a relevant population of postmenopausal women, and so the subject effect is a random variable. The model can be written

$$U_i \sim N(0, \sigma_u^2) \qquad , i = 1, ..., 5 \tag{10.16}$$

$$Y_{ij}|u_i \sim N(a + u_i, \sigma^2) \quad , i = 1, ..., 5, \; j = 1, ..., 16 , \tag{10.17}$$

where u_i is the (unobserved) realized value of U_i, and Y_{ij} is the $\log_{10}$(estrone) measurement to be taken from replicate j on subject i. This model is saying that a randomly chosen subject has a mean value (over measurements on replicate vials) of $\log_{10}$(estrone) that varies around a by the random amount $U_i \sim N(0, \sigma_u^2)$.

There are only five subjects with which to estimate between-subject variability, and so it makes sense to use restricted maximum likelihood (REML, Section 9.2.1) to take into account the loss of a degree of freedom from estimation of the intercept parameter a. The above model is a one-way random-effects ANOVA, and its likelihood $L(a, \sigma^2, \sigma_u^2; \mathbf{y})$ and restricted likelihood can be written in closed form because the marginal distribution of the data is multivariate normal (e.g. see McCulloch *et al.* 2008) of dimension 80.

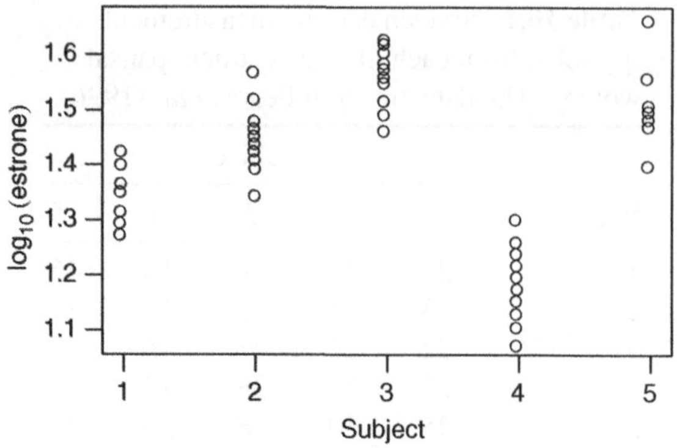

Figure 10.1 $\log_{10}$(estrone) measurements for each of five subjects.

A visual inspection of the data (Figure 10.1) shows that there is, without doubt, considerable between-subject variability. Nonetheless, for completeness, the null hypothesis $H_0 : \sigma_u^2 = 0$ is formally considered here.

Under H_0, the parameter σ_u^2 is on the boundary of the parameter space, and hence the considerations of Section 4.7.3 apply. In particular, as described in greater detail in Example 4.12, the LRT statistic has an asymptotic distribution that is an equal mixture of a zero and a χ_1^2 random variable (Self and Liang 1987, Stram and Lee 1994). However, Pinheiro and Bates (2000) have shown that this provides a crude approximation to the true sampling distribution of the LRT statistic when the number of groups is small. The estrone experiment has just five subjects and a better alternative would be to explore the sampling distribution of the LRT statistic by bootstrapping. A second alternative is available to R users. The R-package `RLRsim` provides function `exactLRT` to numerically evaluate the true finite-sampling distribution of the LRT statistic for testing hypotheses of the form $H_0 : \sigma_u^2 = 0$ in linear mixed models.

To create a bit of variety in the application of SAS, R and ADMB, each of these software tools was used to provide a different piece of the inferential picture. Specifically, `PROC MIXED` was used to produce a contour plot of the profile restricted likelihood for the variance parameters σ^2 and σ_u^2. R function `exactLRT` was used to test $H_0 : \sigma_u^2 = 0$, and the ADMB implementation demonstrates the iterative use of ADMB from within R.

10.4.1 SAS

The SAS code below assumes that dataset `estrone` contains two variables, `person` (a factor variable for subject) and `y` ($\log_{10}$(estrone)). The `MODEL` and `RANDOM` statements in `PROC MIXED` are used to specify the fixed and random components, respectively, of the LMM. Here, the only fixed term in the model is the intercept parameter a, and this is included by default. The `RANDOM INT / SUBJECT=person` statement is used to specify that `person` is the grouping variable, and that u_i (the random effect of `person`) is to be added to the intercept, as specified in Equation (10.17).

`PROC MIXED` uses restricted likelihood by default, but putting the option `METHOD=ML` in the procedure statement will force it to use standard likelihood. The many features of `PROC MIXED` include the `PARMS` statement to calculate the restricted log-likelihood over a grid of values of σ_u^2 and σ^2. This facilitated the production of a contour plot of the profile restricted log-likelihood

$$l_R^*(\sigma^2, \sigma_u^2) = \max_a l_R(a, \sigma^2, \sigma_u^2),$$

where l_R denotes the restricted log-likelihood.

```
*Dataset to hold the restricted lhood points for the contour plot;
ODS OUTPUT ParmSearch=parms;

*REML fit to log10(estrone) data using PROC MIXED;
*NOPROFILE option is used, to prevent partial profiling of lhood;
PROC MIXED DATA=estrone NOPROFILE;
  MODEL y= / SOLUTION;
  RANDOM INT / SUBJECT=person;
  PARMS (0.0 TO 0.2 BY 0.0001) (0.0015 TO 0.0060 BY 0.00005);
RUN;
```

From the output in Figure 10.2, the maximal value of the restricted log-likelihood is 103.03, and the REML estimate is $(\widehat{a}, \widehat{\sigma}^2, \widehat{\sigma}_u^2) = (1.42, 0.00325, 0.0175)$. The approximate variance of $\widehat{\sigma}_u^2$ can be obtained by using the `ASYCOV` procedure

Iteration history			
Iteration	**Evaluations**	**-2 Res Log Like**	**Criterion**
1	2	-206.05467655	0.00000000

Covariance parameter estimates		
Cov Parm	**Subject**	**Estimate**
Intercept	person	0.01749
Residual		0.003254

Solution for fixed effects					
Effect	**Estimate**	**Standard Error**	**DF**	**t Value**	**Pr > |t|**
Intercept	1.4175	0.05949	4	23.83	<.0001

Figure 10.2 Selected output tables from the REML fit to log$_{10}$(estrone) obtained using `PROC MIXED`.

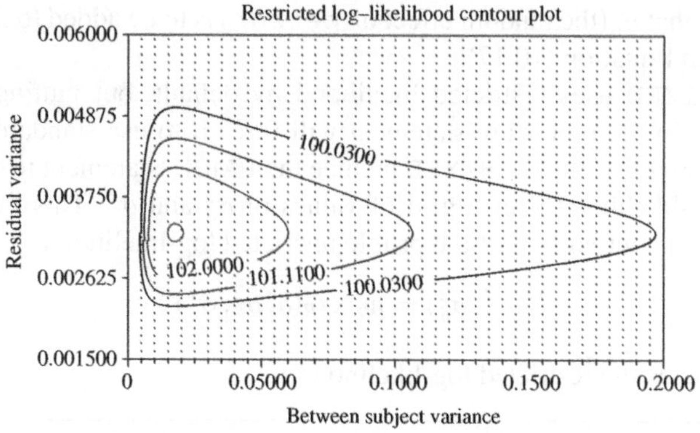

Figure 10.3 Contour plot of the profile restricted log-likelihood from a linear random effects fit to the $\log_{10}$(estrone) data. The contour at 101.11 can be used to obtain (restricted) LR confidence intervals for σ^2 and σ_u^2. The contour at 100.03 is the 95 % (restricted) LR confidence region.

option, however it is preferable that inference regarding σ_u^2 be made using likelihood-ratio methods because the profile likelihood for σ_u^2 is highly asymmetric (see Figure 10.4). This is a strong warning sign that any Wald-based inference using this parameterization will be extremely unreliable (Section 4.3).

The contour plot (Figure 10.3) of the profiled restricted log-likelihood is produced by the following code.

```
AXIS2 LABEL=(ANGLE=90 "Residual variance");
AXIS1 LABEL=("Between subject variance");
TITLE "Restricted log-likelihood contour plot";
PROC GCONTOUR DATA=parms;
  PLOT covP2*covP1=resloglike / LEVELS=(100.03 101.11 102 103)
    NOLEGEND HREF=(0.005 TO 0.2 BY 0.005)
    LHREF=34 AUTOLABEL HAXIS=AXIS1 VAXIS=AXIS2;
  RUN;
QUIT;
```

Likelihood ratio confidence intervals and regions can be visually determined from the contour plot, as demonstrated in Section 3.5.1. For example, the 95 % LRCIs for the individual variance parameters can be visually determined from the contour corresponding to a profile restricted log-likelihood of $103.03 - 0.5\chi^2_{1,0.95} = 101.11$. For σ_u^2, this interval is approximately (0.0055, 0.106). Similarly, the 95 % LR confidence region can be obtained as all those points having a profile restricted log-likelihood of at least 100.03.

The $\log_{10}$(estrone) measurements are iid normal under $H_0 : \sigma_u^2 = 0$, and re-running `PROC MIXED` with the `RANDOM` statement removed calculates the

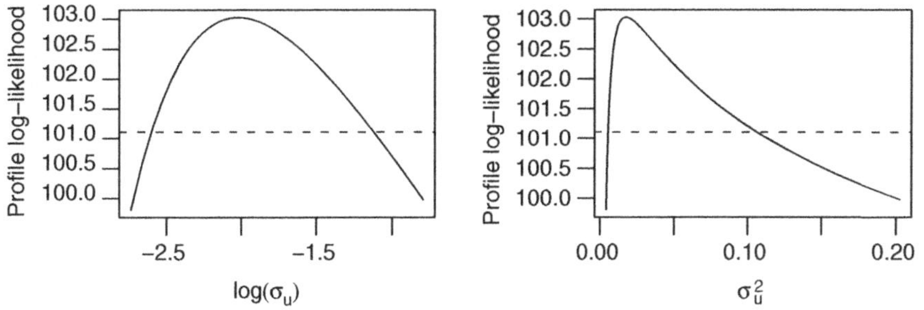

Figure 10.4 Profile restricted log-likelihoods for log σ_u and σ_u^2, for a linear random-effects model fitted to the $\log_{10}$(estrone) data.

maximized restricted log-likelihood under H_0, which is 45.68. The LRT statistic for H_0 is therefore $2(103.03 - 45.68) = 114.70$. Under the null hypothesis, the parameter lives on the boundary of the parameter space, and the usual asymptotic behaviour of the LRT statistic does not apply. In fact, the asymptotic distribution is an equal mixture of a zero and a χ_1^2 distribution (Stram and Lee 1994) (see Example 4.12). Asymptotic results would be very approximate here, especially due to there being only five subjects, though this matters little due to the extreme magnitude of the test statistic, and there is clearly massive evidence against H_0. However, for completeness, an exact test of H_0 is demonstrated in Section 10.4.2.

10.4.2 R

It is assumed that dataframe `estrone.df` contains two variables, $y = \log_{10}$(estrone) and the subject variable `Person`. Since the intercept term is fitted by default, the model expression simply requires the syntax `(1|Person)` to specify that the random effect of `Person` is added to the intercept.

```
> library(lme4)
> REMLfit=lmer(y~(1|Person),data=estrone.df)
> REMLfit

    AIC    BIC logLik deviance REMLdev
 -200.1 -192.9  103.0   -209.9  -206.1
Random effects:
 Groups   Name        Variance  Std.Dev.
 Person   (Intercept) 0.0174942 0.132265
 Residual             0.0032544 0.057047
Number of obs: 80, groups: Person, 5

Fixed effects:
            Estimate Std. Error t value
(Intercept)  1.41751    0.05949   23.83
```

The exact finite-sample p-value for the (restricted) likelihood ratio test of $H_0 : \sigma_u^2 = 0$ is estimated using the `exactRLRT` function in package `RLRsim` (Scheipl *et al.* 2008). This function numerically evaluates the intractable theoretical formula that was derived in Crainiceanu and Ruppert (2004).

```
> library(RLRsim)
> exactRLRT(REMLfit)

        simulated finite sample distribution of RLRT.
        (p-value based on 10000 simulated values)

data:
RLRT = 114.6992, p-value < 2.2e-16
```

At the time of writing, the `lme4` package does not include methods to obtain the profile restricted likelihood, but this is planned for future versions and may now be available.

10.4.3 ADMB

The ADMB template file, `EstroneREMLProf.tpl` is given in Section 10.8.1. The log of the joint density function of the observed data and latent variables, $f(\boldsymbol{y}, \boldsymbol{u}; \boldsymbol{\theta})$, is programmed within the `PROCEDURE_SECTION` of the ADMB template file. From the specification of the model in (10.16) and (10.17), this joint density is

$$\begin{aligned} f(\boldsymbol{y}, \boldsymbol{u}; a, \sigma^2, \sigma_u^2) &= f(\boldsymbol{u}; \sigma_u^2) f(\boldsymbol{y}|\boldsymbol{u}; a, \sigma^2) \\ &= \prod_{i=1}^{5} \left(\frac{1}{\sqrt{2\pi}\sigma_u} \exp\left(-\frac{u_i^2}{2\sigma_u^2} \right) \right. \\ &\quad \times \left. \prod_{j=1}^{16} \frac{1}{\sqrt{2\pi}\sigma} \exp\left(-\frac{(y_{ij} - (a + u_i))^2}{2\sigma^2} \right) \right) . \end{aligned} \tag{10.18}$$

At the time of writing, the built-in likelihood profiling capabilities of ADMB that were demonstrated in Section 1.4.3 are not implemented for latent variable models. In the R code below, function `Profile` is created to call an ADMB executable that, for a given value of σ_u^2, calculates the profiled restricted log-likelihood

$$l_R^*(\sigma_u^2) = \max_{a,\sigma^2} l_R(a, \sigma^2, \sigma_u^2) .$$

Function `plkhci` (introduced in Section 3.4.1) is then used to find a 95 % LR confidence interval for σ_u^2. Note that the ADMB program uses the parameterization $(a, \log \sigma, \log \sigma_u)$, and so the value of `est` specified in the list object x is $\log \widehat{\sigma}_u$.

```
> #Create data file. y is a vector of log10(estrone)
> cat("#np \n",5,"\n #m \n",16,"\n #y \n",y,file="EstroneREMLProf.dat")
>
> #Define function to return profile (negative) log-likelihood
> Profile=function(logsigu) {
+   write(logsigu,"Logsigu.txt")
+   runAD("EstroneREMLProf",argvec="< Logsigu.txt > Out.txt")
+   lhood=(scan("EstroneREMLProf.rep",nmax=1))
+   return(lhood) }

>
> #Find LR interval for logsigu using plkhci function
> library(Bhat)
> x=list(label="logsigu",est=0.5*log(0.01749),low=-3,upp=-1)
> logsiguCI=plkhci(x,Profile,"logsigu")
> logsiguCI
[1] -2.599026 -1.121132

> #Back transform to get LR interval for between subject variance
> exp(2*logsiguCI)
[1] 0.005527317 0.106217707
```

Strong asymmetry in the profiled restricted log-likelihood for σ_u^2 shows the danger of assuming approximate normality of $\widehat{\sigma}_u^2$ (Figure 10.4).

10.5 Nonlinear mixed-effects model

The data (Table 10.2) are measurements of the circumference (mm) of five orange trees at seven different sampling occasions, and are reported in Draper and Smith (1981, p. 524). These data have been heavily used as an example of nonlinear mixed modelling, including by SAS Institute (1999), and within the R-package

Table 10.2 Circumference (mm) of five orange trees. The data are from Draper and Smith (1981).

	Tree				
Age	1	2	3	4	5
118	30	33	30	32	30
484	58	69	51	62	49
664	87	111	75	112	81
1004	115	156	108	167	125
1231	120	172	115	179	142
1372	142	203	139	209	174
1582	145	203	140	214	177

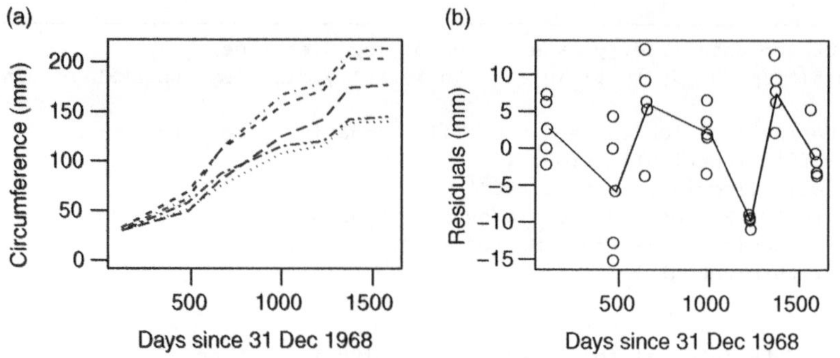

Figure 10.5 (a) Orange tree circumference data with the measurements from each tree connected by a dashed line; (b) residuals from the fit of the model with a random effect of tree on parameter a.

`lme4` where they are provided as dataframe `Orange`. In these examples, the expected size of a tree is modelled as an S-shaped function of age (measured as number of days since 31 Dec 1968), by using a logistic function that is scaled to range between 0 and some upper maximum, a. Specifically, with x_j denoting the tree age on sampling occasion j, $j = 1, ..., 7$, the expected circumference can be expressed as

$$\mu_j = \frac{a}{1 + \exp(-(x_j - b)/c)} \,. \tag{10.19}$$

In this equation, parameter a corresponds to the full-grown (as age goes to infinity) expected circumference, and b corresponds to the age at which the expected circumference is equal to half of a. Parameter $c > 0$ is an inverse growth rate parameter, with smaller values of c corresponding to faster growth.

It is clear from a scatter plot of the data (Figure 10.5a) that the measured circumferences are not independently distributed around the expected values, as each tree appears to have its own underlying growth curve. These individual curves appear to differ primarily in vertical scale, but otherwise have similar shape. This suggests using a model that permits the asymptotic size to vary between tree. The trees can be assumed to be a random sample from the population of all suitable orange trees, and it would therefore be appropriate that their asymptotic sizes be modelled using random effects. Letting subscript i, $i = 1, ..., 5$ denote the tree, a potential model for the measured circumference y_{ij} on tree i at time j would be

$$\begin{aligned} U_i &\sim N(0, \sigma_u^2)\,. \\ Y_{ij}|u_i &\sim N(\mu_{ij}, \sigma^2)\,, \end{aligned} \tag{10.20}$$

where u_i is the (unobserved) realization of U_i, and

$$\mu_{ij} = \frac{a + u_i}{1 + \exp(-(x_j - b)/c)} . \qquad (10.21)$$

Under this model, the asymptotic size of tree i is $a + u_i$. The parameters to be estimated are $\boldsymbol{\theta} = (a, b, c, \sigma^2, \sigma_u^2)$.

A closer inspection of Figure 10.5a shows that the measured circumferences on the five trees appear to be correlated at each sampling time. For example, all five trees had a noticeably higher size increment between measurements taken at ages of 1231 and 1371 days than between the ages of 1004 and 1231. This may well be due to seasonal growth factors, but in the absence of knowledge about the conditions under which the trees were reared (but see Exercise 10.4), an alternative model for including this feature of the data would be to include a random effect of sampling occasion. For example,

$$\begin{aligned} U_i &\sim N(0, \sigma_u^2) \\ V_j &\sim N(0, \sigma_v^2) \\ Y_{ij}|u_i, v_j &\sim N(\mu_{ij}, \sigma^2) , \end{aligned}$$

where v_j is the (unobserved) realization of V_j, and

$$\mu_{ij} = \frac{a + u_i + v_j}{1 + \exp(-(x_j - b)/c)} .$$

This model adds an extra parameter σ_v^2, and is an example of a crossed-effects NLMM. Note that this model does not take into account the temporal order of the sampling – this could be accommodated by imposing an autoregressive structure on v_j, $j = 1, ..., 7$ (see Exercise 10.5).

The model with a random tree effect is fitted using the SAS procedure `NLMIXED` in Section 10.5.1. This procedure uses Gauss-Hermite quadrature and can accommodate only one random effect. The help file for function `nlmer` in the R package `lme4` includes example code for fitting the random tree-effect model to these data, but (at the time of writing) it gave an erroneous estimate of σ_u^2, though with addition of the optional quadrature argument `nAGQ=2` it did produce results similar to those produced by `NLMIXED`. ADMB is used in Section 10.5.2 to fit the model having both random tree and day effects.

10.5.1 SAS

Data set `Orange` contains 35 rows and three variables, `tree`, `age` and `y`. The required `PROC NLMIXED` code includes programming statements to calculate the expected circumference as specified by Equation (10.21), and a `RANDOM` statement to specify that the value of `u` varies by the subject variable `tree` according to a $N(0, \sigma_u^2)$ distribution. The starting values for the parameters were deduced from inspection of Figure 10.5a.

Parameter	Estimate	StandardError
a	192.05	15.6577
b	727.91	35.2484
c	348.07	27.0797
sigmasq	61.5127	15.8825
sigmasqu	1001.50	649.49

Figure 10.6 MLEs and their standard errors for the random tree-effect nonlinear growth model.

```
PROC NLMIXED DATA=Orange;
  PARMS a=200 b=725 c=350 sigmasq=50 sigmasqu=900;
  ExpSize=(a+u)/(1+exp(-(age-b)/c));
  MODEL y~NORMAL(ExpSize,sigmasq);
  RANDOM u~NORMAL(0,sigmasqu) SUBJECT=tree;
  PREDICT  y-ExpSize OUT=Residuals;
RUN;
```

The maximized value of the log-likelihood was -131.57, and the mean (over the hypothetical population of apple trees) full-grown circumference was estimated to be $\widehat{a} = 192$ mm (Figure 10.6). However, $\widehat{\sigma}_u^2$ is just over 1000, which estimates that the full-grown size has a standard deviation of $\sqrt{1000} \approx 32$ mm between randomly chosen trees.

The residuals from this fit were obtained using the `PREDICT` statement, and are calculated as $y_{ij} - \widehat{\mu}_{ij}$, where $\widehat{\mu}_{ij}$ is obtained from (10.21) by setting $\boldsymbol{u} = (u_1, ..., u_5)$ to the values that maximize $f(\boldsymbol{y}, \boldsymbol{u}; \widehat{\boldsymbol{\theta}})$. The residual plot (Figure 10.5b) confirms the correlation within sampling occasion.

10.5.2 ADMB

Letting $\boldsymbol{u} = (u_1, ..., u_5)$ denote the random tree effects, and $\boldsymbol{v} = (v_1, ..., v_7)$ the random day effects, the ADMB template file (Section 10.8.2) contains programming code to calculate the joint density of $\boldsymbol{y}$, $\boldsymbol{u}$ and $\boldsymbol{v}$,

$$\begin{aligned} f(\boldsymbol{y}, \boldsymbol{u}, \boldsymbol{v}; \boldsymbol{\theta}) &= f(\boldsymbol{u}; \sigma_u^2) f(\boldsymbol{v}; \sigma_v^2) f(\boldsymbol{y}|\boldsymbol{u}, \boldsymbol{v}; a, b, c, \sigma^2) \\ &= \left(\prod_i \frac{1}{\sqrt{2\pi}\sigma_u} \exp\left(-\frac{u_i^2}{2\sigma_u^2}\right)\right) \times \left(\prod_j \frac{1}{\sqrt{2\pi}\sigma_v} \exp\left(-\frac{v_j^2}{2\sigma_v^2}\right)\right) \\ &\quad \times \left(\prod_{i,j} \frac{1}{\sqrt{2\pi}\sigma} \exp\left(\frac{-(y_{ij} - \mu_{ij})^2}{2\sigma^2}\right)\right), \end{aligned} \tag{10.22}$$

where $\boldsymbol{\theta} = (a, b, c, \sigma^2, \sigma_u^2, \sigma_v^2)$.

Table 10.3 ADMB (using Laplace approximation) and simulated ML fits (from Millar (2004)) of the crossed-effects NLMM to the orange tree circumference data. The Monte-Carlo simulation error of the simulated MLE is shown in the right-hand column.

	ADMB		Simulated ML	
	$\widehat{\boldsymbol{\theta}}$	std. error	$\widehat{\boldsymbol{\theta}}$	sim. error
a	196.2	19.4	195.9	0.2
b	748.4	62.3	747.6	0.3
c	352.9	33.3	352.7	0.1
σ^2	28.1	8.2	28.1	0.01
σ_u^2	1061.0	68.8	1059.8	0.5
σ_v^2	109.9	9.1	109.1	0.2

The fit of this model has previously been obtained using simulated maximum likelihood[3] (Millar 2004). The ADMB estimates, obtained using Laplace approximation, were very close to those obtained by simulated likelihood (Table 10.3). Since u_i and v_j appear linearly in μ_{ij}, it is the case that the log of the joint density function is quadratic in $\boldsymbol{u}$ and $\boldsymbol{v}$, and hence the Laplace approximation will be exact, and so there was no need to use importance sampling.

The maximized log-likelihood of the crossed-effects model is $l(\widehat{\boldsymbol{\theta}}) = -125.45$. The LRT statistic for the hypothesis of no effect of sampling occasion, H_0, $\sigma_v^2 = 0$, is therefore $2(-125.45 - (-131.57)) = 12.24$. This is sufficiently large to provide very strong evidence against H_0. Bootstrap simulation of the p-value is left to Exercise 10.3.

10.6 Generalized linear mixed-effects model

Proportion data from a multi-centre clinical trial were presented in Section 8.3.1 where they were analyzed using generalized estimating equations. Here, these data are re-analyzed using a generalized linear mixed-effects model (GLMM). The logistic model is used, that is,

$$p_{ij} = \frac{\exp(\eta_{ij})}{1 + \exp(\eta_{ij})}, \tag{10.23}$$

[3] A computer intensive approach that is based on importance sampling, but with the approximating distribution obtained from a Bayesian fit of the specified model.

where p_{ij} is the probability of a favourable outcome from treatment $j = 1, 2$ in clinic $i = 1, \ldots, 8$, and η_{ij} is the linear predictor. Recall that the inverse of (10.23) is the logit link function

$$\eta_{ij} = \text{logit}(p_{ij}) = \log\left(\frac{p_{ij}}{1 - p_{ij}}\right),$$

which corresponds to the log-odds of a favourable outcome.

It will be assumed that the random effect of clinic is additive in the linear predictor. That is,

$$\eta_{ij} = a + bt_j + u_i, \tag{10.24}$$

where t_j is an indicator variable that equals 1 for the drug treatment and 0 for the control, and u_i are iid $N(0, \sigma_u^2)$. This is a form of a random-intercepts model, where $a_i = a + u_i$ is the random intercept for clinic i. The parameters to be estimated are $\boldsymbol{\theta} = (a, b, \sigma_u^2)$.

The model in (10.24) assumes that the treatment has the same effect on the log-odds of a favourable outcome at all clinics. A convenient way to allow for variability in the treatment effects between clinics would be to introduce an additional clinic-by-treatment interaction random effect. That is,

$$\eta_{ij} = a + bt_j + u_i + v_{ij}. \tag{10.25}$$

where v_{ij} are iid $N(0, \sigma_v^2)$ and are independent of the u_i's. This adds a fourth parameter, σ_v^2, to the model.

The likelihood of the model with only the clinic random effect can be expressed in separable form as a product of one-dimensional integrals (with respect to each u_i), and hence the software packages are able to fit it using Gauss-Hermite quadrature. With the addition of the clinic-by-treatment random effect, the fits use Laplace sampling.

10.6.1 R

At the time of writing, the `lme4` package (version 0.999375-37) was undergoing major redevelopment, and its implementation of Gauss-Hermite quadrature was found to be unstable for this application. Instead, package `glmmML` includes a function of the same name that is able to fit GLMMs of the form in (10.24) where there is a single random effect that is additive in the linear predictor, that is, a random intercept model. The fit below used Gauss-Hermite quadrature (`method="ghq"`) and was not sensitive to the choice of the number of quadrature points (specified using the `d=` option), and the code below implicitly uses the default of eight. Considerable between-clinic variability was detected, $\widehat{\sigma}_u^2 = 1.401^2 \approx 1.96$.

```
> y=c(11,10,16,22,14,7,2,1,6,0,1,0,1,1,4,6)
> n=c(36,37,20,32,19,19,16,17,17,12,11,10,5,9,6,7)
> clinic=as.factor(rep(1:8,rep(2,8)))
> trmt=as.factor(rep(c("Drug","Control"),8))
>
> library(glmmML)
> glmmML(y/n~trmt,cluster=clinic,family="binomial",weight=n,
+        method="ghq")

               coef se(coef)      z Pr(>|z|)
(Intercept) -1.1976   0.5564 -2.152   0.0314
trmtDrug     0.7385   0.3004  2.458   0.0140

Scale parameter in mixing distribution:  1.401 gaussian
Std. Error:                              0.4259
Residual deviance: 35.65 on 13 degrees of freedom   AIC: 41.65
```

The observations are modelled as binomial, and so it is possible to compute the residual deviance, as twice the difference between the log-likelihoods of the saturated model and the fitted mixed-effects model. The deviance is seen to be 35.65 on 13 d.o.f. However, it would extremely dubious to compare this deviance against a χ^2_{13} distribution for the purpose of assessing goodness of fit because the GLMM model is not nested within the saturated model, since there are no random effects in the saturated model. Thus, the deviance has no justification as a likelihood ratio test statistic here. Moreover, concerns over sparsity of the data are warranted, because six of the sixteen observations have a value of zero or one.

A parametric bootstrap was used to obtain an approximate sampling distribution of the deviance. That is, new binomial data were generated from the model specified in Equation (10.23) and (10.24), using parameter values $(\widehat{a}, \widehat{b}, \widehat{\sigma}_u^2) = (-1.20, 0.739, 1.96)$. The realized deviance of 35.65 was found to be typical of the simulated deviances (Figure 10.7), and hence provides no evidence of lack of fit.

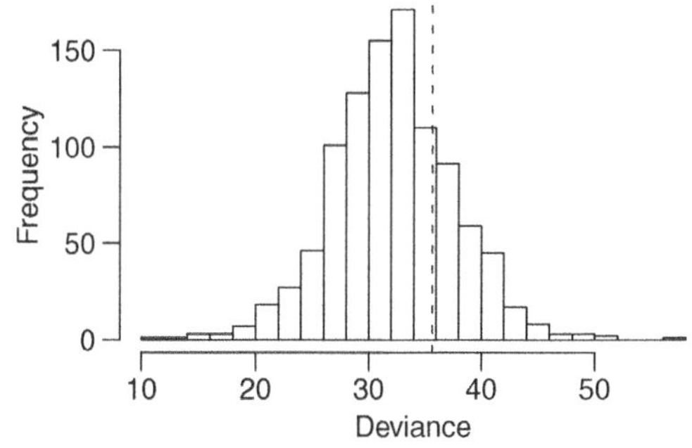

Figure 10.7 Simulated deviances from a parametric bootstrap of the random clinic-effects model of the multi-centre clinical trial data. The deviance of the fitted model is shown by a vertical dashed line.

Analysis of deviance (that is, likelihood ratio tests) should be more reasonable for the purpose of testing the null hypothesis of no treatment effect, $H_0 : b = 0$, notwithstanding that it may be very approximate due to the limited amount of data. Fitting the mixed-effects model without the `trmt` covariate yields a deviance of 41.91. The increase in deviance is 6.26, corresponding to a p-value of 0.012 when judged against a χ^2_1 distribution. In this case, the LR p-value is very close to the 0.014 Wald p-value given in the `glmmML` output.

The `lmer` function in package `lme4` was stable for the purpose of fitting the model with clinic-by-treatment interaction via Laplace approximation.

```
subj=factor(1:16)
lmer(y/n~trmt+(1|clinic)+(1|subj),family="binomial",weight=n)
```

The log-likelihood from this fit was compared to that of the (Laplace approximation) fit of the model with a random clinic effect only (from `glmmML`). To two decimal places, the (Laplace-approximated) log-likelihood is unchanged by addition of the interaction term, and hence this term is clearly not required (Table 10.4).

10.6.2 SAS

With the data entered into dataset `infection` using the SAS code from Section 8.3.1, the model in (10.24) is fitted via Gauss-Hermite quadrature using the following additional code.

```
PROC GLIMMIX DATA=infection METHOD=QUAD;
      MODEL y/n = trmt / S DIST=BINOMIAL;
      RANDOM INT / SUBJECT=clinic;
RUN;
```

An additional `RANDOM` statement with `SUBJECT=clinic*trmt` will fit the additional random interaction term using Laplace approximation.

10.6.3 ADMB

The ADMB template file in Section 10.8.3 contains code to calculate the joint density function from the model with the random effect of clinic,

$$
\begin{aligned}
f(\boldsymbol{y}, \boldsymbol{u}; \boldsymbol{\theta}) &= f(\boldsymbol{u}; \sigma_u^2) f(\boldsymbol{y}|\boldsymbol{u}; a, b) \\
&= \left(\prod_i \frac{1}{\sqrt{2\pi}\sigma_u} \exp\left(-\frac{u_i^2}{2\sigma_u^2}\right) \right) \times \left(\prod_{i,j} \binom{n_{ij}}{y_{ij}} p_{ij}^{y_{ij}} (1 - p_{ij})^{n_{ij} - y_{ij}} \right),
\end{aligned}
$$

where p_{ij} is given by (10.23), and $\boldsymbol{\theta} = (a, b, \sigma_u^2)$.

Table 10.4 The calculated MLEs and their standard errors from fitting models (10.24) and (10.25) to the multi-centre clinical trials data. $\widehat{l(\boldsymbol{\theta})}$ denotes the maximum of the approximation to the likelihood. The fits using GREML and importance sampling were implemented in ADMB, and the other fits were obtained using each of R, SAS and ADMB.

	Clinic effect only						With clinic-by-trmt interaction			
	Laplace		Quadrature		GREML		Laplace		Imp. Sampling	
	$\widehat{\theta}$	s.e.	$\widehat{\theta}$	s.e.	$\widehat{\theta}$	s.e.	$\widehat{\theta}$	s.e.	$\widehat{\theta}$	s.e.
a	−1.20	0.56	−1.20	0.56	−1.21	0.60	−1.20	0.56	−1.20	0.56
b	0.74	0.32	0.74	0.30	0.75	0.30	0.74	0.32	0.75	0.32
σ_u^2	1.93	1.17	1.96	1.19	2.36	1.50	1.93	1.17	1.96	1.20
σ_v^2							0.01	0.15	0.01	0.16
$\widehat{l(\boldsymbol{\theta})}$	−37.10		−37.02				−37.10		−37.01	

The template file in Section 10.8.3 implements the generalized restricted maximum likelihood (GREML) fit using integrated likelihood, as described in Section 9.3.2. This varies from the conventional ML fit by specifying regression coefficients a and b as variables over which to integrate. As anticipated, the estimated value of $\widehat{\sigma}_u^2$ is increased compared to that from using standard likelihood (Table 10.4), due to the effective adjustment for the degrees-of-freedom lost by estimation of a and b.

It should be noted that the GREML estimates of regression coefficients a and b were obtained by refitting the model with σ_u^2 fixed at the estimated value of $\widehat{\sigma}_u^2$. This is the approach recommended by Lee *et al.* (2006) and Noh and Lee (2007) and will be reasonable provided that σ_u^2 is at least approximately parameter orthogonal (Section 11.6.1) to the regression coefficients.

ADMB was also used to improve upon the fit obtained from the Laplace approximation of the model with random interaction term, by using importance sampling (Table 10.4) with $m = 2000$ sampling points.

10.6.4 GLMM vs GEE

In Section 8.3.1 these data were analyzed using a generalized estimating equation (GEE). In particular, using the GEE with exchangeable correlation structure (on the log-odds ratios), the GEE estimates of the regression coefficients were found to be $\widetilde{a} = -0.8740$ and $\widetilde{b} = 0.5532$ (Figure 8.6). These differ substantially from the MLEs obtained via the GLMM model (Table 10.4), but this is not necessarily a cause for concern, because the two approaches require very different interpretation.

The GEE is a population-level (i.e. marginal) model that is derived from modelling the expected value and variance of $\boldsymbol{Y}$. The GLMM is a subject-specific (i.e.

conditional) model that is specified using the latent random variables, $\boldsymbol{U}$, and the conditional distribution of $\boldsymbol{Y}$ given the (unobserved) realized values of the latent variables. Neuhaus, Kalbfleisch and Hauck (1991) present a theoretical comparison of the GEE and GLMM approaches when applied to binary data, and for the random intercepts type of model used here, they found that the population-level model will necessarily shrink covariate effects towards zero. Thus, the difference in the estimated values, $\widehat{b} = 0.74$ and $\widetilde{b} = 0.55$ should not be interpreted to mean that the GLMM estimated greater efficacy of the treatment.

Simulation can be used to assess the behaviour of the GLMM at the population-level, and this will allow a meaningful comparison with the results from the GEE approach. Note that Equation (10.23) can be written

$$p_i(t) = \frac{\exp(a + bt + u_i)}{1 + \exp(a + bt + u_i)}, \tag{10.26}$$

where $t = 0$ and $t = 1$ correspond to the control and drug treatments, respectively. The hypothetical population of clinics can be simulated by generating iid values of u_i distributed $N(0, \sigma_u^2)$. This was done one million times (i.e. for $i = 1, ..., 1\,000\,000$ hypothetical clinics) using the estimates from the GREML fit $(\widehat{a}, \widehat{b}, \widehat{\sigma}_u^2) = (-1.21, 0.75, 2.36)$ as the parameter values. In addition, for purposes of display, (10.26) was evaluated on a fine grid of t values between 0 and 1, rather than just at the values 0 and 1. The GLMM population-level curve was then calculated as the mean of these one million curves,

$$\overline{p}(t) = \frac{\sum_{i=1}^{10^6} p_i(t)}{10^6}.$$

The GLMM population-level curve was barely distinguishable from the GEE population curve obtained from (10.26) using $\widetilde{a} = -0.8740$ and $\widetilde{b} = 0.5532$ (Figure 10.8), and setting $u_i = 0$. That is, these two methodologies are giving near

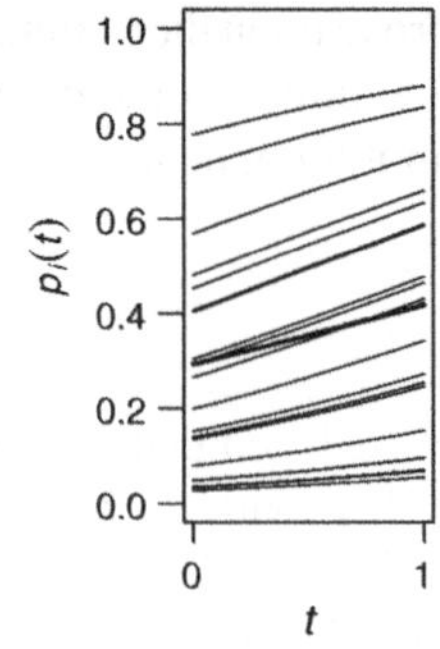

Figure 10.8 The GLMM population-level curve (thick solid line) and the GEE curve (thick dashed line) are so similar as to be indistinguishable. The first 20 random curves $p_i(t), i = 1, ..., 20$ from the GLMM are shown as fine solid lines.

identical fits at the population level. The GLMM population curve was also simulated using the standard MLEs rather than the GREML estimates, but the difference was negligible.

The choice of methodology to use depends largely on the research questions. For example, if between-clinic variability is of interest then of course the GLMM approach would be required. It between-clinic variability is not of interest then a GEE model may be preferred, particularly if there were doubts that the assumed normal distribution for u_i was reasonable.

10.7 State-space model for count data

The data $y_t, t = 1, ..., 52$, are a time series of the weekly counts of new cases of acute haemorrhagic conjunctivitus in the Chiba prefecture of Japan during 1987 (Figure 10.9). Kashiwagi and Yanagimoto (1992) modelled y_t as independent Poisson observations with mean λ_t, where λ_t followed the state-space equation

$$\log(\lambda_t) = \log(\lambda_{t-1}) + u_t \; , \quad t = 2, ..., 52 \; , \tag{10.27}$$

where u_t are iid $N(0, \sigma_u^2)$. Note that, given $\boldsymbol{u} = (u_2, ..., u_{52})$, the values of $\lambda_2, ..., \lambda_{52}$ can be calculated from any given value of λ_1. Thus, this model has just two fixed parameters, λ_1 and σ_u^2. In the terminology of state-space models, it is the quantities $\lambda_t, t = 1, ..., 52$ that are the underlying states. Kashiwagi and Yanagimoto (1992) applied recursive numerical integration to evaluate the likelihood and hence obtain the MLEs of λ_1 and σ_u^2. Here, ADMB is used.

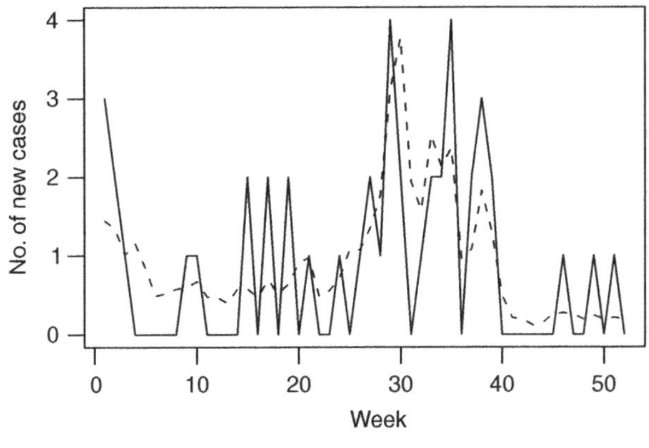

Figure 10.9 Weekly incidence (solid line) of new cases of acute haemorrhagic conjunctivitus in Chiba-prefecture in Japan during 1987. The dotted line shows the fitted values.

The ADMB template file in Section 10.8.4 contains code to calculate the joint density

$$f(\boldsymbol{y},\boldsymbol{u};\boldsymbol{\theta}) = f(\boldsymbol{u};\sigma_u^2)f(\boldsymbol{y}|\boldsymbol{u};\lambda_1)$$
$$= \left(\prod_{t=2}^{52}\frac{1}{\sqrt{2\pi}\sigma_u}\exp\left(-\frac{u_i^2}{2\sigma_u^2}\right)\right)\times\left(\prod_{t=1}^{52}\frac{e^{-\lambda_t}\lambda_t^{y_t}}{y_t!}\right).$$

The template file takes advantage of the partial separability that arises from the property that the likelihood can be evaluated using recursive integration of dimension two or less.[4]

The MLE is calculated to be $\widehat{\boldsymbol{\theta}} = (\widehat{\lambda}_1, \widehat{\sigma}_u^2) = (1.44, 0.126)$ and $l(\widehat{\boldsymbol{\theta}}) = -66.07$. The predicted values of $\widehat{\lambda}_i$, $i = 2, ..., 52$ (Figure 10.9) are the values that maximize $f(\boldsymbol{y},\boldsymbol{u};\widehat{\boldsymbol{\theta}})$, and are found in the parameter file produced by ADMB.

10.8 ADMB template files

When an ADMB template file includes random effects, it is currently necessary to use the `raneff` option when creating the executable. For example,

```
makeAD("MyADMBprogram",raneff=T)
```

10.8.1 One-way linear random-effects model using REML

Features to note:

- The `PROCEDURE_SECTION` programs the log of the joint density function given in (10.18).
- For the purposes of calculating a profile likelihood for σ_u^2, the value of $\log\sigma_u$ is input as a command-line argument. To maximize over σ_u^2, the `LOCAL_CALCS` section should be removed and `log_sigma_u` declared as an `init_bounded_number` in the `PARAMETER_SECTION`.
- This code implements a REML fit via integrated likelihood (Section 9.3), which can be regarded as treating a like a latent variable, in the sense that it is integrated from the likelihood. A standard ML fit would simply require changing the declaration of a to an `init_number`.
- The third argument in the specification of the random effects vector, $\boldsymbol{u}$, specifies it is to be used in phase 2 of the optimization (see Section 5.4). That is,

[4] This allows for efficient ML estimation of state-space models for lengthy time series. The state-space model herein was applied to a simulated time series of length 2000. Maximization of the Laplace approximation required less than 5 min of computing time on a PC.

a fit is achieved with $\boldsymbol{u} = \mathbf{0}$ in phase 1, and this is then used as the starting point for full optimization.

```
DATA_SECTION
  init_int np                 //Number of people
  init_int m                  //Number of measurements on each
  init_matrix y(1,np,1,m)     // Data matrix (5 by 16)
  number log_sigma_u
 LOCAL_CALCS
  cin>>log_sigma_u;           //Read text file named in command line
 END_CALCS

PARAMETER_SECTION
  init_bounded_number log_sigma(-10,10)  // log(residual s.e)
  random_effects_number a                // Intercept parameter
  random_effects_vector u(1,np,2)        // Random effects vectors
  objective_function_value neg_log_joint

PROCEDURE_SECTION
  int i,j;
  const double pi=3.14159265359;
  dvariable pred, f; // f is the integrand
  dvariable sigma = exp(log_sigma);
  dvariable sigma_u=exp(log_sigma_u);
  f=0;
  for(i=1;i<=np;i++)
  {
    //Random effect contribution to joint density
    f=f-0.5*log(2*pi)-log_sigma_u-0.5*(square(u(i)/sigma_u));
    pred=a+u[i];
    for(j=1;j<=m;j++)
    {
      //Data contribution to joint density
      f=f-0.5*log(2*pi)-log_sigma-0.5*square((y(i,j)-pred)/sigma);
    }
  }
  neg_log_joint=-f;

REPORT_SECTION
  report << neg_log_joint << endl;
```

10.8.2 Nonlinear crossed mixed-effects model

```
DATA_SECTION
  init_int ntree                    //Number of trees
  init_int nday                     //Number of days
  init_matrix y(1,ntree,1,nday)     // Response vector
  init_vector t(1,nday)             // Days of measurement
```

```
PARAMETER_SECTION
  init_number a                              //  Asymptotic size
  init_number b                                 // Age at mid size
  init_number c                                 // Growth rate paramter
  init_bounded_number log_sigma(-5,5,1)         // log(residual s.e)
  init_bounded_number log_sigma_u(-5,5,2)       // log(tree effect s.e)
  init_bounded_number log_sigma_v(-5,5,2)       // log(day effect s.e)
  random_effects_vector u(1,ntree,2)     // Random tree effects
  random_effects_vector v(1,nday,2)          // Random day effect
  objective_function_value neg_log_joint
  sdreport_number sigmasq
  sdreport_number sigmasq_u
  sdreport_number sigmasq_v

PROCEDURE_SECTION
  int i,j;
  const double pi=3.14159265359;
  dvariable pred, f; // f is the integrand
  sigmasq = exp(2*log_sigma);
  sigmasq_u = exp(2*log_sigma_u);
  sigmasq_v = exp(2*log_sigma_v);
  f = 0.0;
  for(i=1;i<=ntree;i++)
  {
    for(j=1;j<=nday;j++)
    {
      pred = (a+u(i)+v(j))/(1+exp(-(t(j)-b)/c));
      f = f-0.5*log(2*pi)-log_sigma-0.5*square(y(i,j)-pred)/sigmasq;
    }
  }
// Random effect contribution from u
  for(i=1;i<=ntree;i++)
  {
    f = f-0.5*log(2*pi)-log_sigma_u-0.5*square(u(i))/sigmasq_u;
  }
// Random effects contribution from v
  for(j=1;j<=nday;j++)
  {
    f = f-0.5*log(2*pi)-log_sigma_v-0.5*square(v(j))/sigmasq_v;
  }
  neg_log_joint=-f;
```

10.8.3 Generalized linear mixed model using GREML

```
DATA_SECTION
  init_int nc                    // Number of clinics
  init_int m                     // Number of obs per clinic
  init_matrix fav(1,nc,1,m)          // Favourable counts
  init_matrix unfav(1,nc,1,m)    // Unfavourable counts
```

```
PARAMETER_SECTION
//  init_vector coefs(1,2)          // For standard ML fit
  init_number log_sigmau            // log(clinic effect s.e.)
  sdreport_number sigmau2
  random_effects_vector coefs(1,2)  // For GREML fit
  random_effects_vector u(1,nc)     // Random effects
  objective_function_value f

PROCEDURE_SECTION
  int i,j;
  dvariable eta,prob,a,b;
  const double pi=3.14159265359;
  a=coefs[1]; b=coefs[2];
  sigmau2=exp(2*log_sigmau);
  f=0.0;
  for(i=1;i<=nc;i++)
  {
    f=f+0.5*log(2*pi)+0.5*square(u[i])/sigmau2+log_sigmau;
    for(j=1;j<=m;j++)
    {
      eta=a+(2-j)*b+u[i];
      prob=exp(eta)/(1+exp(eta));
      f=f-fav(i,j)*log(prob)-unfav(i,j)*log(1-prob);
      f=f-lgamma(fav(i,j)+unfav(i,j)+1);
      f=f+lgamma(fav(i,j)+1)+lgamma(unfav(i,j)+1);
    }
  }
```

10.8.4 State-space model for count data

```
DATA_SECTION
  init_int n              //Number of observations
  init_vector y(1,n)      //Number of new cases of conjunctivitus

PARAMETER_SECTION
  init_bounded_number log_lambda1(-5,6)
  init_bounded_number log_sigma(-5,5,2)    // NB: Phase 2 parameter
  sdreport_number lambda1                  // Report its s.e.
  sdreport_number sigma2                   // Report its s.e
  random_effects_vector log_lambda(2,n,2)  // Latent state vector
  objective_function_value f

PROCEDURE_SECTION
  int i;
  dvariable hack; // Spare variable for use when i=1
  lambda1=exp(log_lambda1);
  sigma2=exp(2*log_sigma);
  f=0.0;
  i=1;
  indiv(hack,log_lambda1,log_sigma,i);
  i=2;
  indiv(log_lambda1,log_lambda[2],log_sigma,i);
  for(i=3;i<=n;i++)
  { indiv(log_lambda[i-1],log_lambda[i],log_sigma,i); }
```

```
SEPARABLE_FUNCTION void indiv(const dvariable& u0,
                   const dvariable& ui,const dvariable& lsig, int i);
  const double pi=3.14159265359;
  f=f+mfexp(ui) - y[i]*ui +lgamma(y[i]+1);  // Condtl Poisson term
  if(i>1) f=f+0.5*log(2*pi)+0.5*square((ui-u0)/exp(lsig)) +lsig;
```

10.9 Exercises

10.1 The estrone data in Table 10.1 were obtained by assaying batches of four vials each (Gail *et al.* 1996). For each subject, the first four vials comprise batch 1, the next four comprise batch 2, and so on.

1. Using restricted maximum likelihood, fit a nested LMM to the $\log_{10}$(estrone) data, with random grouping factors subject and batch (nested within subject).
2. Which model do you prefer, the model with nested random effects or the model with only the subject random effect?

10.2 For the nested LMM in Exercise 10.1, conduct a bootstrap simulation to evaluate the sampling distribution of the LRT statistic for testing the hypothesis of no batch effect, and compare to the actual LRT statistic.

10.3 For the orange tree data, conduct a bootstrap simulation to evaluate the sampling distribution of the LRT statistic for the hypothesis of no day effect, and compare to the actual LRT statistic.

10.4 Instead of adding a random day effect to the tree-effect model, fit a seasonal model to the orange tree data. Assume a northern hemisphere climate which causes the rate of growth to be at its minimum on 31 December, and at its greatest on 30 June. Specifically, let the (inverse) growth rate parameter vary with calendar day $k = 1, ..., 365$, according to the sinusoidal curve

$$c_k = c + \delta \cos(2\pi k/365) .$$

Here the parameters to be estimated are $\boldsymbol{\theta} = (a, b, c, \delta, \sigma^2, \sigma_u^2)$.

10.5 Use ADMB to fit an AR(1) error term to the nonlinear model of orange tree circumference.

10.6 Taking advantage of separability, and making use of the `integrate` function, write an R function to calculate the log-likelihood for model (10.24) applied to the binomial data from the multi-centre clinical trial in Section 10.6. Next, use `optim` to find the MLE.

10.7 Write an ADMB template file that specifies model (10.24) (using standard likelihood) in separable form. Run this in ADMB using the Gauss-Hermite quadrature option.

Part III

THEORETICAL FOUNDATIONS

11

Cramér-Rao inequality and Fisher information

... the only way of understanding nature and making predictions is to study the laws of change and formulate appropriate rules of action. – Calyampudi Radhakrishna Rao

11.1 Introduction

Parts I and II of this book presented maximum likelihood as a natural methodology for estimation and inference. This chapter, and the next, develop the theory under which maximum likelihood can be regarded as an optimal methodology.

In the general case, the sampling distribution of MLEs is intractable to determine explicitly, but the theory is able to approximate the sampling distribution for large sample sizes. Moreover, the approximating distribution has desirable properties, including unbiasedness, and variance equalling the best that can possibly be achieved by any unbiased estimator (should one exist). For this reason, MLEs are said to be *asymptotically unbiased* and *asymptotically efficient.*

It is the Cramér-Rao (CR) inequality that establishes the minimum variance that can be achieved by unbiased estimators, and the key concept in the statement of this inequality is the *expected Fisher information.* Fisher information is a central concept in the theory of ML inference, and much of this chapter is aimed at providing familiarity with Fisher information and its properties.

The CR inequality and (expected) Fisher information are initially presented in the one-dimensional case, $\theta \in \mathbb{R}$ (Section 11.2), in which case the Fisher information is a scalar quantity. Section 11.3 extends the inequality to unbiased estimators

---

Maximum Likelihood Estimation and Inference: With Examples in R, SAS and ADMB, First Edition. Russell B. Millar.

of $\zeta = g(\theta)$, and Sections 11.4 and 11.5 provide alternative formulae for calculation of Fisher information. In the general case, $\boldsymbol{\theta} \in \mathbb{R}^s$, the Fisher information is an $s \times s$ matrix (Section 11.6). Examples are presented in Section 11.7.

11.1.1 Notation

The notation $E_{\boldsymbol{\theta}}$ denotes expectation over $\boldsymbol{Y} \in \mathbb{R}^n$ for a given $\boldsymbol{\theta}$, under the distribution with density $f(\boldsymbol{y}; \boldsymbol{\theta})$. That is, it is expectation under repetition of an experiment that generates $\boldsymbol{Y}$ according to $f(\boldsymbol{y}; \boldsymbol{\theta})$. Note that $E_{\boldsymbol{\theta}}$ is being used as an abbreviation for the more explicit but cumbersome notation $E_{\boldsymbol{Y};\boldsymbol{\theta}}$. Similarly, $\text{var}_{\boldsymbol{\theta}}$ will be used to denote the variance under repetition of the experiment.

It will be assumed that the model is a collection of distributions that are absolutely continuous with respect to some sigma-finite measure. However, familiarity with measure theory is not required, and the notation will omit these details for simplicity. The notation employed will be for that of a continuous distribution, and so, for example, the expected value of a real-valued function $h(\boldsymbol{Y})$ will be written

$$E_{\boldsymbol{\theta}}[h(\boldsymbol{Y})] = \int h(\boldsymbol{y}) f(\boldsymbol{y}; \boldsymbol{\theta}) d\boldsymbol{y}.$$

If $\boldsymbol{Y}$ has a discrete distribution, then $f(\boldsymbol{y}; \boldsymbol{\theta})$ denotes a probability mass function and it is implicitly understood that

$$\int h(\boldsymbol{y}) f(\boldsymbol{y}; \boldsymbol{\theta}) d\boldsymbol{y} \equiv \sum_{\boldsymbol{y}} h(\boldsymbol{y}) f(\boldsymbol{y}; \boldsymbol{\theta}).$$

11.2 The Cramér-Rao inequality for $\theta \in \mathbb{R}$

Definition 11.1 Expected Fisher information, $\theta \in \mathbb{R}$. Let $\boldsymbol{Y}$ have a distribution with density function $f(\boldsymbol{y}; \theta), \theta \in \mathbb{R}$. Assuming that $\frac{\partial f(\boldsymbol{y};\theta)}{\partial \theta}$ exists, the expected Fisher information about θ contained in $\boldsymbol{Y}$ is

$$\mathcal{I}(\theta) = E_\theta \left[\left(\frac{\partial l(\theta; \boldsymbol{Y})}{\partial \theta} \right)^2 \right] \tag{11.1}$$

$$= \int \left(\frac{\partial \log f(\boldsymbol{y}; \theta)}{\partial \theta} \right)^2 f(\boldsymbol{y}; \theta) d\boldsymbol{y}. \tag{11.2}$$

Expected Fisher information can appear to be an unusual beast at first sight. It may help to write (11.1) more explicitly as

$$\mathcal{I}(\theta_0) = E_{\theta_0} \left[\left(\left. \frac{\partial l(\theta; \boldsymbol{Y})}{\partial \theta} \right|_{\theta=\theta_0} \right)^2 \right], \tag{11.3}$$

where $\mathcal{I}(\theta_0)$ is the expected Fisher information evaluated at the fixed parameter value θ_0.

Example 11.1. Binomial. Let Y be distributed Bin(n, p). To within a constant the likelihood is $L(p; y) = p^y(1-p)^{n-y}$ and the log-likelihood is

$$l(p; y) = y \log p + (n - y)\log(1 - p), \tag{11.4}$$

and so

$$\begin{aligned}\frac{\partial l(p; y)}{\partial p} &= \frac{y}{p} - \frac{n-y}{1-p}\\ &= \frac{y - np}{p(1-p)}.\end{aligned}$$

The expected Fisher information for p provided by observing Y is therefore

$$\begin{aligned}\mathcal{I}(p) &= E_p\left[\left(\frac{Y - np}{p(1-p)}\right)^2\right]\\ &= \frac{E_p[(Y-np)^2]}{p^2(1-p)^2}\\ &= \frac{np(1-p)}{p^2(1-p)^2}\\ &= \frac{n}{p(1-p)}.\end{aligned}$$

□

Further examples of the calculation of $\mathcal{I}(\boldsymbol{\theta})$ are given in Section 11.7.

In Equation (11.3), the derivative of the log-likelihood can be denoted

$$s(\theta, \boldsymbol{Y}) = \frac{\partial l(\theta; \boldsymbol{Y})}{\partial \theta}, \tag{11.5}$$

where $s(\theta, \boldsymbol{Y})$ is commonly known as the *score function* when regarded as a function of θ (for fixed $\boldsymbol{Y}$), or the *score statistic* when regarded as a function of $\boldsymbol{Y}$ (for fixed θ). The score statistic can be regarded as a random variable that is obtained from a transformation of $\boldsymbol{Y}$ from $\mathbb{R}^n$ to $\mathbb{R}$. The expected Fisher information is the second moment of this random variable.

Before stating the Cramér-Rao inequality, a formal definition of an unbiased estimator is needed.

Definition 11.2 Unbiased estimator. The estimator $T(\boldsymbol{Y})$ of $\theta \in \mathbb{R}$ is an *unbiased estimator* of θ if $E_{\boldsymbol{\theta}}[T(\boldsymbol{Y})] = \theta$ for all $\theta \in \Theta$.

Theorem 11.1 Cramér-Rao inequality for estimation of $\boldsymbol{\theta} \in \boldsymbol{\Theta} \subset \mathbb{R}$. Let $\boldsymbol{Y} = (Y_1, ..., Y_n)$ be a sample from a distribution with density function $f(\boldsymbol{y}; \theta)$, and let $T(Y)$ be any unbiased estimator of $\theta \in \Theta \subset \mathbb{R}$. If $\text{var}_\theta(T) < \infty$ and $\int T(y)f(y;\theta)dy$ and $\int f(y;\theta)dy$ can be differentiated with respect to θ under the integral sign, then

$$\text{var}_\theta(T) \geq \frac{1}{\mathcal{I}(\theta)} \tag{11.6}$$

provided that $0 < \mathcal{I}(\theta) < \infty$. The right-hand side of (11.6) is called the *Cramér-Rao lower bound* for unbiased estimation of θ.

Proof of CR inequality. It is required to show that $\text{var}_\theta(T)\mathcal{I}(\theta) \geq 1$. Letting f be abbreviated notation for $f(y, \theta)$,

$$\begin{aligned}\text{var}_\theta(T)\mathcal{I}(\theta) &= \int (T(y)-\theta)^2 f dy \int \left(\frac{\partial \log f}{\partial\theta}\right)^2 f dy \\ &\geq \left(\int (T(y)-\theta)\left(\frac{\partial \log f}{\partial\theta}\right) f dy\right)^2,\end{aligned}$$

by the Cauchy-Schwarz inequality (Section 13.5.2). The above expression can be written

$$\begin{aligned}\left(\int (T(y)-\theta)\frac{f'}{f} f dy\right)^2 &= \left(\int (T(y)-\theta)f' dy\right)^2 \\ &= \left(\int T(y)f' dy + \theta \int f' dy\right)^2,\end{aligned} \tag{11.7}$$

where $f' \equiv f'(y;\theta)$ denotes the derivative of $f(y;\theta)$ with respect to θ. Now, the integral $\int f(y;\theta)dy$ can be differentiated under the integral sign, and so

$$\begin{aligned}\theta \int f' dy &= \theta \frac{\partial \int f dy}{\partial\theta} \\ &= \theta \frac{\partial 1}{\partial\theta} = 0,\end{aligned} \tag{11.8}$$

by virtue of the fact that

$$\int f(y;\theta)dy = 1 \text{ for all } \theta \in \Theta.$$

Similarly,

$$\begin{aligned}\int T(y)f' dy &= \frac{\partial \int T(y) f dy}{\partial\theta} \\ &= \frac{\partial\theta}{\partial\theta} = 1,\end{aligned} \tag{11.9}$$

since T is an unbiased estimator of θ. Substituting (11.8) and (11.9) into (11.7) gives the required result. □

Example 11.1 continued. For $Y \sim \text{Bin}(n, p)$, the CR inequality states that, for any unbiased estimator $T(Y)$ of p,

$$\text{var}_p(T(Y)) \geq \frac{1}{\mathcal{I}(p)} = \frac{p(1-p)}{n}.$$

In this particular case, the MLE $\widehat{p} = Y/n$ is an unbiased estimator of p, and has variance that attains the bound. □

The Cramér-Rao inequality does *not* imply the existence of an unbiased estimator of θ that will achieve the lower bound, nor even the existence of *any* unbiased estimator of θ (e.g. see Example 11.2 and Box 11.1).

In practice, the CR inequality is of limited utility because the construction of unbiased estimators that achieve the lower bound is only possible within the class of exponential family models (Chapter 7) and subject to θ being an appropriate parameterization (Pawitan 2001, p. 223), notwithstanding that this class of models does include the normal linear models. More generally, all unbiased estimators (if any) of θ could have variance that exceeds the lower bound of the CR inequality. (It might then be possible to apply the Rao-Blackwell method to find one having the smallest variance – see Pawitan (2001, p. 225) for details.)

The Cramér-Rao inequality is next extended to estimators of functions of $\theta \in \mathbb{R}$ (Section 11.3), and to functions of $\boldsymbol{\theta} \in \Theta \subset \mathbb{R}^s$ (Section 11.6) .

11.3 Cramér-Rao inequality for functions of θ

Here, it is desired to obtain a version of the Cramér-Rao inequality for unbiased estimators of $\zeta = g(\theta)$ where $g : \Theta \to \mathbb{R}$ is assumed differentiable with derivative denoted $g'(\theta) = \frac{\partial \zeta}{\partial \theta}$.

If $g(\theta)$ is invertible then the model can be re-parameterized as a function of ζ. To distinguish models parameterized by θ from those parameterized by ζ, a temporary modification of notation will be employed by introducing subscripts to the log-likelihood, so that l_θ and l_ζ denote the log-likelihood functions for θ and ζ respectively. That is, $l_\zeta(\zeta; \mathbf{y}) = l_\theta(g^{-1}(\zeta); \mathbf{y}) = l_\theta(\theta; \mathbf{y})$. From the chain rule of differentiation

$$\begin{aligned}\frac{\partial l_\zeta(\zeta; \mathbf{y})}{\partial \zeta} &= \frac{\partial l_\theta(\theta; \mathbf{y})}{\partial \theta} \frac{\partial \theta}{\partial \zeta} \\ &= \frac{\partial l_\theta(\theta; \mathbf{y})}{\partial \theta} \left(\frac{\partial \zeta}{\partial \theta} \right)^{-1} = \frac{1}{g'(\theta)} \frac{\partial l_\theta(\theta; \mathbf{y})}{\partial \theta}.\end{aligned}$$

The expected Fisher information about ζ from observing $\boldsymbol{y}$ can therefore be obtained as

$$\begin{aligned}\mathcal{I}(\zeta) &= E_\zeta\left[\left(\frac{\partial l_\zeta(\zeta;\boldsymbol{Y})}{\partial\zeta}\right)^2\right]\\ &= \frac{1}{g'(\theta)^2}E_\theta\left[\left(\frac{\partial l_\theta(\theta;\boldsymbol{Y})}{\partial\theta}\right)^2\right] = \frac{\mathcal{I}(\theta)}{g'(\theta)^2}.\end{aligned} \tag{11.10}$$

Example 11.2. Log-odds. Suppose that Y is binomially distributed Bin(n, p), $0 < p < 1$, and that the quantity of interest is the log-odds, $\zeta = g(p) = \text{logit}(p) = \log(p/(1-p))$. The inverse of g is

$$p = g^{-1}(\zeta) = \frac{e^\zeta}{1+e^\zeta},$$

and $\mathcal{I}(\zeta)$ can be obtained from working with the binomial likelihood in (11.4), but expressed as a function of ζ (see Exercise 11.3). Alternatively, using (11.10), the expected Fisher information about ζ from Y is

$$\begin{aligned}\mathcal{I}(\zeta) &= \left(\frac{\partial\log(p/(1-p))}{\partial p}\right)^{-2}\mathcal{I}(\theta)\\ &= \left(\frac{1}{p(1-p)}\right)^{-2}\frac{n}{p(1-p)}\\ &= np(1-p).\end{aligned}$$

The CR inequality for unbiased estimators $T(Y)$ of the log-odds is therefore

$$\text{var}_p(T) \geq \frac{1}{\mathcal{I}(\zeta)} = \frac{1}{np(1-p)}.$$

However, it can be shown that no unbiased estimator of the log-odds exists – see Box 11.1. □

For the estimation problem in Example 11.2, if $T(Y)$ is an unbiased estimator of the log-odds then

$$E_p(T(Y)) = \sum_{y=0}^{n}\binom{n}{y}T(y)p^y(1-p)^{n-y} = \log(p/(1-p)) \quad \text{for all } p,\ 0<p<1.$$

Note that $E_p(T(Y)) \leq \max_y T(y)$, yet $g(p) = \log(p/(1-p))$ is unbounded as $p \to 1$. Hence no unbiased estimator of the log-odds can exist.

Box 11.1

It is immediate from (11.10) that the Cramér-Rao lower bound for unbiased estimators $T(\boldsymbol{Y})$ of $\zeta = g(\theta)$ is $g'(\theta)^2 \mathcal{I}(\theta)^{-1}$. This lower bound also holds more generally for differentiable functions $g(\theta)$ that are not necessarily invertible, and is stated below.

Theorem 11.2 If $T(\boldsymbol{Y})$ is an unbiased estimator of the differentiable function $g(\theta)$, and the conditions of Theorem 11.1 hold, then

$$\mathrm{var}_\theta(T) \geq \frac{(g'(\theta))^2}{\mathcal{I}(\theta)} \tag{11.11}$$

where $g'(\theta) = \frac{\partial g}{\partial \theta}$ evaluated at θ.

Theorem 11.2 can be proved directly by minor modification to the proof of Theorem 11.1 (Exercise 11.1).

11.4 Alternative formulae for $\mathcal{I}(\theta)$

The next lemma provides alternative formulae for calculation of $\mathcal{I}(\theta)$. It is often the case that calculating Fisher information using the form in (11.2) can be cumbersome compared to use of the alternative (11.14) that is given below.

Lemma 11.1. If $\frac{\partial f(\boldsymbol{y};\theta)}{\partial \theta}$ exists and $\int f(\boldsymbol{y};\theta)d\boldsymbol{y}$ can be differentiated (with respect to θ) under the integral sign then

$$E_\theta\left[\frac{\partial l(\theta;\boldsymbol{Y})}{\partial \theta}\right] = 0, \tag{11.12}$$

from which it immediately follows that

$$\mathcal{I}(\theta) = \mathrm{var}_\theta\left[\frac{\partial l(\theta;\boldsymbol{Y})}{\partial \theta}\right]. \tag{11.13}$$

Furthermore, if $\frac{\partial^2 f(\boldsymbol{y};\theta)}{\partial \theta^2}$ exists and $\int f(\boldsymbol{y};\theta)d\boldsymbol{y}$ can be twice differentiated under the integral sign, then

$$\mathcal{I}(\theta) = -E_\theta\left[\frac{\partial^2 l(\theta;\boldsymbol{Y})}{\partial \theta^2}\right]. \tag{11.14}$$

Proof. Note that

$$E_\theta\left[\frac{\partial l(\theta; Y)}{\partial \theta}\right] = E_\theta\left[\frac{\partial \log f(Y; \theta)}{\partial \theta}\right]$$
$$= \int \frac{f'(y; \theta)}{f(y; \theta)} f(y; \theta) dy$$
$$= \int f'(y; \theta) dy$$
$$= \frac{\partial \int f(y; \theta) dy}{\partial \theta} = \frac{\partial 1}{\partial \theta} = 0.$$

To show (11.14), application of the chain rule of differentiation gives

$$\frac{\partial^2 l(\theta; y)}{\partial \theta^2} \equiv \frac{\partial}{\partial \theta} \frac{f'(y; \theta)}{f(y; \theta)}$$
$$= \frac{f''(y; \theta)}{f(y; \theta)} - \left(\frac{f'(y; \theta)}{f(y; \theta)}\right)^2.$$

The first term on the right hand side vanishes upon taking its expectation (since it is the double derivative (with respect to θ) of $\int f(y; \theta) dy$), which gives the required result.

Example 11.1 continued. For $Y \sim \text{Bin}(n, p)$,

$$\frac{\partial l(p; y)}{\partial p} = \frac{y}{p} - \frac{n - y}{1 - p},$$

and so the second derivative of the log-likelihood is

$$\frac{\partial^2 l(p; y)}{\partial p^2} = -\frac{y}{p^2} - \frac{n - y}{(1 - p)^2}. \tag{11.15}$$

Taking expectations gives

$$\mathcal{I}(p) = -E_p\left[\frac{\partial^2 l(p; Y)}{\partial p^2}\right]$$
$$= E_p\left[\frac{Y}{p^2} + \frac{n - Y}{(1 - p)^2}\right]$$
$$= \frac{n}{p} + \frac{n}{1 - p}$$
$$= \frac{n}{p(1 - p)}.$$

□

11.5 The iid data case

When the observations are iid then it would seem natural that the information from $Y_1, \ldots, Y_n$ will be n times the information from a single observation. Indeed, if Y_i are iid, then

$$\frac{\partial l(\theta; \boldsymbol{Y})}{\partial \theta} = \sum_{i=1}^{n} \frac{\partial \log f(Y_i; \theta)}{\partial \theta},$$

where $\frac{\partial \log f(Y_i;\theta)}{\partial \theta}$, $i = 1, \ldots, n$ are also iid random variables. Hence

$$\begin{aligned}\mathcal{I}(\theta) &= \text{var}_\theta \left[\frac{\partial l(\theta; \boldsymbol{Y})}{\partial \theta}\right] \\ &= \sum_{i=1}^{n} \text{var}_\theta \left[\frac{\partial \log f(Y_i; \theta)}{\partial \theta}\right] = n\,\mathcal{I}_1(\theta),\end{aligned}$$

where $\mathcal{I}_1(\theta)$ denotes the expected Fisher information from one observation.

Example 11.1 continued. A Bin(n, p) experiment is equivalent to observing n iid Bernoulli trials. $\mathcal{I}_1(p)$, the information from a single Bernoulli trial, is $1/(p(1-p))$. □

11.6 The multi-dimensional case, $\boldsymbol{\theta} \in \Theta \subset \mathbb{R}^s$

The notation, concepts and results of Sections 11.2–11.5 extend in a natural way to the multi-parameter case where $\boldsymbol{\theta} \in \Theta \subset \mathbb{R}^s$. In particular, assuming the partial differentiability of $f(\boldsymbol{y}; \boldsymbol{\theta})$ with respect to each θ_i, $i = 1, \ldots, s$, the score statistic is now a s-dimensional random vector

$$s(\boldsymbol{\theta}; \boldsymbol{Y}) = \frac{\partial l(\boldsymbol{\theta}; \boldsymbol{Y})}{\partial \boldsymbol{\theta}} = \begin{pmatrix} \frac{\partial l(\boldsymbol{\theta};\boldsymbol{Y})}{\partial \theta_1} \\ \cdot \\ \cdot \\ \cdot \\ \frac{\partial l(\boldsymbol{\theta};\boldsymbol{Y})}{\partial \theta_s} \end{pmatrix}.$$

Definition 11.3 Expected Fisher information, $\boldsymbol{\theta} \in \mathbb{R}^s$. The expected Fisher information matrix, $\mathcal{I}(\boldsymbol{\theta})$, is defined to be the $s \times s$ matrix with (i, j) element

$$\mathcal{I}(\boldsymbol{\theta})_{ij} = E_{\boldsymbol{\theta}} \left[\frac{\partial l(\boldsymbol{\theta}; \boldsymbol{Y})}{\partial \theta_i} \frac{\partial l(\boldsymbol{\theta}; \boldsymbol{Y})}{\partial \theta_j}\right]. \tag{11.16}$$

That is, $\mathcal{I}(\boldsymbol{\theta})$ is the expectation of the outer product of the vector score statistic with itself

$$\mathcal{I}(\boldsymbol{\theta}) = E_{\boldsymbol{\theta}}\left[\frac{\partial l(\boldsymbol{\theta};\boldsymbol{Y})}{\partial \boldsymbol{\theta}}\frac{\partial l(\boldsymbol{\theta};\boldsymbol{Y})}{\partial \boldsymbol{\theta}}^T\right]. \tag{11.17}$$

From (11.17), $\mathcal{I}(\boldsymbol{\theta})$ is necessarily a non-negative definite matrix, and for well-posed models it will be positive definite, and therefore non-singular (i.e. invertible). In practice, if $\mathcal{I}(\boldsymbol{\theta})$ is singular then it will often be due to the model parameters being non-identifiable, as can arise in linear regression (Section 11.7.3) when the design matrix is not of full rank.

Box 11.2

The multi-parameter version of the Cramér-Rao inequality is stated analogously to Theorem 11.2, but with the lower bound expressed using a quadratic form.

Theorem 11.3 Multiparameter Cramér-Rao inequality, $\boldsymbol{\theta} \in \Theta \subset \mathbb{R}^s$. If $T(\boldsymbol{Y})$ is an unbiased estimator of the differentiable function $g(\boldsymbol{\theta}) : \Theta \to \mathbb{R}$ then

$$\text{var}_{\boldsymbol{\theta}}(T) \geq g'(\boldsymbol{\theta})^T \mathcal{I}(\boldsymbol{\theta})^{-1} g'(\boldsymbol{\theta}), \tag{11.18}$$

where $g'(\boldsymbol{\theta}) = \frac{\partial g(\boldsymbol{\theta})}{\partial \boldsymbol{\theta}}$ is the column vector with ith element equal to $\partial g(\boldsymbol{\theta})/\partial \theta_i$.

The proof of Theorem 11.3 requires an extension of the Cauchy-Schwarz inequality and is not provided here. It can be found in Lehmann (1983, Section 2.7 therein).

Example 11.2. CR bound for θ_i. If $g(\boldsymbol{\theta}) = \theta_i$, then $g'(\boldsymbol{\theta})$ is the s-dimensional vector having element i equal to unity, and all others equal to zero. Hence, the lower bound on the variance of unbiased estimators of θ_i is $[\mathcal{I}(\boldsymbol{\theta})^{-1}]_{ii}$, the ith diagonal element of the inverse Fisher information matrix. □

11.6.1 Parameter orthogonality

Let the parameter vector $\boldsymbol{\theta}$ be partitioned as $\boldsymbol{\theta} = (\boldsymbol{\psi}, \boldsymbol{\lambda})$ where $\boldsymbol{\theta} \in \mathbb{R}^s$, $\boldsymbol{\psi} \in \mathbb{R}^r$ and $\boldsymbol{\lambda} \in \mathbb{R}^{s-r}$. The information matrix $\mathcal{I}(\boldsymbol{\theta})$ can then be partitioned as

$$\mathcal{I}(\boldsymbol{\theta}) = \mathcal{I}(\boldsymbol{\psi}, \boldsymbol{\lambda}) = \begin{pmatrix} \mathcal{I}_{\psi\psi} & \mathcal{I}_{\psi\lambda} \\ \mathcal{I}_{\lambda\psi} & \mathcal{I}_{\lambda\lambda} \end{pmatrix}.$$

The $r \times r$ matrix $\mathcal{I}_{\psi\psi} \equiv \mathcal{I}_{\psi\psi}(\theta)$ is

$$\mathcal{I}_{\psi\psi} = E_\theta \left[\frac{\partial l}{\partial \psi} \frac{\partial l}{\partial \psi}^T \right],$$

where $\partial l / \partial \psi$ is the r-dimensional vector score statistic containing the partial derivatives of the log-likelihood with respect to the elements of ψ. Similarly, $\mathcal{I}_{\psi\lambda} \equiv \mathcal{I}_{\psi\lambda}(\theta)$ is the $r \times (s-r)$ matrix

$$\mathcal{I}_{\psi\lambda} = E_\theta \left[\frac{\partial l}{\partial \psi} \frac{\partial l}{\partial \lambda}^T \right],$$

and analogously for $\mathcal{I}_{\lambda\psi}$ and $\mathcal{I}_{\lambda\lambda}$. Note that $\mathcal{I}_{\lambda\psi} = \mathcal{I}_{\psi\lambda}^T$.

If all elements of $\mathcal{I}_{\psi\lambda}$ are zero, denoted $\mathcal{I}_{\psi\lambda} = \mathbf{0}$, then ψ and λ are said to be (information) orthogonal parameters. Then $\mathcal{I}(\theta)$ has block diagonal form, and

$$\mathcal{I}(\theta)^{-1} = \begin{pmatrix} \mathcal{I}_{\psi\psi}^{-1} & \mathbf{0} \\ \mathbf{0} & \mathcal{I}_{\lambda\lambda}^{-1} \end{pmatrix}.$$

It follows that the Cramér-Rao lower bound for unbiased estimators of differentiable functions $g(\psi) : \mathbb{R}^r \to \mathbb{R}$ is $g'(\psi)^T \mathcal{I}_{\psi\psi}^{-1}(\theta) g'(\psi)$. Note that this lower bound is the same as would be obtained if the value of λ were known, in which case the parameter vector would be ψ alone.

In the general case where ψ and λ need not be orthogonal, the Cramér-Rao lower bound for unbiased estimators of $g(\psi)$ is greater (or at best equal) when λ is an unknown parameter, compared to the case when λ is known. Essentially, this is saying that ψ is harder to estimate in the presence of the additional unknown parameters in λ (see the gamma distribution example in Section 11.7.6, and Exercise 11.10).

11.6.2 Alternative formulae for $\mathcal{I}(\theta)$

Analogously to Section 11.5, if the data are iid then it is also the case that $\mathcal{I}(\theta) = n\,\mathcal{I}_1(\theta)$ in the multi-parameter case, where $\mathcal{I}_1(\theta)$ is the information matrix for a single observation. The alternative formulae for the calculation of expected Fisher information extend to the multi-parameter case in a natural way, and are stated below.

Lemma 11.2. If the first-order partial derivatives (with respect to θ_i, $i = 1, \ldots, s$) of $f(y;\theta)$ exist, and $\int f(y;\theta) dy$ can be differentiated (with respect to each θ_i) under the integral sign, then

$$E_\theta \left[\frac{\partial l(\theta; Y)}{\partial \theta} \right] = \mathbf{0}, \tag{11.19}$$

from which it immediately follows that $\mathcal{I}(\boldsymbol{\theta})$ is the $s \times s$ variance matrix of the vector score statistic. That is,

$$\mathcal{I}(\boldsymbol{\theta})_{ij} = \mathrm{cov}_{\boldsymbol{\theta}}\left(\frac{\partial l}{\partial \theta_i}, \frac{\partial l}{\partial \theta_j}\right). \tag{11.20}$$

Furthermore, if all second-order partial derivatives of $f(y; \boldsymbol{\theta})$ exist, and $\int f(\mathbf{y}; \boldsymbol{\theta}) d\mathbf{y}$ can be twice differentiated under the integral sign, then

$$\mathcal{I}(\boldsymbol{\theta})_{ij} = -E_{\boldsymbol{\theta}}\left[\frac{\partial^2 l(\boldsymbol{\theta}; \mathbf{Y})}{\partial \theta_i \partial \theta_j}\right]. \tag{11.21}$$

That is, $\mathcal{I}(\boldsymbol{\theta})$ is the negative of the expected value of the $s \times s$ Hessian matrix of second derivatives of the log-likelihood.

11.6.3 Fisher information for re-parameterized models

Suppose that the model is re-parameterized using $\boldsymbol{\zeta} = g(\boldsymbol{\theta})$, where $g : \Theta \to \mathbb{R}^s$ is differentiable and invertible. Denote

$$g(\boldsymbol{\theta}) = \begin{pmatrix} g_1(\boldsymbol{\theta}) \\ \cdot \\ \cdot \\ \cdot \\ g_s(\boldsymbol{\theta}) \end{pmatrix}$$

where each co-ordinate $g_i : \Theta \to \mathbb{R}$ has derivative $g_i'(\boldsymbol{\theta}) = \left(\frac{\partial g_i}{\partial \theta_1}, \ldots, \frac{\partial g_i}{\partial \theta_s}\right)^T$. Then the Fisher information matrix $\mathcal{I}(\boldsymbol{\zeta})$ for $\boldsymbol{\zeta}$ can be obtained from the Fisher information matrix $\mathcal{I}(\boldsymbol{\theta})$ using the natural extension of Equation (11.10). That is,

$$\mathcal{I}(\boldsymbol{\zeta}) = \mathbf{G}(\boldsymbol{\theta})^{-T} \mathcal{I}(\boldsymbol{\theta}) \mathbf{G}(\boldsymbol{\theta})^{-1}, \tag{11.22}$$

where $\mathbf{G}(\boldsymbol{\theta})^{-T}$ denotes the inverse of the transpose (or equivalently, transpose of the inverse) of $\mathbf{G}(\boldsymbol{\theta})$, and $\mathbf{G}(\boldsymbol{\theta})$ is the $s \times s$ Jacobian matrix of derivatives

$$\mathbf{G}(\boldsymbol{\theta}) = \begin{pmatrix} g_1'(\boldsymbol{\theta})^T \\ \cdot \\ \cdot \\ \cdot \\ g_s'(\boldsymbol{\theta})^T \end{pmatrix}.$$

Application of (11.22) is demonstrated in Section 11.7.1.

11.7 Examples of Fisher information calculation

11.7.1 Normal(μ, σ^2)

Let $Y_1, ..., Y_n$ be iid $N(\mu, \sigma^2)$. Here, note that the information calculations are made with respect to parameters $(\theta_1, \theta_2) = (\mu, \sigma^2)$. Equation (11.22) is then used to obtain the information matrix for parameters $(\zeta_1, \zeta_2) = (\mu, \sigma)$.

Now, the contribution to the log-likelihood from a single data point y is

$$l(\boldsymbol{\theta}; y) = l(\mu, \sigma^2; y) = -\frac{1}{2}\log(2\pi\sigma^2) - \frac{1}{2\sigma^2}(y-\mu)^2,$$

and so

$$\begin{aligned}\frac{\partial l}{\partial \mu} &= \frac{1}{\sigma^2}(y-\mu),\\ \frac{\partial l}{\partial \sigma^2} &= \frac{-1}{2\sigma^2} + \frac{1}{2\sigma^4}(y-\mu)^2.\end{aligned}$$

The elements of $\mathcal{I}_1(\boldsymbol{\theta}) = \mathcal{I}_1(\mu, \sigma^2)$ are

$$\begin{aligned}[\mathcal{I}_1]_{11} &= E_{\boldsymbol{\theta}}\left[\left(\frac{\partial l}{\partial \mu}\right)^2\right]\\ &= \frac{1}{\sigma^4}E_{\boldsymbol{\theta}}\left[(Y-\mu)^2\right] = \frac{1}{\sigma^2},\\ [\mathcal{I}_1]_{12} &= E_{\boldsymbol{\theta}}\left[\frac{\partial l}{\partial \mu}\frac{\partial l}{\partial \sigma^2}\right]\\ &= \frac{1}{2\sigma^4}E_{\boldsymbol{\theta}}\left[-(Y-\mu) + \frac{1}{\sigma^2}(Y-\mu)^3\right] = 0,\\ [\mathcal{I}_1]_{22} &= E_{\boldsymbol{\theta}}\left[\left(\frac{\partial l}{\partial \sigma^2}\right)^2\right]\\ &= \frac{1}{4\sigma^4}E_{\boldsymbol{\theta}}\left[1 - \frac{2}{\sigma^2}(Y-\mu)^2 + \frac{1}{\sigma^4}(Y-\mu)^4\right]\\ &= \frac{1}{4\sigma^4}\left[1 - \frac{2\sigma^2}{\sigma^2} + \frac{3\sigma^4}{\sigma^4}\right] = \frac{1}{2\sigma^4}.\end{aligned}$$

Alternatively, using the second derivative calculations of formula (11.21),

$$\begin{aligned}[\mathcal{I}_1]_{11} &= -E_{\boldsymbol{\theta}}\left[\frac{\partial^2 l}{\partial \mu^2}\right]\\ &= -E_{\boldsymbol{\theta}}\left[\frac{-1}{\sigma^2}\right] = \frac{1}{\sigma^2},\end{aligned}$$

$$[\mathcal{I}_1]_{12} = -E_{\boldsymbol{\theta}}\left[\frac{\partial^2 l}{\partial\mu\partial\sigma^2}\right]$$
$$= \frac{1}{\sigma^4}E_{\boldsymbol{\theta}}[(Y-\mu)] = 0,$$
$$[\mathcal{I}_1]_{22} = -E_{\boldsymbol{\theta}}\left[\frac{\partial^2 l}{\partial(\sigma^2)^2}\right]$$
$$= \frac{-1}{2\sigma^4} + \frac{1}{\sigma^6}E_{\boldsymbol{\theta}}\left[(Y-\mu)^2\right] = \frac{1}{2\sigma^4}.$$

Thus, the information matrix for a single observation is

$$\mathcal{I}_1(\mu,\sigma^2) = \begin{pmatrix} 1/\sigma^2 & 0 \\ 0 & 1/(2\sigma^4) \end{pmatrix},$$

and note that μ and σ^2 are orthogonal parameters. For a sample of size n,

$$\mathcal{I}(\mu,\sigma^2) = n\begin{pmatrix} 1/\sigma^2 & 0 \\ 0 & 1/(2\sigma^4) \end{pmatrix},$$

and

$$\mathcal{I}(\mu,\sigma^2)^{-1} = \frac{1}{n}\begin{pmatrix} \sigma^2 & 0 \\ 0 & 2\sigma^4 \end{pmatrix}.$$

That is, the CR lower bounds for the variances of unbiased estimators of μ and σ^2 are $[\mathcal{I}^{-1}]_{11} = \sigma^2/n$ and $[\mathcal{I}^{-1}]_{22} = 2\sigma^4/n$ respectively. The MLE of μ is $\overline{Y}$ which is unbiased and has variance that equals the CR lower bound.

As a consequence of the parameter orthogonality, the CR lower bound for each parameter does not depend on whether the other parameter is known. For example, if σ^2 were known, one would simply calculate $\mathcal{I}(\mu) = n/\sigma^2$, and the bound would be $\mathcal{I}(\mu)^{-1} = \sigma^2/n$. If μ were known, then $\mathcal{I}(\sigma^2)^{-1} = 2\sigma^4/n$ and in this case the MLE of σ^2 would be $\widehat{\sigma}^2 = \frac{1}{n}\sum_{i=1}^n (Y_i-\mu)^2$. This is an unbiased estimator of σ^2 and has variance that equals the CR lower bound.

In practice, μ will not be known and it is then the case that no unbiased estimator of σ^2 can achieve the CR lower bound (see Pawitan 2001, p. 223). The MLE of σ^2 is $\widehat{\sigma}^2 = \frac{n-1}{n}S^2$ where S^2 is the familiar unbiased sample variance. The estimator S^2 has distribution (Seber and Lee 2003, p. 48)

$$S^2 = \frac{\sum_{i=1}^n (Y_i - \overline{Y})^2}{n-1} \sim \frac{\sigma^2\chi^2_{n-1}}{n-1}.$$

Since $E[\chi^2_{n-1}] = n - 1$, S^2 is unbiased for σ^2, and $\widehat{\sigma}^2$ has bias $\frac{-\sigma^2}{n}$. Also,

$$\begin{aligned}\text{var}_{\boldsymbol{\theta}}(S^2) &= \left(\frac{\sigma^2}{n-1}\right)^2 \text{var}(\chi^2_{n-1}) \\ &= \left(\frac{\sigma^2}{n-1}\right)^2 2(n-1) = \frac{2\sigma^4}{n-1},\end{aligned}$$

which is greater than the CR lower bound. Also,

$$\begin{aligned}\text{var}_{\boldsymbol{\theta}}(\widehat{\sigma}^2) &= \left(\frac{n-1}{n}\right)^2 \text{var}_{\boldsymbol{\theta}}(S^2) \\ &= \left(\frac{n-1}{n}\right)\frac{2\sigma^4}{n},\end{aligned}$$

which is smaller than the CR lower bound. Of course, this does not contradict the CR inequality because it does not apply to biased estimators.

For large n, the bias of $\widehat{\sigma}^2$ tends to zero and it is said to be asymptotically unbiased. The limiting variance (as $n \to \infty$) of both S^2 and $\widehat{\sigma}^2$ is the CR lower bound, and hence they are both asymptotically efficient.

Finally, if this model were re-parameterized using $\boldsymbol{\zeta} = (\mu, \sigma) = (\theta_1, \sqrt{\theta_2})$, then $\mathcal{I}(\boldsymbol{\zeta})$ could be obtained by re-doing the information calculations, or from using Equation (11.22) as follows,

$$\begin{aligned}\mathcal{I}(\boldsymbol{\zeta}) = \mathcal{I}(\mu, \sigma) &= \begin{pmatrix} 1 & 0 \\ 0 & \frac{1}{2\sqrt{\theta_2}} \end{pmatrix}^{-T} \mathcal{I}(\boldsymbol{\theta}) \begin{pmatrix} 1 & 0 \\ 0 & \frac{1}{2\sqrt{\theta_2}} \end{pmatrix}^{-1} \\ &= n \begin{pmatrix} 1 & 0 \\ 0 & 2\sigma \end{pmatrix} \begin{pmatrix} 1/\sigma^2 & 0 \\ 0 & 1/(2\sigma^4) \end{pmatrix} \begin{pmatrix} 1 & 0 \\ 0 & 2\sigma \end{pmatrix} \\ &= n \begin{pmatrix} 1/\sigma^2 & 0 \\ 0 & 2/\sigma^2 \end{pmatrix}.\end{aligned}$$

11.7.2 Exponential family distributions

Let $Y_1, \ldots, Y_n$ be independent with distribution belonging to the exponential (dispersion) family (Section 7.2.1) such that the log-density of y_i can be written in the form

$$\log f(y_i; \psi_i, \phi_i) = \frac{y\psi_i - b(\psi_i)}{a(\phi_i)} + c(y, \phi_i). \tag{11.23}$$

For each i, it will assumed that $a(\phi_i) = \phi / w_i$ for some $\phi > 0$ and known values w_i. In general, ϕ is a parameter to be estimated, but in some familiar cases it can be considered known (e.g. see Example 7.2). Furthermore, ψ_i is assumed to be a differentiable function of parameters β_j, $j = 1, \ldots, p$. The parameters to be estimated are $\boldsymbol{\theta} = (\beta_1, \ldots, \beta_p, \phi)$, and $\mathcal{I}(\boldsymbol{\beta})$ will be used to denote the upper-left $p \times p$ sub-matrix of $\mathcal{I}(\boldsymbol{\theta})$ corresponding to information calculations with respect to $\boldsymbol{\beta}$ only.

Letting $l_i = l(\boldsymbol{\beta}, \phi; y_i)$ denote the contribution to the log-likelihood from observation i,

$$\begin{aligned}\frac{\partial l(\boldsymbol{\beta}, \phi; \mathbf{y})}{\partial \beta_j} &= \sum_{i=1}^{n} \frac{\partial l_i}{\partial \beta_j} = \sum_{i=1}^{n} \frac{\partial l_i}{\partial \psi_i} \frac{\partial \psi_i}{\partial \beta_j} \\ &= \sum_{i=1}^{n} \frac{(y_i - b'(\psi_i))}{a(\phi_i)} \frac{\partial \psi_i}{\partial \beta_j}. \end{aligned} \tag{11.24}$$

It is more intuitive to re-express (11.24) in terms of the mean and variance of Y_i. This can be accomplished by noting that, from Exercise 11.8,

$$\mu_i = E_{\boldsymbol{\theta}}[Y_i] = b'(\psi_i), \tag{11.25}$$

and

$$\mathrm{var}_{\boldsymbol{\theta}}(Y_i) = b''(\psi_i) a(\phi). \tag{11.26}$$

From (11.25) it follows that $\partial \mu_i / \partial \psi_i = b''(\psi_i)$, and hence

$$\begin{aligned}\frac{\partial \psi_i}{\partial \beta_j} &= \frac{\partial \psi_i}{\partial \mu_i} \frac{\partial \mu_i}{\partial \beta_j} \\ &= \frac{1}{b''(\psi_i)} \frac{\partial \mu_i}{\partial \beta_j} \\ &= \frac{a(\phi_i)}{\mathrm{var}_{\boldsymbol{\theta}}(Y_i)} \frac{\partial \mu_i}{\partial \beta_j}.\end{aligned}$$

Substituting this into Equation (11.24) gives

$$\frac{\partial l(\boldsymbol{\beta}, \phi; \mathbf{y})}{\partial \beta_j} = \sum_{i=1}^{n} \frac{(y_i - \mu_i)}{\mathrm{var}_{\boldsymbol{\theta}}(Y_i)} \frac{\partial \mu_i}{\partial \beta_j}. \tag{11.27}$$

The (j, k)'th element of $\mathcal{I}(\boldsymbol{\beta})$ is

$$\begin{aligned}[\mathcal{I}(\boldsymbol{\beta})]_{jk} &= E_{\boldsymbol{\theta}}\left[\frac{\partial l(\boldsymbol{\beta}, \phi; \boldsymbol{Y})}{\partial \beta_j} \frac{\partial l(\boldsymbol{\beta}, \phi; \boldsymbol{Y})}{\partial \beta_k}\right] \\ &= \sum_{i=1}^{n} \frac{E_{\boldsymbol{\theta}}[(Y_i - \mu_i)^2]}{\mathrm{var}_{\boldsymbol{\theta}}(Y_i)^2} \frac{\partial \mu_i}{\partial \beta_j} \frac{\partial \mu_i}{\partial \beta_k} \\ &= \sum_{i=1}^{n} \frac{\frac{\partial \mu_i}{\partial \beta_j} \frac{\partial \mu_i}{\partial \beta_k}}{\mathrm{var}_{\boldsymbol{\theta}}(Y_i)}. \end{aligned} \tag{11.28}$$

This can be written in matrix notation as

$$\mathcal{I}(\boldsymbol{\beta}) = \mathbf{G}^T\mathbf{V}^{-1}\mathbf{G}, \tag{11.29}$$

where $\mathbf{V}$ is the $n \times n$ diagonal matrix with diagonal entry i equal to $\text{var}_{\boldsymbol{\theta}}(Y_i)$, and $\mathbf{G}$ is the $n \times p$ Jacobian matrix with (i, j) element equal to $\partial\mu_i/\partial\beta_j$.

Differentiation of (11.24) with respect to ϕ results in a summation of terms in which y_i appears only through the multiplicative term $(y_i - b'(\psi_i))$ in the numerator. These terms therefore have expectation of zero, and it follows that $\mathcal{I}(\boldsymbol{\theta})$ has block diagonal form

$$\mathcal{I}(\boldsymbol{\theta}) = \begin{pmatrix} \mathcal{I}(\boldsymbol{\beta}) & \mathbf{0} \\ \mathbf{0} & \mathcal{I}_{\phi\phi}(\boldsymbol{\theta}) \end{pmatrix}, \tag{11.30}$$

where $\mathcal{I}_{\phi\phi}(\boldsymbol{\theta})$ denotes Fisher information with respect to ϕ.

11.7.3 Linear regression model

Let each $Y_i, i = 1, \ldots, n$ be independently distributed as $N(\boldsymbol{x}_i^T\boldsymbol{\beta}, \sigma^2)$ where $\boldsymbol{x}_i = (x_{i1}, \ldots, x_{ip})^T$ is a p-dimensional vector of known covariates and $\boldsymbol{\beta} \in \mathbb{R}^p$ and σ^2 are parameters to be estimated. The normal distribution is a member of the exponential (dispersion) family (Example 7.1), where $\phi = \sigma^2$. Since $x_{ij} = \partial\mu_i/\partial\beta_j$, it follows from (11.29) that

$$\mathcal{I}(\boldsymbol{\beta}) = \frac{\mathbf{X}^T\mathbf{X}}{\sigma^2},$$

where $\mathbf{X}$ is the $n \times p$ design matrix having $\boldsymbol{x}_i^T$ in row i.

Analogously to the iid $N(\mu, \sigma^2)$ case in Section 11.7.1, $\mathcal{I}_{\sigma^2\sigma^2}(\boldsymbol{\theta}) = n/2\sigma^4$. From (11.30), the expected Fisher information is the $(p+1) \times (p+1)$ matrix

$$\mathcal{I}(\boldsymbol{\beta}, \sigma^2) = \begin{pmatrix} \frac{\mathbf{X}^T\mathbf{X}}{\sigma^2} & \mathbf{0} \\ \mathbf{0} & \frac{n}{2\sigma^4} \end{pmatrix}. \tag{11.31}$$

The likelihood equations for $\widehat{\boldsymbol{\beta}}$ are given by setting the partial derivative in (11.27) to zero, for $j = 1, \ldots, p$. (11.27). Using matrix notation, this is

$$\frac{\partial l(\boldsymbol{\beta}, \sigma^2)}{\partial\boldsymbol{\beta}} = \frac{1}{\sigma^2}(\mathbf{X}^T\boldsymbol{y} - \mathbf{X}^T\mathbf{X})\boldsymbol{\beta} = 0, \tag{11.32}$$

which is sometimes also called the normal equation. Assuming that the design matrix is of full rank, the MLE is

$$\widehat{\boldsymbol{\beta}} = (\mathbf{X}^T\mathbf{X})^{-1}\mathbf{X}^T\boldsymbol{y}, \tag{11.33}$$

which is also the usual least squares estimator. This estimator has variance matrix $\sigma^2(\mathbf{X}^T\mathbf{X})^{-1}$, and therefore for any linear function $g(\boldsymbol{\beta}) \in \mathbb{R}$, the unbiased estimator $g(\widehat{\boldsymbol{\beta}})$ has variance that attains the CR lower bound.

11.7.4 Nonlinear regression model

Let each Y_i, $i = 1, \ldots, n$ be independently distributed as $N(\mu_i(\boldsymbol{\beta}), \sigma^2)$, where each μ_i is differentiable with respect to $\boldsymbol{\beta} = (\beta_1, \ldots, \beta_p)$. It follows immediately from (11.28) and (11.30) that

$$\mathcal{I}(\boldsymbol{\beta}, \sigma^2) = \begin{pmatrix} \mathcal{I}(\boldsymbol{\beta}) & \mathbf{0} \\ \mathbf{0} & \frac{n}{2\sigma^4} \end{pmatrix}, \tag{11.34}$$

where $\mathcal{I}(\boldsymbol{\beta})$ has (j, k)'th element

$$[\mathcal{I}(\boldsymbol{\beta})]_{jk} = \frac{1}{\sigma^2} \sum_{i=1}^{n} \frac{\partial \mu_i}{\partial \beta_j} \frac{\partial \mu_i}{\partial \beta_k}.$$

11.7.5 Generalized linear model with canonical link function

In the notation of Section 7.2.2, $\eta = g(\mu)$ where η is the linear predictor and g is the link function. The canonical link function is given by taking g to be the inverse of the function b' in (11.25). With this choice, $\mu = b'(\psi)$, and it follows that

$$\psi = (b')^{-1}(\mu) = g(\mu) = \eta.$$

That is, ψ is equal to the linear predictor. Therefore $\partial\psi_i/\partial\beta_j = x_{ij}$, which does not depend on $\boldsymbol{\beta}$. From (11.24), it follows that the Hessian matrix, $\partial^2 l/\partial\boldsymbol{\beta}^2$, does not depend on y_i. This establishes that the expected and observed Fisher information (Section 12.3.1) are identical when the canonical link is used in a GLM.

11.7.6 Gamma(α, β)

Let $Y_1, \ldots, Y_n$ be iid from a Gamma(α, β) distribution with density function

$$f(y; \alpha, \beta) = \frac{y^{\alpha-1} e^{-y/\beta}}{\beta^\alpha \Gamma(\alpha)}, \qquad y > 0.$$

It can be shown (Exercise 11.7) that

$$\mathcal{I}_1(\alpha, \beta) = \begin{pmatrix} \Psi'(\alpha) & 1/\beta \\ 1/\beta & \alpha/\beta^2 \end{pmatrix}, \tag{11.35}$$

where $\Psi(\alpha)$ denotes the digamma function, $\Psi(\alpha) = \Gamma'(\alpha)/\Gamma(\alpha)$. Thus,

$$\mathcal{I}_1(\alpha, \beta)^{-1} = \frac{\beta^2}{\alpha\Psi'(\alpha) - 1} \begin{pmatrix} \alpha/\beta^2 & -1/\beta \\ -1/\beta & \Psi'(\alpha) \end{pmatrix}.$$

and for a sample of size n the CR lower bounds for the variances of unbiased estimators of α and β are therefore $\alpha/(n(\alpha\Psi'(\alpha) - 1))$ and $\beta^2\Psi'(\alpha)/(n(\alpha\Psi'(\alpha) - 1))$, respectively. Since the CR lower bound is the inverse of information, it can be said that $n(\alpha\Psi'(\alpha) - 1)/\alpha$ is the information for α when β is unknown, and similarly, that $n(\alpha\Psi'(\alpha) - 1)/\beta^2\Psi'(\alpha)$ is the information for β when α is unknown.

Compare the above to the case where α (the shape parameter) is known. Then $\mathcal{I}(\beta) = n\alpha/\beta^2$ and so the CR lower bound for estimation of β is $\beta^2/(n\alpha)$. In this case the MLE is $\widehat{\beta} = \overline{Y}/\alpha$ and it is unbiased with variance equal to the CR lower bound. Note that $\beta^2\Psi'(\alpha)/(n(\alpha\Psi'(\alpha) - 1)) \geq \beta^2/(n\alpha)$. This is suggesting that β is harder to estimate when α is not known.

In this example, it is seen that the CR lower bound for estimation of one parameter is reduced when it it is assumed that the value of the other parameter is known. It can be shown that this phenomenon holds more generally (Exercise 11.10).

11.8 Exercises

11.1 Prove (11.11) via slight modification to the proof of the CR inequality.

11.2 Let $Y_1, \ldots, Y_n$ be iid Pois(λ).

1. Verify that $\mathcal{I}(\lambda) = n/\lambda$.
2. Suppose that it is of interest to estimate $g(\lambda) = P(Y = 0) = \exp(-\lambda)$. Verify that the CR lower bound for unbiased estimation of $g(\lambda)$ is $\lambda\exp(-2\lambda)/n$.

11.3 Using either Equation (11.1) or (11.14), calculate $\mathcal{I}(\zeta)$ for a binomial experiment with n trials and log-odds ζ. That is, using the log-likelihood

$$l(\zeta; y) = y(\zeta - \log(1 + e^{\zeta})) - (n - y)\log(1 + e^{\zeta}).$$

11.4 Let Y be from a geometric distribution with density function

$$f(y) = p^y(1 - p)\,,\, y = 0, 1, 2, ...$$

where $0 < p < 1$. Show that $\mathcal{I}_1(p) = \frac{1}{p(1-p)^2}$.
You may find it helpful to utilize the knowledge that $E_p[Y] = \frac{p}{1-p}$.

11.5 Let Y be from the distribution with density

$$f(y; \beta) = \frac{\beta}{(y + \beta)^2}, \quad y \geq 0, \quad \beta > 0.$$

Show that $\mathcal{I}_1(\beta) = 1/(3\beta^2)$.

11.6 Let Y have a logistic(θ) distribution with density

$$f(y,\theta) = \frac{\exp(y-\theta)}{(1+\exp(y-\theta))^2}, \quad y \in \mathbb{R}, \quad \theta \in \mathbb{R}.$$

Show that $\mathcal{I}_1(\theta) = 1/3$.

11.7 Verify the Fisher information matrix, $\mathcal{I}_1(\alpha, \beta)$, in (11.35) for $Y \sim$ Gamma(α, β).

11.8 Let Y have an exponential (dispersion) family distribution with density $f(y; \psi, \phi)$ given by Equation (11.23).

1. Using the identities $E_{\boldsymbol{\theta}}[\frac{\partial l}{\partial \theta}] = 0$ and $E_{\boldsymbol{\theta}}[\frac{\partial^2 l}{\partial \theta^2}] + E_{\boldsymbol{\theta}}[(\frac{\partial l}{\partial \theta})^2] = 0$, show that

$$E_{\boldsymbol{\theta}}[Y] = b'(\psi) \quad \text{and} \quad \text{var}_{\boldsymbol{\theta}}(Y) = b''(\psi)a(\phi).$$

2. The binomial proportion $Y \sim \text{Bin}(n, p)/n$ has a distribution belonging to the exponential family (Example 7.2). Hence, use the above result to obtain the mean and variance of Y.

11.9 Repeat the information calculations in Section 11.7.2 using $[\mathcal{I}(\boldsymbol{\beta})]_{jk} = -E_{\boldsymbol{\theta}}\left[\frac{\partial^2 l}{\partial \beta_j \partial \beta_k}\right]$.

11.10 Suppose that $\boldsymbol{\theta} = (\boldsymbol{\psi}, \boldsymbol{\lambda})$ where $\boldsymbol{\theta} \in \mathbb{R}^s$, $\boldsymbol{\psi} \in \mathbb{R}^r$ and $\boldsymbol{\lambda} \in \mathbb{R}^{s-r}$. Denote the expected Fisher information as

$$\mathcal{I}(\boldsymbol{\theta}) = \mathcal{I}(\boldsymbol{\psi}, \boldsymbol{\lambda}) = \begin{pmatrix} \mathcal{I}_{\psi\psi} & \mathcal{I}_{\psi\lambda} \\ \mathcal{I}_{\lambda\psi} & \mathcal{I}_{\lambda\lambda} \end{pmatrix},$$

where $\mathcal{I}_{\psi\psi}$ is $r \times r$, $\mathcal{I}_{\lambda\lambda}$ is $(s-r) \times (s-r)$, and $\mathcal{I}_{\psi\lambda} = \mathcal{I}_{\lambda\psi}^T$ is $r \times (s-r)$. Assume that $\mathcal{I}(\boldsymbol{\theta})$ is positive definite, with inverse denoted by

$$\mathcal{I}(\boldsymbol{\theta})^{-1} = \mathcal{I}(\boldsymbol{\psi}, \boldsymbol{\lambda})^{-1} = \begin{pmatrix} \mathcal{I}^{\psi\psi} & \mathcal{I}^{\psi\lambda} \\ \mathcal{I}^{\lambda\psi} & \mathcal{I}^{\lambda\lambda} \end{pmatrix}.$$

1. By expanding $\mathcal{I}(\boldsymbol{\psi}, \boldsymbol{\lambda})^{-1}\mathcal{I}(\boldsymbol{\psi}, \boldsymbol{\lambda}) = \mathbf{I}_s$ (the $s \times s$ identity matrix), show that

$$\mathcal{I}^{\psi\psi} = \left(\mathcal{I}_{\psi\psi} - \mathcal{I}_{\psi\lambda}\mathcal{I}_{\lambda\lambda}^{-1}\mathcal{I}_{\lambda\psi}\right)^{-1}.$$

2. Hence, show that $\mathcal{I}^{\psi\psi} \geq \mathcal{I}_{\psi\psi}^{-1}$ where $\geq$ is defined using the Löwner ordering. That is, for any vector $\boldsymbol{h} \in \mathbb{R}^r$, show that

$$\boldsymbol{h}^T\mathcal{I}^{\psi\psi}\boldsymbol{h} \geq \boldsymbol{h}^T\mathcal{I}_{\psi\psi}^{-1}\boldsymbol{h}.$$

This establishes that the CR lower bound for unbiased estimation of $\boldsymbol{\psi}$ is higher (or equal) in the presence of additional parameters $\boldsymbol{\lambda}$, compared to the case where $\boldsymbol{\lambda}$ is known.

Hint: If $\mathbf{A}$ and $\mathbf{B}$ are $r \times r$ matrices that are symmetric and positive definite, and

$$\boldsymbol{h}^T\mathbf{A}\boldsymbol{h} \geq \boldsymbol{h}^T\mathbf{B}\boldsymbol{h} \quad \text{for all } \boldsymbol{h} \in \mathbb{R}^r,$$

then

$$\boldsymbol{h}^T\mathbf{A}^{-1}\boldsymbol{h} \leq \boldsymbol{h}^T\mathbf{B}^{-1}\boldsymbol{h} \quad \text{for all } \boldsymbol{h} \in \mathbb{R}^r.$$

12

Asymptotic theory and approximate normality

... *if the difference between n and n* $-$ 1 *ever matters to you, then you are probably up to no good anyway* ... – Press *et al.* (2007)

12.1 Introduction

The central-limit theorem for maximum likelihood estimators is derived in Section 12.2 for the case of iid data and $\theta \in \mathbb{R}$. This is obtained in three steps, via Lemmas 12.1, 12.2 and Theorem 12.1. In plain language, the first lemma shows that, for sufficiently large sample size, the likelihood will be greater when evaluated at the true unknown parameter value θ_0 than at any other pre-specified parameter value. The second lemma uses this result to establish that MLEs are consistent estimators, at least in cases where the MLE is unique (with high probability). Finally, Theorem 12.1 uses a Taylor series expansion of the derivative of the log-likelihood function (i.e. the score function) to obtain the asymptotic normality of a $\sqrt{n}$-standardized sequence of maximum likelihood estimators. The result is stated for multi-parameter models, $\boldsymbol{\theta} \in \mathbb{R}^s$ (Section 12.2.2), and generalized to functions $g(\boldsymbol{\theta}) : \mathbb{R}^s \rightarrow \mathbb{R}^p$ (Section 12.2.3), misspecified models (Section 12.2.4), M-estimators (Section 12.2.5), and to the non-iid data case (Section 12.2.6).

Section 12.3 provides a translation of the central-limit result into a more pragmatic interpretation based on approximate normality, and shows how this can be put to good use. Examples are used to caution that approximate normality of the MLE does not in any way guarantee that the MLE inherits the usual properties of

Maximum Likelihood Estimation and Inference: With Examples in R, SAS and ADMB, First Edition. Russell B. Millar.

the normal distribution. For example, although the distribution of the MLE may be well approximated by a normal, it may not have an expected value (Example 12.6), or may not exist with some small positive probability (Example 12.2).

Section 12.4 looks at the construction of asymptotically correct hypothesis tests and confidence intervals/regions. These are asymptotically correct in the sense that they have approximately the desired level or coverage for sufficiently large sample size. Section 12.5 presents the theory of likelihood ratio tests and confidence intervals, and Section 12.6 provides brief coverage of the lesser-used Rao-score statistic.

Throughout this chapter, n will be used to denote sample size, and will be used as a subscript on estimators. For example, T_n in Definition 12.1 denotes an estimator based on a sample of size n. In particular, $\widehat{\theta}_n$ in Lemma 12.2 and Theorem 12.1 denotes the ML estimator obtained as a root of the likelihood equation from a sample of size n.

12.2 Consistency and asymptotic normality

The central-limit theorem for maximum likelihood estimators requires that they be *consistent*, that is, that the MLE converges to the true parameter value in probability. The formal definition of a consistent estimator is given below. In this definition and in Lemma 12.1, the parameter is assumed to be in $\mathbb{R}^s$. Restriction to the scalar-parameter case is required for Lemma 12.2 and Theorem 12.1.

Definition 12.1 Consistent estimator. Let $T_n \equiv T_n(Y_1, \ldots, Y_n)$ denote a sequence (indexed by sample size $n = 1, 2, 3, ...$) of estimators of $g(\boldsymbol{\theta}) \in \mathbb{R}$. The estimator (sequence) T_n is said to be consistent if for every $\boldsymbol{\theta}$ in Θ

$$T_n \rightarrow_{P_\theta} g(\boldsymbol{\theta}),$$

where $\rightarrow_{P_\theta}$ denotes convergence in probability when $\boldsymbol{\theta}$ is the true parameter, i.e. for every $\epsilon > 0$, $P_{\boldsymbol{\theta}}(|T_n - g(\boldsymbol{\theta})| > \epsilon) \rightarrow 0$ as $n \rightarrow \infty$. An estimator that is not consistent is said to be inconsistent.

The following regularity conditions are required in the statement of Lemma 12.1.

R1: The observations $\boldsymbol{Y} = (Y_1, \ldots, Y_n)$ are iid from a distribution having density function $f(y; \boldsymbol{\theta})$ within the parametric statistical model indexed by $\boldsymbol{\theta}$, $\{f(y; \boldsymbol{\theta}) : \boldsymbol{\theta} \in \Theta\}$.

R2: The statistical model is identifiably parameterized. That is, the statistical distributions corresponding to the density functions $f(y; \boldsymbol{\theta})$, $\boldsymbol{\theta} \in \Theta$ are distinct.

R3: The distributions in the statistical model have common support. That is, the set $\mathbf{S} = \{y : f(y; \boldsymbol{\theta}) > 0\}$ does not depend on $\boldsymbol{\theta}$.

The following lemma shows that, for sufficiently large sample size, the likelihood of the true parameter value, $\boldsymbol{\theta}_0$, is greater than that of any other fixed $\boldsymbol{\theta}$ (with high probability). In this lemma, use of the $\boldsymbol{\theta}_0$ subscript in the notation $P_{\boldsymbol{\theta}_0}$ and $\rightarrow_{P_{\boldsymbol{\theta}_0}}$ is used to explicitly denote that the probabilistic statements are being made with regard to the sampling distribution of $\boldsymbol{Y} = (Y_1, \ldots, Y_n)$ when $\boldsymbol{\theta}_0$ is the true parameter value. For ease of notation, this explicit subscribing is dropped in later sections, but is retained in notation for the expectation and variance, $E_{\boldsymbol{\theta}_0}$ and $\mathrm{var}_{\boldsymbol{\theta}_0}$, where appropriate.

Lemma 12.1. $\boldsymbol{\theta}_0$ is best.
If regularity conditions R1-R3 hold, then

$$P_{\boldsymbol{\theta}_0}\big(L(\boldsymbol{\theta}_0; \boldsymbol{Y}) > L(\boldsymbol{\theta}; \boldsymbol{Y})\big) \rightarrow 1 \ \ as \ \ n \rightarrow \infty, \tag{12.1}$$

for any fixed $\boldsymbol{\theta}$ not equal to the true parameter value $\boldsymbol{\theta}_0$.

Proof. The statement in (12.1) is

$$P_{\boldsymbol{\theta}_0}\left(\prod_{i=1}^{n} f(Y_i; \boldsymbol{\theta}_0) > \prod_{i=1}^{n} f(Y_i; \boldsymbol{\theta})\right) \rightarrow 1,$$

and this can be re-expressed as

$$P_{\boldsymbol{\theta}_0}\left(\sum_{i=1}^{n} \log\left(\frac{f(Y_i; \boldsymbol{\theta})}{f(Y_i; \boldsymbol{\theta}_0)}\right) < 0\right) \rightarrow 1,$$

or equivalently,

$$P_{\boldsymbol{\theta}_0}\left(\frac{1}{n}\sum_{i=1}^{n} \log\left(\frac{f(Y_i; \boldsymbol{\theta})}{f(Y_i; \boldsymbol{\theta}_0)}\right) < 0\right) \rightarrow 1. \tag{12.2}$$

Now, the term on the left-hand side of the inequality in (12.2) is the average of n iid terms, and by the weak law of large numbers (Example 13.8), it converges in probability to its expected value. That is,

$$\frac{1}{n}\sum_{i=1}^{n} \log\left(\frac{f(Y_i; \boldsymbol{\theta})}{f(Y_i; \boldsymbol{\theta}_0)}\right) \rightarrow_{P_{\boldsymbol{\theta}_0}} E_{\boldsymbol{\theta}_0}\left[\log\left(\frac{f(Y; \boldsymbol{\theta})}{f(Y; \boldsymbol{\theta}_0)}\right)\right].$$

The negative of the log function is convex, and application of Jensen's inequality (Section 13.5.1) gives

$$E_{\boldsymbol{\theta}_0}\left[\log\left(\frac{f(Y; \boldsymbol{\theta})}{f(Y; \boldsymbol{\theta}_0)}\right)\right] < \log E_{\boldsymbol{\theta}_0}\left[\frac{f(Y; \boldsymbol{\theta})}{f(Y; \boldsymbol{\theta}_0)}\right]. \tag{12.3}$$

Now, $E_{\boldsymbol{\theta}_0}$ denotes expectation under the distribution with density function $f(y, \boldsymbol{\theta}_0)$, and so (12.3) reduces to

$$\log E_{\boldsymbol{\theta}_0}\left[\frac{f(Y;\boldsymbol{\theta})}{f(Y;\boldsymbol{\theta}_0)}\right] = \log \int \frac{f(y;\boldsymbol{\theta})}{f(y;\boldsymbol{\theta}_0)} f(y;\boldsymbol{\theta}_0)dy$$
$$= \log 1 = 0.$$

Thus, it has been shown that $\frac{1}{n}\sum_{i=1}^{n}\log\left(\frac{f(Y_i;\boldsymbol{\theta})}{f(Y_i;\boldsymbol{\theta}_0)}\right)$ converges in probability to a negative quantity, from which it follows that it is negative with probability tending to unity (see Exercise 13.3), which establishes (12.2). □

The next lemma will assume the following additional regularity condition.

R4: The parameter space Θ is a subset of $\mathbb{R}$, and θ_0 is not on the boundary of Θ. That is, the interval $(\theta_0 - a, \theta_0 + a)$ is entirely contained within Θ for all sufficiently small $a > 0$. (This regularity condition will always be satisfied if Θ is an open subset of $\mathbb{R}$.)

Lemma 12.2. Consistency of $\widehat{\theta}_n$.
Assume that regularity conditions R1 - R4 hold, and that $f(y;\theta)$ is differentiable with respect to θ, with derivative $f'(y;\theta)$. Then, with probability tending to one as $n \to \infty$, the likelihood equation

$$\frac{\partial \log L(\theta;\boldsymbol{Y})}{\partial\theta} \equiv \frac{\partial l(\theta;\boldsymbol{Y})}{\partial\theta}$$
$$= \frac{\partial \sum_{i=1}^{n}\log f(Y_i;\theta)}{\partial\theta}$$
$$= \sum_{i=1}^{n}\frac{f'(Y_i;\theta)}{f(Y_i;\theta)} = 0,$$

has a root $\widehat{\theta}_n \equiv \widehat{\theta}_n(\boldsymbol{Y})$ that tends to the true value θ_0 in probability.

Proof. For any sufficiently small choice of a (see condition R4), consider the shape of the log-likelihood on the interval $(\theta_0 - a, \theta_0 + a)$. By virtue of the differentiability of the likelihood, and hence log-likelihood, it is necessarily the case that the log-likelihood has a turning point (i.e. a root of the likelihood equation) in the interval $(\theta_0 - a, \theta_0 + a)$ whenever the log-likelihood is higher at θ_0 then at both $\theta_0 - a$ and $\theta_0 + a$. By Lemma 12.1, the probability of this tends to one because it is the intersection of the event $\{l(\theta_0;\boldsymbol{Y}) > l(\theta_0 - a;\boldsymbol{Y})\}$ and the event $\{l(\theta_0;\boldsymbol{Y}) > l(\theta_0 + a;\boldsymbol{Y})\}$, both of which have probability tending to one as $n \to \infty$ (see Exercise 12.1)

If the likelihood equation has multiple roots in $(\theta_0 - a, \theta_0 + a)$ then take $\widehat{\theta}_n$ to be the root closest to θ_0 (this removes the dependence of $\widehat{\theta}_n$ on a.) This establishes that, for arbitrarily small choice of a, and with probability $\to 1$, there exists a root

$\widehat{\theta}_n$ such that $|\widehat{\theta}_n - \theta_0| < a$. That is, $P_{\theta_0}(|\widehat{\theta}_n - \theta_0| > a) \to 0$, which establishes that $\widehat{\theta}_n$ converges in probability to θ_0. □

Lemma 12.2 comes very close to establishing the existence of a consistent sequence of roots $\widehat{\theta}_n$ of the likelihood equation, but stops just short of this. Note that a root of the likelihood equation need not exist for every possible outcome $\mathbf{y}$. However, a root does exist with probability tending to unity, and so a consistent sequence of estimators $\widehat{\theta}_n^*$ can be obtained as

$$\widehat{\theta}_n^* = \begin{cases} \widehat{\theta}_n, & \text{if a root exists} \\ \widetilde{\theta}, & \text{otherwise,} \end{cases} \tag{12.4}$$

where $\widetilde{\theta}$ is any arbitrary value in the parameter space Θ. As in the proof of Lemma 12.2, if the likelihood equation has multiple roots then $\widehat{\theta}_n$ in (12.4) is taken to be the root closest to θ_0. If a unique root exists with probability tending to one as $n \to \infty$, then consistency of $\widehat{\theta}_n^*$ will be assured for any choice of root of the likelihood equation.

The next two examples demonstrate the strange behaviour that the likelihood can exhibit on subsets of the sample space having zero probability or probability tending to zero.

Example 12.1. No root of iid $N(0, \sigma^2)$ likelihood equation. Let $Y_1, ..., Y_n$ be iid $N(0, \sigma^2)$ with parameter space Θ given by the open interval defined by $\sigma > 0$. If all Y_i are identically zero then there is no root of the likelihood equation in Θ and the likelihood is unbounded as σ is made arbitrarily small. This misbehaviour of the likelihood is vacuous in practice, because the outcome $Y_i = 0, i = 1, \ldots, n$, is a zero-probability event. □

Example 12.2. No root of binomial likelihood equation for the log-odds. Let $B_1, ..., B_n$ be iid Bernoulli with $P(B_i = 1) = p,\ 0 < p < 1$, with the model parameterized by the log-odds, $\zeta = \text{logit}(p) = \log(p/(1-p)) \in \Theta = \mathbb{R}$. Letting $Y_n = \sum B_i$ denote the Bin(n, p) distributed sum of the n Bernoulli observations, the likelihood for ζ given the observation y_n is

$$\begin{aligned} L(\zeta; y_n) &\propto p^{y_n}(1-p)^{n-y_n} \\ &= \left(\frac{e^\zeta}{1+e^\zeta}\right)^{y_n} \left(\frac{1}{1+e^\zeta}\right)^{n-y_n} \\ &= \frac{e^{\zeta y_n}}{(1+e^\zeta)^n}. \end{aligned}$$

If y_n is other than 0 or n then the likelihood has a root at $\widehat{\zeta}_n = \text{logit}(\widehat{p}_n)$ where $\widehat{p}_n = y_n/n$. However, if y_n is 0 or n then this likelihood does not possess a root in $\mathbb{R}$. For example, if $y_n = 0$ then $L(\zeta; 0) = (1 + e^\zeta)^{-n}$ is monotone decreasing on $\mathbb{R}$. However, since $0 < p < 1$, the probability that Y_n equals 0 or n tends to zero as n increases. Thus, with probability tending to one, the root $\widehat{\zeta}_n$ exists, and this sequence of roots converges in probability to the true value $\zeta_0 = \text{logit}(p_0)$, where p_0 is the true value of p. □

12.2.1 Asymptotic normality, $\theta \in \mathbb{R}$

Establishing the asymptotic normality of maximum likelihood estimators requires further regularity conditions on the statistical model.

R5: The integral $\int f(y;\theta)dy$ can be twice differentiated (with respect to θ) under the integral sign.

R6: $0 < \mathcal{I}_1(\theta) < \infty$, where $\mathcal{I}_1(\theta)$ is the expected Fisher information about θ from a single observation of Y.

R7: The third derivative of $f(y;\theta)$ exists and satisfies

$$\left| \frac{\partial^3 \log f(y;\theta)}{\partial\theta^3} \right| \leq M(y),$$

for all $y \in \mathbf{S}$ (the support set) and all θ in some interval $(\theta_0 - c, \theta_0 + c)$, and $E_{\theta_0}[M(Y)] < \infty$. (This cumbersome condition is required to be able to ignore the remainder term of the Taylor series expansion used in the proof of Theorem 12.1 below.)

Theorem 12.1 Asymptotic normality of $\widehat{\theta}_n \in \mathbb{R}$.
If θ_0 is the true parameter value and regularity conditions R1–R7 hold, then any consistent sequence $\widehat{\theta}_n \equiv \widehat{\theta}_n(Y_1, \ldots, Y_n)$ of roots of the likelihood equation satisfies

$$\sqrt{n}(\widehat{\theta}_n - \theta_0) \rightarrow_D N\left(0, \frac{1}{\mathcal{I}_1(\theta_0)}\right). \tag{12.5}$$

Proof. Expand the score function $l'(\theta) = \frac{\partial l(\theta;Y)}{\partial\theta}$ about θ_0 using a Taylor series with Lagrange's form of remainder (Apostol 1967), and evaluate at the root of the likelihood equation, $\widehat{\theta}_n$. This gives

$$\begin{aligned} 0 &= l'(\widehat{\theta}_n) \\ &= l'(\theta_0) + (\widehat{\theta}_n - \theta_0)l''(\theta_0) + \frac{1}{2}(\widehat{\theta}_n - \theta_0)^2 l'''(\theta_n^*), \end{aligned} \tag{12.6}$$

where θ_n^* is some value between θ_0 and $\widehat{\theta}_n$. After some re-arrangement, this gives

$$\sqrt{n}(\widehat{\theta}_n - \theta_0) = \frac{(1/\sqrt{n})l'(\theta_0)}{-(1/n)l''(\theta_0) - (1/2n)(\widehat{\theta}_n - \theta_0)l'''(\theta_n^*)}. \tag{12.7}$$

The required result is obtained after showing that

$$\frac{1}{\sqrt{n}}l'(\theta_0) \rightarrow_D N(0, \mathcal{I}_1(\theta_0)), \tag{12.8}$$

and

$$-\frac{1}{n}l''(\theta_0) \rightarrow_P \mathcal{I}_1(\theta_0), \tag{12.9}$$

and

$$\frac{1}{n}(\widehat{\theta}_n - \theta_0)l'''(\theta_n^*) \rightarrow_P 0. \tag{12.10}$$

From (12.9) and (12.10), application of Slutsky's theorem (Section 13.6.3) establishes that the denominator in (12.7) converges in probability to $\mathcal{I}_1(\theta_0)$. That is,

$$-\frac{1}{n}l''(\theta_0) - \frac{1}{2n}(\widehat{\theta}_n - \theta_0)l'''(\theta_n^*) \rightarrow_P \mathcal{I}_1(\theta_0).$$

Thus, (12.7) can be expressed as A_n/B_n where A_n converges in distribution to $N(0, \mathcal{I}_1(\theta_0))$ and $1/B_n$ converges in probability to $1/\mathcal{I}_1(\theta_0)$. A second application of Slutsky's theorem then gives

$$\sqrt{n}(\widehat{\theta}_n - \theta_0) \rightarrow_D \frac{1}{\mathcal{I}_1(\theta_0)}N(0, \mathcal{I}_1(\theta_0)) = N\left(0, \frac{1}{\mathcal{I}_1(\theta_0)}\right),$$

as required.

The following arguments establish (12.8)–(12.10). In particular, (12.8) and (12.9) make use of the fact that the expected Fisher information $\mathcal{I}_1(\theta_0)$ can be calculated as the variance of $l'(\theta_0; Y)$, or as the negative of the expected value of $l''(\theta_0; Y)$ (see Equation (11.13) and (11.14), respectively).

- The left-hand side of (12.8) can be written

  $$\begin{aligned}\frac{1}{\sqrt{n}}l'(\theta_0) &= \frac{1}{\sqrt{n}}\sum_{i=1}^{n} l'(\theta_0, Y_i)\\ &= \sqrt{n}\left(\frac{1}{n}\sum_{i=1}^{n} l'(\theta_0, Y_i)\right),\end{aligned}$$

  where the terms in the summation are iid with mean 0 (from Lemma 11.1). It follows from the central limit theorem (Example 13.6) that

  $$\begin{aligned}\frac{1}{\sqrt{n}}l'(\theta_0) &\rightarrow_D N\left(0, \mathrm{var}_{\theta_0}\left(l'(\theta_0, Y)\right)\right)\\ &= N(0, \mathcal{I}_1(\theta_0)).\end{aligned} \tag{12.11}$$

- The left-hand side of (12.9) is

$$-\frac{1}{n}l''(\theta_0) = -\frac{1}{n}\sum_{i=1}^{n} l''(\theta_0, Y_i),$$

where each term in the summation is iid. It follows from the weak law of large numbers that

$$\begin{aligned} -\frac{1}{n}l''(\theta_0) &\to_P -E_{\theta_0}[l''(\theta_0, Y)] \\ &= \mathcal{I}_1(\theta_0). \end{aligned} \qquad (12.12)$$

- Finally, to show (12.10), first note that $\theta_0 - c < \theta_n^* < \theta_0 + c$ with probability tending to 1, since $\widehat{\theta}_n$ converges in probability to θ_0. Hence, from condition R7,

$$\left|\frac{1}{n}l'''(\theta_n^*)\right| = \frac{1}{n}\left|\sum_{i=1}^{n} l'''(\theta_n^*, Y_i)\right| \leq \frac{1}{n}\sum_{i=1}^{n} M(Y_i),$$

with probability tending to 1 also. By the weak law of large numbers

$$\frac{1}{n}\sum_{i=1}^{n} M(Y_i) \to_P E_{\theta_0}[M(Y)] < \infty,$$

and hence $l'''(\theta_n^*)/n$ is bounded in probability. That is, there exists $B < \infty$ (take B to be any value greater than $E_{\theta_0}[M(Y)]$) such that

$$P\left(\left|\frac{1}{n}l'''(\theta_n^*)\right| > B\right) \to 0 \text{ as } n \to \infty. \qquad (12.13)$$

Since $\widehat{\theta}_n - \theta_0 \to_P 0$, Equation (12.10) follows (see Exercise 13.4). □

Example 12.3. Asymptotic distribution of the binomial proportion. Let $B_1, ..., B_n$ be iid Bernoulli with $P(B_i = 1) = p,\ 0 < p < 1$, and let $\widehat{p}_n = \frac{1}{n}\sum B_i$ denote the ML estimator of p. The expected Fisher information from a single Bernoulli observation is $\mathcal{I}_1(p) = 1/(p(1-p))$ (Example 11.1), and so, with p_0 denoting the true unknown value of p, Theorem 12.1 gives

$$\sqrt{n}(\widehat{p}_n - p_0) \to_D N\left(0, \frac{1}{\mathcal{I}_1(p_0)}\right) = N(0, p_0(1-p_0)).$$

Figure 13.2 provides a graphical demonstration of this convergence, for $p_0 = 0.5$ and for n taking the values 10, 100, 1000 and 10000. □

Example 12.2 continued. Binomial model indexed by log-odds. From Example 11.2, $\mathcal{I}_1(\zeta) = p(1-p)$ is the expected Fisher information about $\zeta = \text{logit}(p)$ from a single Bernoulli observation. From Theorem 12.1, with $\zeta_0 = \text{logit}(p_0)$ denoting the true value of ζ,

$$\sqrt{n}(\widehat{\zeta}_n - \zeta_0) \to_D N\left(0, \frac{1}{p_0(1-p_0)}\right), \tag{12.14}$$

where $\widehat{\zeta}_n = \text{logit}(\widehat{p}_n)$ is the MLE, provided that $\widehat{p}_n$ is not equal to 0 or 1. For $\widehat{\zeta}_n$ to be a properly defined estimator it is enough to set $\widehat{\zeta}_n$ to any arbitrary real number when $\widehat{p}_n$ equals 0 or 1. These events have probability tending to zero as $n \to \infty$ (since $0 < p_0 < 1$), and hence the estimator $\widehat{\zeta}_n$ so defined is a consistent estimator, and (12.14) holds. □

12.2.2 Asymptotic normality: $\boldsymbol{\theta} \in \mathbb{R}^s$

Extending the consistency and asymptotic normality results to the multiparameter case, $\boldsymbol{\theta} \in \Theta \subset \mathbb{R}^s$ is reasonably straightforward (but extremely tedious) and so only the results are stated below.

Recall from Section 11.6 that the $s \times s$-dimensional expected Fisher information matrix $\mathcal{I}(\boldsymbol{\theta})$ has (i, j) element

$$\begin{aligned}\mathcal{I}(\boldsymbol{\theta})_{ij} &= E_{\boldsymbol{\theta}}\left[\frac{\partial l(\boldsymbol{\theta};\boldsymbol{Y})}{\partial \theta_i}\frac{\partial l(\boldsymbol{\theta};\boldsymbol{Y})}{\partial \theta_j}\right] \\ &= \text{cov}_{\boldsymbol{\theta}}\left(\frac{\partial l(\boldsymbol{\theta};\boldsymbol{Y})}{\partial \theta_i}, \frac{\partial l(\boldsymbol{\theta};\boldsymbol{Y})}{\partial \theta_j}\right),\end{aligned}$$

and provided the likelihood can be twice differentiated under the integral sign, it can also be obtained as

$$\mathcal{I}(\boldsymbol{\theta})_{ij} = -E_{\boldsymbol{\theta}}\left[\frac{\partial^2 l(\boldsymbol{\theta};\boldsymbol{Y})}{\partial \theta_i \partial \theta_j}\right]. \tag{12.15}$$

Theorem 12.2 Asymptotic normality of $\widehat{\boldsymbol{\theta}}_n \in \mathbb{R}^s$. Under appropriate conditions (similar to those of the one-dimensional parameter case), any consistent sequence $\widehat{\boldsymbol{\theta}}_n \equiv \widehat{\boldsymbol{\theta}}_n(Y_1, \ldots, Y_n)$ of roots of the likelihood equations satisfies

$$\sqrt{n}(\widehat{\boldsymbol{\theta}}_n - \boldsymbol{\theta}_0) \to_D N_s\left(\mathbf{0}, \mathcal{I}_1(\boldsymbol{\theta}_0)^{-1}\right), \tag{12.16}$$

where $\mathcal{I}_1(\boldsymbol{\theta})$ is the expected Fisher information (matrix) about $\boldsymbol{\theta}_0$ from one observation.

The convergence result in (12.16) can be stated in normalized form, as

$$\mathcal{I}(\boldsymbol{\theta}_0)^{\frac{1}{2}}(\widehat{\boldsymbol{\theta}}_n - \boldsymbol{\theta}_0) \to_D N_s(\mathbf{0}, \mathbf{I}_s), \tag{12.17}$$

where $\mathcal{I}(\boldsymbol{\theta}_0)^{\frac{1}{2}}$ denotes the positive definite matrix satisfying $\mathcal{I}(\boldsymbol{\theta}_0) = \mathcal{I}(\boldsymbol{\theta}_0)^{\frac{1}{2}}\,\mathcal{I}(\boldsymbol{\theta}_0)^{\frac{1}{2}}$ (see Appendix A.4 of Seber and Lee 2003), and $\mathbf{I}_s$ is the s-dimensional identity matrix. Note that $\mathcal{I}(\boldsymbol{\theta}_0)^{\frac{1}{2}} = \sqrt{n}\,\mathcal{I}_1(\boldsymbol{\theta}_0)^{\frac{1}{2}}$.

Example 12.4. IID $N(\mu, \sigma^2)$ model. The expected Fisher information matrix about $\boldsymbol{\theta} = (\mu, \sigma^2)$ from a single observation $Y \sim N(\mu, \sigma^2)$ was obtained in Section 11.7.1. Hence, letting $(\overline{Y}_n, \widehat{\sigma}_n^2)$ denote the ML estimator from an iid sample of size n,

$$\sqrt{n}\left(\begin{pmatrix}\overline{Y}_n \\ \widehat{\sigma}_n^2\end{pmatrix} - \begin{pmatrix}\mu_0 \\ \sigma_0^2\end{pmatrix}\right) \to_D N_2\left(\begin{pmatrix}0 \\ 0\end{pmatrix}, \begin{pmatrix}\sigma_0^2 & 0 \\ 0 & 2\sigma_0^4\end{pmatrix}\right),$$

where (μ_0, σ_0^2) is the true parameter value.

Of course, for this model it is not necessary to utilize the above asymptotic normality since the exact distribution of $(\overline{Y}_n, \widehat{\sigma}_n^2)$ is known (Seber and Lee 2003, p. 48). In particular, $\overline{Y}_n$ and $\widehat{\sigma}_n^2$ are independent, and

$$\begin{aligned}\sqrt{n}(\overline{Y}_n - \mu_0) &\sim N(0, \sigma_0^2) \\ \sqrt{n}(\widehat{\sigma}_n^2 - \sigma_0^2) &\sim \sqrt{n}\sigma_0^2\left(\frac{\chi_{n-1}^2}{n} - 1\right).\end{aligned} \tag{12.18}$$

The convergence in distribution of $\sqrt{n}(\widehat{\sigma}_n^2 - \sigma_0^2)$ to $N(0, 2\sigma_0^4)$ can be established directly from (12.18) (Exercise 13.8). □

Example 12.5. Orthogonal parameters. Let the parameter vector $\boldsymbol{\theta}$ be partitioned as $\boldsymbol{\theta} = (\boldsymbol{\psi}, \boldsymbol{\lambda})$ where $\boldsymbol{\psi} \in \mathbb{R}^r$ and $\boldsymbol{\lambda} \in \mathbb{R}^{s-r}$, and assume that $\boldsymbol{\psi}$ and $\boldsymbol{\lambda}$ are orthogonal (Section 11.6.1). That is, the information matrix $\mathcal{I}_1(\boldsymbol{\theta})$ is block diagonal

$$\mathcal{I}_1(\boldsymbol{\theta}) = \mathcal{I}_1(\boldsymbol{\psi}, \boldsymbol{\lambda}) = \begin{pmatrix}\mathcal{I}_{1,\boldsymbol{\psi}\boldsymbol{\psi}} & \mathbf{0} \\ \mathbf{0} & \mathcal{I}_{1,\boldsymbol{\lambda}\boldsymbol{\lambda}}\end{pmatrix},$$

where $\mathcal{I}_{1,\boldsymbol{\psi}\boldsymbol{\psi}}$ is the $r \times r$ matrix

$$\mathcal{I}_{1,\boldsymbol{\psi}\boldsymbol{\psi}} = E_{\boldsymbol{\theta}}\left[\frac{\partial l(\boldsymbol{\theta}; Y)}{\partial \boldsymbol{\psi}}\frac{\partial l(\boldsymbol{\theta}; Y)}{\partial \boldsymbol{\psi}}^T\right],$$

and analogously for the $(s-r) \times (s-r)$ matrix $\mathcal{I}_{1,\boldsymbol{\lambda}\boldsymbol{\lambda}}$. Then

$$\mathcal{I}_1(\boldsymbol{\theta})^{-1} = \begin{pmatrix}\mathcal{I}_{1,\boldsymbol{\psi}\boldsymbol{\psi}}^{-1} & \mathbf{0} \\ \mathbf{0} & \mathcal{I}_{1,\boldsymbol{\lambda}\boldsymbol{\lambda}}^{-1}\end{pmatrix},$$

and it follows that

$$\sqrt{n}(\widehat{\boldsymbol{\psi}}_n - \boldsymbol{\psi}_0) \to_D N_r(\mathbf{0}, \mathcal{I}_{1,\boldsymbol{\psi}\boldsymbol{\psi}}^{-1}) \tag{12.19}$$

$$\sqrt{n}(\widehat{\boldsymbol{\lambda}}_n - \boldsymbol{\lambda}_0) \to_D N_{s-r}(\mathbf{0}, \mathcal{I}_{1,\boldsymbol{\lambda}\boldsymbol{\lambda}}^{-1}). \tag{12.20}$$

Note that the asymptotic variance of $\sqrt{n}(\widehat{\boldsymbol{\lambda}}_n - \boldsymbol{\lambda}_0)$ in (12.20) is the same as would be obtained if $\boldsymbol{\psi}$ were known. That is, by virtue of parameter orthogonality, the asymptotic distribution of $\widehat{\boldsymbol{\lambda}}_n$ does not depend on whether $\boldsymbol{\psi}$ is known or unknown. Similarly, the asymptotic distribution of $\widehat{\boldsymbol{\psi}}_n$ does not depend on whether $\boldsymbol{\lambda}$ is known or unknown. □

Example 12.4 continued. In the $N(\mu, \sigma^2)$ model, parameters μ and σ^2 are orthogonal, and hence, for example, the asymptotic distribution of $\sqrt{n}(\widehat{\sigma}_n^2 - \sigma_0^2)$ is unchanged if μ were known. This can be verified directly (see Exercise 13.8), since $\widehat{\sigma}_n^2 = \sum(Y_i - \mu)^2/n$ when μ is known. □

12.2.3 Asymptotic normality of $g(\boldsymbol{\theta}) \in \mathbb{R}^p$

Consider inference about $g(\boldsymbol{\theta}) : \mathbb{R}^s \to \mathbb{R}^p$, where $g(\boldsymbol{\theta})$ is differentiable at $\boldsymbol{\theta}_0 \in \mathbb{R}^s$ with $p \times s$ Jacobian matrix $\mathbf{G}(\boldsymbol{\theta}_0)$. Then, application of the delta theorem (Section 13.6.4) to (12.16) gives the convergence result

$$\begin{aligned} \sqrt{n}\left(g(\widehat{\boldsymbol{\theta}}_n) - g(\boldsymbol{\theta}_0)\right) &\to_D \mathbf{G}(\boldsymbol{\theta}_0)N_s(\mathbf{0}, \mathcal{I}_1(\boldsymbol{\theta}_0)^{-1}) \\ &= N_p(\mathbf{0}, \mathbf{G}(\boldsymbol{\theta}_0)\mathcal{I}_1(\boldsymbol{\theta}_0)^{-1}\mathbf{G}(\boldsymbol{\theta}_0)^T). \end{aligned} \tag{12.21}$$

Example 12.6. Coefficient of variation. For Y_i distributed iid $N(\mu, \sigma^2)$, the asymptotic convergence of the MLE was given in Example 12.4. Now, suppose that the asymptotic distribution of the ML estimator of the coefficient of variation, σ/μ, is of interest. Here $g(\mu, \sigma^2) = \sigma/\mu = \sqrt{\sigma^2}/\mu$ and so

$$g'(\mu, \sigma^2) \equiv \mathbf{G}(\mu, \sigma^2) = \left(-\frac{\sigma}{\mu^2}, \frac{1}{2\sigma\mu}\right).$$

Therefore,

$$\begin{aligned} \sqrt{n}\left(\frac{\widehat{\sigma}_n}{\overline{Y}_n} - \frac{\sigma_0}{\mu_0}\right) &\to_D N\left(0, \left(-\frac{\sigma_0}{\mu_0^2}, \frac{1}{2\sigma_0\mu_0}\right)\begin{pmatrix}\sigma_0^2 & 0\\ 0 & 2\sigma_0^4\end{pmatrix}\begin{pmatrix}-\frac{\sigma_0}{\mu_0^2}\\ \frac{1}{2\sigma_0\mu_0}\end{pmatrix}\right) \\ &= N\left(0, \frac{\sigma_0^2}{\mu_0^2}\left(\frac{\sigma_0^2}{\mu_0^2} + \frac{1}{2}\right)\right). \end{aligned}$$

Although the statistic $\sqrt{n}\left(\frac{\widehat{\sigma}_n}{\overline{Y}_n} - \frac{\sigma_0}{\mu_0}\right)$ converges in distribution, it can be shown that its expectation and variance do not exist. □

12.2.4 Asymptotic normality under model misspecification

Suppose that the statistical model is misspecified by working with the density function $\widetilde{f}(\boldsymbol{y};\boldsymbol{\theta})$ rather than the correct density function $f(\boldsymbol{y};\boldsymbol{\theta})$. Define $\widetilde{\boldsymbol{\theta}}_0 \in \Theta$ to be the parameter value such $\widetilde{f}(\boldsymbol{y};\widetilde{\boldsymbol{\theta}}_0)$ is closest to the true density function $f(\boldsymbol{y};\boldsymbol{\theta}_0)$.[1] That is

$$\Delta\big(\widetilde{f}(y;\widetilde{\boldsymbol{\theta}}_0),\, f(y;\boldsymbol{\theta}_0)\big) = \min_{\boldsymbol{\theta}\in\Theta} \Delta\big(\widetilde{f}(y;\boldsymbol{\theta}),\, f(y;\boldsymbol{\theta}_0)\big),$$

where it is assumed that $\widetilde{\boldsymbol{\theta}}_0$ exists and is unique. In some cases it may be that $\widetilde{\boldsymbol{\theta}}_0 = \boldsymbol{\theta}_0$ but this is not so in general.

Given observations $\boldsymbol{y}$, the likelihood function being maximized, $\widetilde{l}(\boldsymbol{\theta}) = \log \widetilde{f}(\boldsymbol{y};\boldsymbol{\theta})$, $\boldsymbol{\theta} \in \Theta$, is not the true likelihood function. However, by virtue of the definition of $\widetilde{\boldsymbol{\theta}}_0$, and subject to appropriate regularity conditions, it follows (see Pawitan 2001, pp. 370–374 for details) that

$$E_{\boldsymbol{\theta}_0}\left[\left.\frac{\partial \widetilde{l}(\boldsymbol{\theta};\boldsymbol{Y})}{\partial \boldsymbol{\theta}}\right|_{\boldsymbol{\theta}=\widetilde{\boldsymbol{\theta}}_0}\right] = \boldsymbol{0} \tag{12.22}$$

where $E_{\boldsymbol{\theta}_0}$ continues to denote expectation under the true model. Letting $\widetilde{\boldsymbol{\theta}}_n$ denote the MLE under maximization of $\widetilde{l}$, it can be shown that $\widetilde{\boldsymbol{\theta}}_n$ is a consistent estimator of $\widetilde{\boldsymbol{\theta}}_0$ (Pawitan 2001), and is asymptotically normal.

For $\theta \in \mathbb{R}$, the asymptotic normality of $\widetilde{\theta}_n$ can be established by repeating the steps in the proof of Theorem 12.1, but now using a Taylor series expansion of $\widetilde{l}'(\widetilde{\theta}_n)$ around $\widetilde{\theta}_0$. It is no longer the case that $\mathrm{var}_{\theta_0}\big(\widetilde{l}'(\widetilde{\theta}_0, Y)\big)$ and $-E_{\theta_0}\big[\widetilde{l}''(\widetilde{\theta}_0, Y)\big]$ are necessarily equal, and so using (12.11) and (12.12) in place of (12.8) and (12.9) gives

$$\sqrt{n}(\widetilde{\theta}_n - \widetilde{\theta}_0) \to_D N\left(0, \frac{\mathrm{var}_{\theta_0}\big(\widetilde{l}'(\widetilde{\theta}_0, Y)\big)}{E_{\theta_0}\big[\widetilde{l}''(\widetilde{\theta}_0, Y)\big]^2}\right). \tag{12.23}$$

In the multi-parameter case, $\boldsymbol{\theta} \in \mathbb{R}^s$, the convergence result becomes

$$\sqrt{n}(\widetilde{\boldsymbol{\theta}}_n - \widetilde{\boldsymbol{\theta}}_0) \to_D N_s\left(\boldsymbol{0}, \mathbf{A}_1^{-1}\mathbf{B}_1\mathbf{A}_1^{-1}\right), \tag{12.24}$$

where $\mathbf{A}_1$ and $\mathbf{B}_1$ are the symmetric $s \times s$-dimensional matrices

$$\mathbf{A}_1 = -E_{\boldsymbol{\theta}_0}\big[\widetilde{l}''(\widetilde{\boldsymbol{\theta}}_0, Y)\big], \tag{12.25}$$

[1] The difference is quantified using the Kullback-Leibler divergence (Kullback 1959), $\Delta(f_1, f_2) = \int f_1(y) \log\left(\frac{f_1(y)}{f_2(y)}\right) dy$.

and

$$\mathbf{B}_1 = \text{var}_{\boldsymbol{\theta}_0}\left(\widetilde{l}'(\boldsymbol{\theta}_0, Y)\right). \tag{12.26}$$

Here, the subscript 1 on $\mathbf{A}_1$ and $\mathbf{B}_1$ is used to denote that the above expectation and variance are with respect to a single observation, Y.

12.2.5 Asymptotic normality of M-estimators

The previous section introduced the notion that asymptotically normal estimators can be obtained when the model is incorrectly specified. This can be used to great advantage, especially in situations where there is difficulty in specifying the correct form of the likelihood. Useful estimators can be obtained by solving equations that possess similar properties to a likelihood equation. Indeed, Example 12.7 shows that asymptotically normal estimators of θ can be obtained from just the correct specification of the mean and variance of Y. The name *M-estimator* was coined by Huber (1964), and alludes to the fact that these estimators are similar in construction to maximum likelihood estimators.

Specifically, suppose that $\boldsymbol{\theta} \in \Theta \subset \mathbb{R}^s$, and for any given value of y, let $\mathcal{U}(\boldsymbol{\theta}; y) : \Theta \to \mathbb{R}^s$ be a function that satisfies the zero-mean property

$$E_{\boldsymbol{\theta}}[\mathcal{U}(\boldsymbol{\theta}; Y)] = \mathbf{0} \quad \text{for all} \quad \boldsymbol{\theta} \in \Theta. \tag{12.27}$$

Here, $\mathcal{U}(\boldsymbol{\theta}; y)$ acts as a surrogate for the score function $\partial l(\boldsymbol{\theta}; y)/\partial \boldsymbol{\theta}$. For an iid sample $\mathbf{y} = (y_1, \ldots, y_n)$, an estimating function can be defined as

$$U_n(\boldsymbol{\theta}; \mathbf{y}) = \sum_{i=1}^{n} \mathcal{U}(\boldsymbol{\theta}; y_i),$$

and analogously to the likelihood equation, the estimating equation is defined to be

$$U_n(\boldsymbol{\theta}; \mathbf{y}) = \mathbf{0}. \tag{12.28}$$

The solution $\widetilde{\boldsymbol{\theta}}_n$ to the estimating equation is the M-estimator.

Subject to appropriate regularity conditions (see Pawitan 2001, p. 405), it can be shown that

$$\sqrt{n}(\widetilde{\boldsymbol{\theta}}_n - \boldsymbol{\theta}_0) \to_D N_s\left(\mathbf{0}, \mathbf{A}_1^{-1}\mathbf{B}_1\mathbf{A}_1^{-T}\right), \tag{12.29}$$

where $\mathbf{A}_1$ and $\mathbf{B}_1$ are the $s \times s$-dimensional matrices

$$\mathbf{A}_1 = -E_{\boldsymbol{\theta}_0}\left[\frac{\partial \mathcal{U}(\boldsymbol{\theta}_0, Y)}{\partial \theta}\right], \tag{12.30}$$

and

$$\mathbf{B}_1 = \text{var}_{\boldsymbol{\theta}_0}(\mathcal{U}(\boldsymbol{\theta}_0, Y)). \tag{12.31}$$

Note that $\mathbf{A}_1$ need not be symmetric, and the notation $\mathbf{A}_1^{-T}$ is used to denote the transpose of $\mathbf{A}_1^{-1}$, or equivalently, the inverse of $\mathbf{A}_1^T$.

Example 12.7. Suppose that $\theta \in \mathbb{R}$ and Y_i are iid with mean $\mu(\theta)$, where μ is a differentiable and invertible function of the parameter of interest, θ. The estimating function

$$\mathcal{U}(\theta; Y) = (Y - \mu),$$

satisfies the zero-mean property, and the estimating equation is

$$U_n(\theta, \mathbf{Y}) = \sum_{i=1}^{n} (Y_i - \mu) = 0.$$

This estimating equation is satisfied when $\mu = \overline{Y}_n$, and hence the M-estimator of θ is the value $\widetilde{\theta}_n$ such that $\mu(\widetilde{\theta}_n) = \overline{Y}_n$.

Now, if the variance of Y_i is σ^2, then (12.30) and (12.31) are the scalar quantities

$$\mathbf{A}_1 = -\frac{\partial \mu}{\partial \theta}, \quad \text{and} \quad \mathbf{B}_1 = \sigma^2.$$

From (12.29), it follows that

$$\begin{aligned}\sqrt{n}(\widetilde{\theta}_n - \theta_0) \to_D \; & N\left(0, \frac{\sigma^2}{\left(\partial\mu/\partial\theta\right)^2}\right) \\ = \; & N\left(0, \left(\frac{\partial \theta}{\partial \mu}\right)^2 \sigma^2\right).\end{aligned}$$

In this example, note that the above convergence result could be obtained more directly from the central limit theorem (see Equation 13.8), followed by application of the delta theorem to the function $\theta = g(\mu)$. □

Heyde (1997) gives a rigorous presentation of M-estimators and their properties, and considers the optimality of various classes of M-estimators. In particular, it can be shown that the log-likelihood is the optimal M-estimator (Theorem 12.3). For convenience, the proof of this theorem assumes that the estimating function is of standardized form, whereby $\mathbf{A}_1$ and $\mathbf{B}_1$ are equal. This can be assumed without loss of generality (Exercise 12.3). In standardized form, the asymptotic variance in (12.29) reduces to the inverse of $\mathrm{var}_{\boldsymbol{\theta}}(\mathcal{U})$, or equivalently, the inverse of $-E_{\boldsymbol{\theta}}[\partial \mathcal{U}/\partial \boldsymbol{\theta}]$.

Example 12.7 continued. The standardized form of $\mathcal{U}$ is

$$\mathcal{U}^{(s)} = \frac{\partial \mu}{\partial \theta} \frac{(Y - \mu)}{\sigma^2},$$

which has variance

$$\operatorname{var}\left(\mathcal{U}^{(s)}\right) = \left(\frac{\partial \mu}{\partial \theta}\right)^2 \frac{1}{\sigma^2}.$$

Note that this is equal to

$$-E\left[\partial \mathcal{U}^{(s)}/\partial \theta\right] = -E\left[\frac{\partial^2 \mu}{\partial \theta^2}\left(\frac{Y-\mu}{\sigma^2}\right) - \left(\frac{\partial \mu}{\partial \theta}\right)^2 \frac{1}{\sigma^2}\right].$$

□

Theorem 12.3 Suppose that $\theta \in \mathbb{R}$, and that Y_i are iid with density function $f(y, \theta)$, and let $\mathcal{U}(\theta; y)$ be an estimating function that satisfies the zero-mean property in (12.27). Then, the asymptotic variance of $\sqrt{n}(\widetilde{\theta}_n - \theta_0)$ is minimized when $\mathcal{U}(\theta; y)$ is the score function $\frac{\partial l(\theta;y)}{\partial \theta}$.

Proof: Without loss of generality it can be assumed that $\mathcal{U}$ is in standardized form, so that the asymptotic variance of $\sqrt{n}(\widetilde{\theta}_n - \theta_0)$ is $1/\operatorname{var}_{\theta_0}(\mathcal{U}) = -1/E_{\theta_0}[\partial \mathcal{U}/\partial \theta]$. Thus, the theorem is proved by showing that $\operatorname{var}_{\theta_0}(\mathcal{U})$ is maximized when $\mathcal{U}$ is chosen to be the score function, or equivalently, that

$$\operatorname{var}_{\theta_0}(\mathcal{U}) \leq \operatorname{var}_{\theta_0}\left(\frac{\partial l(\theta; Y)}{\partial \theta}\right), \tag{12.32}$$

where the right-hand side of (12.32) is the expected Fisher information from one observation. Now, since

$$E_\theta[\mathcal{U}] = \int \mathcal{U}(\theta; y) f(y; \theta) dy = 0 \quad \text{for all } \theta \in \Theta,$$

it follows that its derivative (with respect to θ) is also zero. That is (assuming differentiation under the integral sign is valid), for all $\theta \in \Theta$,

$$\begin{aligned}
\frac{\partial E_\theta[\mathcal{U}]}{\partial \theta} &= \frac{\partial}{\partial \theta} \int \mathcal{U} f dy \\
&= \int \frac{\partial \mathcal{U}}{\partial \theta} f dy + \int \mathcal{U} \frac{\partial f}{\partial \theta} dy \\
&= E_\theta\left[\frac{\partial \mathcal{U}}{\partial \theta}\right] + \int \mathcal{U} \frac{\partial f}{\partial \theta} dy = 0,
\end{aligned}$$

from which it follows that

$$E_\theta\left[\frac{\partial \mathcal{U}}{\partial \theta}\right] = -\int \mathcal{U} \frac{\partial f}{\partial \theta} dy.$$

Therefore,

$$\begin{aligned}
\text{var}_{\theta_0}(\mathcal{U}) = -E_{\theta_0}\left[\frac{\partial \mathcal{U}}{\partial \theta}\right] &= \int \mathcal{U}\frac{\partial f}{\partial \theta}dy \\
&= \int \mathcal{U}\frac{\partial l}{\partial \theta}f dy \\
&= \text{cov}_{\theta_0}\left(\mathcal{U}, \frac{\partial l}{\partial \theta}\right) \leq \sqrt{\text{var}_{\theta_0}(\mathcal{U})\text{var}_{\theta_0}\left(\frac{\partial l}{\partial \theta}\right)},
\end{aligned}$$

from application of the Cauchy-Schwarz inequality (Section 13.5.2). The inequality in (12.32) follows immediately. □

12.2.6 The non-iid case

The asymptotic convergence results can be generalized to cases where the data are not iid. For example, in the case of independent but non-identically distributed data, the asymptotic normality of $\sqrt{n}(\widehat{\boldsymbol{\theta}}_n - \boldsymbol{\theta}_0)$ will be maintained provided that the log-likelihood $l(\boldsymbol{\theta}) = \sum_{i=1}^{n} l(\boldsymbol{\theta}; y_i)$ is not overly dominated by any individual $l(\boldsymbol{\theta}; y_i)$ terms. This condition is stated formally in (12.33).

The expected Fisher information from observing $\boldsymbol{Y} = (Y_1, \ldots, Y_n)$ is

$$\mathcal{I}(\boldsymbol{\theta}) = \sum_{i=1}^{n} \mathcal{I}_i(\boldsymbol{\theta}),$$

where $\mathcal{I}_i(\boldsymbol{\theta})$ is the information from observation of Y_i. Under the condition that the averaged Fisher information matrix converges element-wise, i.e.

$$\overline{\mathcal{I}}_n(\boldsymbol{\theta}) = \frac{\mathcal{I}(\boldsymbol{\theta})}{n} \to \overline{\mathcal{I}}(\boldsymbol{\theta}), \tag{12.33}$$

for some positive definite matrix $\overline{\mathcal{I}}(\boldsymbol{\theta})$, it can be shown (Hoadley 1971) that

$$\sqrt{n}(\widehat{\boldsymbol{\theta}}_n - \boldsymbol{\theta}_0) \to_D N_s\left(\mathbf{0}, \overline{\mathcal{I}}(\boldsymbol{\theta}_0)^{-1}\right). \tag{12.34}$$

Note that in the iid case, $\mathcal{I}_1(\boldsymbol{\theta}_0) = \overline{\mathcal{I}}(\boldsymbol{\theta}_0)$.

12.3 Approximate normality

Here, the n subscript will be removed from the $\widehat{\boldsymbol{\theta}}_n$ notation, since n is now taken to be fixed.

For sufficiently large n, the convergence result of (12.34) can be interpreted as saying

$$\sqrt{n}(\widehat{\boldsymbol{\theta}} - \boldsymbol{\theta}_0) \,\dot\sim\, N_s\left(\mathbf{0}, \overline{\mathcal{I}}(\boldsymbol{\theta}_0)^{-1}\right), \tag{12.35}$$

where $\stackrel{.}{\sim}$ denotes 'approximately distributed'. Dividing both sides by $\sqrt{n}$, and adding $\boldsymbol{\theta}_0$, gives

$$\widehat{\boldsymbol{\theta}} \stackrel{.}{\sim} N_s\left(\boldsymbol{\theta}_0, \mathcal{I}(\boldsymbol{\theta}_0)^{-1}\right). \tag{12.36}$$

Equation (12.36) is a pragmatic statement of the approximate normality of MLEs. Specifically, it states that the distribution of $\widehat{\boldsymbol{\theta}}$ (under repetition of the experiment) is approximately equal to that of a multivariate normal with mean $\boldsymbol{\theta}_0$ and variance matrix given by the inverse of the expected Fisher information from observing $\boldsymbol{Y} = (Y_1, \ldots, Y_n)$.

Example 12.4 continued. For the iid $N(\mu, \sigma^2)$ model, if follows from (12.36) that for n sufficiently large,

$$\begin{pmatrix} \overline{Y} \\ \widehat{\sigma}^2 \end{pmatrix} \stackrel{.}{\sim} N_2\left(\begin{pmatrix} \mu_0 \\ \sigma_0^2 \end{pmatrix}, \frac{1}{n}\begin{pmatrix} \sigma_0^2 & 0 \\ 0 & 2\sigma_0^4 \end{pmatrix}\right). \tag{12.37}$$

□

Loosely speaking, it can be said that, for sufficiently large n, $\widehat{\boldsymbol{\theta}}$ is approximately an unbiased estimator of $\boldsymbol{\theta}_0$ with variance matrix $\mathcal{I}(\boldsymbol{\theta}_0)^{-1}$. By virtue of the delta theorem (Section 13.6.4), for any differentiable function $g(\boldsymbol{\theta}) : \Theta \to \mathbb{R}$, $g(\widehat{\boldsymbol{\theta}})$ is also an approximately unbiased estimator of $g(\boldsymbol{\theta}_0)$, with approximate variance equal to the Cramér-Rao lower bound in (11.18). Consequently, maximum likelihood estimators are said to be asymptotically efficient.

In practice, ML estimators typically have finite expected value and variance if the model is sensibly parameterized, notwithstanding the potential need to correct for possible ill-behaviour that may occur with probability tending to zero as sample size increases (e.g. Equation 12.4). They typically have bias that decreases at a rate of $1/n$ rate, denoted $O(n^{-1})$. From (12.36), the variance is also of order $O(n^{-1})$. The mean-squared error of $\widehat{\theta}$ is therefore

$$\text{mse}(\widehat{\theta}) = \text{bias}(\widehat{\theta})^2 + \text{var}(\widehat{\theta}) = O(n^{-2}) + O(n^{-1}).$$

Thus, the overall performance of MLEs (as measured by the mean-squared error) is typically dominated by their variance for large n.

Box 12.1

It is important to remember that 'approximately distributed' should be interpreted as saying that the distribution of $\widehat{\boldsymbol{\theta}}$ is approximately that of a normal random variable (or vector). However, there is no absolute guarantee that the expected value or variance of $\widehat{\boldsymbol{\theta}}$ even exist (Example 12.6), or that a root of the likelihood exists

(Example 12.2) So, for example, saying that $\widehat{\boldsymbol{\theta}}$ is approximately unbiased has to be recognized as a statement about its approximating distribution.

12.3.1 Estimation of the approximate variance

In practice, calculation of the expected Fisher information may be difficult or intractable and a more pragmatic alternative is required. Since the expected Fisher information is the negative of the expected Hessian matrix of the log-likelihood, it is natural to approximate it using the negative of the Hessian evaluated using the observed data $\mathbf{y}$. This approximation is called the observed Fisher information and will be denoted by $\mathbf{O}(\boldsymbol{\theta})$, having (i, j) element

$$\mathbf{O}_{ij}(\boldsymbol{\theta}) = -\frac{\partial^2 l(\boldsymbol{\theta}; \mathbf{y})}{\partial\theta_i \partial\theta_j}. \tag{12.38}$$

An alternative approximation can be based on the representation of expected Fisher information as the variance of the score function (see Exercise 12.10), but this is not widely used.

In general, it is usually most convenient to evaluate the observed Fisher information at $\widehat{\boldsymbol{\theta}}$, but it could also be evaluated at a hypothesized value $\boldsymbol{\theta}_0$ specified under a null hypothesis. Then, the approximate normality of the MLE can be expressed as

$$\widehat{\boldsymbol{\theta}} \dot\sim N_s\left(\boldsymbol{\theta}_0, \mathbf{O}(\boldsymbol{\theta}_0)^{-1}\right), \tag{12.39}$$

or

$$\widehat{\boldsymbol{\theta}} \dot\sim N_s\left(\boldsymbol{\theta}_0, \mathbf{O}(\widehat{\boldsymbol{\theta}})^{-1}\right). \tag{12.40}$$

Example 12.8. Observed information from iid $Y_i \sim N(\mu, \sigma^2)$. Using the calculations in Section 11.7.1, it follows that the observed Fisher information from observing n iid $N(\mu, \sigma^2)$ observations is

$$\mathbf{O}(\mu, \sigma^2) = \begin{pmatrix} n/\sigma^2 & \sum(y_i - \mu)/\sigma^4 \\ \sum(y_i - \mu)/\sigma^4 & -n/(2\sigma^4) + \sum(y_i - \mu)^2/\sigma^6 \end{pmatrix}.$$

When evaluated at the MLE, this simplifies to

$$\mathbf{O}(\overline{y}, \widehat{\sigma}^2) = n \begin{pmatrix} 1/\widehat{\sigma}^2 & 0 \\ 0 & 1/(2\widehat{\sigma}^4) \end{pmatrix}.$$

In particular, from (12.40), this gives

$$\overline{Y} \dot\sim N(\mu, \widehat{\sigma}^2/n). \tag{12.41}$$

In this example, $\mathcal{I}(\widehat{\boldsymbol{\theta}}) = \mathbf{O}(\widehat{\boldsymbol{\theta}})$, but this does not hold in general. □

Not only is it more convenient to use the observed information in place of the expected information, but it has also been shown that it generally results in a better normal approximation (Efron and Hinkley 1978). The argument for favouring use of the observed information is that it quantifies the information contained in the actual data that were realized, whereas the expected information quantifies the average information under repetition of the experiment (e.g. see Exercise 12.9).

The approximation in (12.40) is by far the most widely used, and is the form used throughout Part II of this text, with the exception of Section 8.3 where generalized estimating equations (GEEs) were presented. These are a form of M-estimator, and their approximate normality is considered in the next section.

12.3.2 Approximate normality of M-estimators

Let $\mathbf{y}$ be observed from the distribution with density $f(\mathbf{y};\boldsymbol{\theta}_0)$, and let $\widetilde{\boldsymbol{\theta}} \in \mathbb{R}^s$ denote the M-estimator obtained as the solution to the s-dimensional system of equations

$$U(\boldsymbol{\theta};\mathbf{y}) = \sum_{i=1}^{n} \mathcal{U}_i(\boldsymbol{\theta}; y_i) = \mathbf{0},$$

where the functions $\mathcal{U}_i$ are assumed to satisfy the zero-mean property in (12.27).

For sufficiently large n, and subject to appropriate regularity conditions, the approximate version of the asymptotic normality result given by Equation (12.29) is

$$\widetilde{\boldsymbol{\theta}} \dot{\sim} N_s\left(\boldsymbol{\theta}_0, \mathbf{A}^{-1}\mathbf{B}\mathbf{A}^{-T}\right), \tag{12.42}$$

where $\mathbf{A}$ and $\mathbf{B}$ are the $s \times s$ matrices

$$\mathbf{A} = -E_{\boldsymbol{\theta}_0}\left[U'(\boldsymbol{\theta}_0, \mathbf{Y})\right], \tag{12.43}$$

and

$$\mathbf{B} = \text{var}_{\boldsymbol{\theta}_0}(U(\boldsymbol{\theta}_0, \mathbf{Y})). \tag{12.44}$$

Matrix $\mathbf{A}$ can be estimated using its empirical counterpart,

$$\widehat{\mathbf{A}} = -U'(\widetilde{\boldsymbol{\theta}}, \mathbf{y}), \tag{12.45}$$

and assuming that Y_i are independent, $\mathbf{B}$ is commonly estimated using the empirical estimate

$$\widehat{\mathbf{B}} = \sum_{i=1}^{n} \mathcal{U}_i(\widetilde{\boldsymbol{\theta}}; y_i)\mathcal{U}_i(\widetilde{\boldsymbol{\theta}}; y_i)^T. \tag{12.46}$$

For more detail, see Section 2.2 of Hardin and Hilbe (2003), and Exercise 12.10. The estimated variance

$$\widehat{\text{var}}(\widetilde{\boldsymbol{\theta}}) = \widehat{\mathbf{A}}^{-1}\widehat{\mathbf{B}}\widehat{\mathbf{A}}^{-T}, \tag{12.47}$$

is commonly called the sandwich estimator of variance because matrix $\widehat{\mathbf{B}}$ is sandwiched between $\widehat{\mathbf{A}}^{-1}$ and its transpose. It is also sometimes called the robust variance estimator, or empirical variance estimator due to the empirical estimation of $\mathbf{B}$.

It was seen in Section 12.2.4 that an asymptotic variance of sandwich form arose in the context of using maximum likelihood with a misspecified model. This suggests that using a sandwich estimator of variance rather than the inverse observed Fisher information would give robustness against misspecification of the model. However, the sandwich estimator is not generally used in this context, because it can have considerably greater sampling variability than the inverse Fisher information, thereby leading to confidence intervals with inferior coverage probability (Kauermann and Carroll 2001).

The following example features an estimation equation that is of great practical relevance. It is widely used to extend generalized linear models to grouped data, and its application is demonstrated in Section 8.3.1.

Example 12.9. Generalized estimating equations. The form of estimating equation used throughout Chapter 8 can be written as the s-dimensional system of equations

$$U(\boldsymbol{\theta}; \boldsymbol{y}) = \dot{\boldsymbol{\mu}}^T \mathbf{V}^{-1}(\boldsymbol{y} - \boldsymbol{\mu}) = \mathbf{0}, \tag{12.48}$$

where $\boldsymbol{\mu} \equiv \boldsymbol{\mu}(\boldsymbol{\theta}) = E_{\boldsymbol{\theta}}[\boldsymbol{Y}] = (\mu_1, \ldots, \mu_n)^T$ depends on parameters $\boldsymbol{\theta} \in \mathbb{R}^s$, and $\dot{\boldsymbol{\mu}}$ denotes the $n \times s$ Jacobian matrix with (i, j) element

$$\dot{\boldsymbol{\mu}}_{ij} = \frac{\partial \mu_i}{\partial \theta_j}.$$

Matrix $\mathbf{V} \equiv \mathbf{V}(\boldsymbol{\theta})$ is $n \times n$ and positive definite, and may also depend on $\boldsymbol{\theta}$, and on correlation parameters $\boldsymbol{\alpha}$. Ideally, $\mathbf{V}$ is equal, or proportional, to $\text{var}_{\boldsymbol{\theta}}(\boldsymbol{Y})$, but the approximate normality result in (12.42) does not require this, and hence $\mathbf{V}$ has a general interpretation as a 'working' variance matrix (see the application in Section 8.3).

When Y_i, $i = 1, \ldots, n$ are independent then $\mathbf{V}$ is a diagonal matrix and the s-dimensional estimating equation has the form used by Wedderburn (1974), as demonstrated in Section 8.2. That is,

$$\sum_{i=1}^{n} \frac{\frac{\partial \mu_i}{\partial \theta_j}(y_i - \mu_i)}{v_i} = 0, \qquad j = 1, \ldots, s, \tag{12.49}$$

where v_i is the ith diagonal element of $\mathbf{V}$.

In the GEE context of Liang and Zeger (1986), the data have the form of independent multivariate observations $\boldsymbol{y}_i = (y_{i1}, \ldots, y_{in_i})^T$, $i = 1, \ldots, m$ (Section 8.3) and the s-dimensional estimating equation can be written in the form

$$\sum_{i=1}^{m} \dot{\boldsymbol{\mu}}_i^T \mathbf{V}_i^{-1}(\boldsymbol{y}_i - \boldsymbol{\mu}_i) = \mathbf{0}\,,$$

where $\boldsymbol{\mu}_i = (E_{\boldsymbol{\theta}}[Y_{i1}], \ldots, E_{\boldsymbol{\theta}}[Y_{in_i}])^T$, $\dot{\boldsymbol{\mu}}_i$ is its $n_i \times s$ Jacobian, and $\mathbf{V}_i$ is the assumed variance matrix of $\boldsymbol{Y}_i = (Y_{i1}, \ldots, Y_{in_i})^T$.

Now, $\dot{\boldsymbol{\mu}}$ and $\mathbf{V}$ do not depend on $\boldsymbol{Y}$, and so it follows immediately that the estimating function $U(\boldsymbol{\theta}; \boldsymbol{y})$ in (12.48) satisfies the zero-mean property (12.27). Moreover, matrices $\mathbf{A}$ and $\mathbf{B}$ are

$$\mathbf{A} = \dot{\boldsymbol{\mu}}^T \mathbf{V}^{-1} \dot{\boldsymbol{\mu}}, \tag{12.50}$$

and

$$\mathbf{B} = \dot{\boldsymbol{\mu}}^T \mathbf{V}^{-1} \text{var}_{\boldsymbol{\theta}}(\boldsymbol{Y}) \mathbf{V}^{-1} \dot{\boldsymbol{\mu}}, \tag{12.51}$$

and so the approximate variance matrix of the M-estimator $\widetilde{\boldsymbol{\theta}}$ is

$$\mathbf{A}^{-1}\mathbf{B}\mathbf{A}^{-T} = \left(\dot{\boldsymbol{\mu}}^T \mathbf{V}^{-1} \dot{\boldsymbol{\mu}}\right)^{-1} \dot{\boldsymbol{\mu}}^T \mathbf{V}^{-1} \text{var}_{\boldsymbol{\theta}}(\boldsymbol{Y}) \mathbf{V}^{-1} \dot{\boldsymbol{\mu}} \left(\dot{\boldsymbol{\mu}}^T \mathbf{V}^{-1} \dot{\boldsymbol{\mu}}\right)^{-1} \tag{12.52}$$

When $\mathbf{V} = \text{var}_{\boldsymbol{\theta}}(\boldsymbol{Y})$ then this simplifies to

$$\mathbf{A}^{-1}\mathbf{B}\mathbf{A}^{-T} = \mathbf{B}^{-1} = \left(\dot{\boldsymbol{\mu}}^T \mathbf{V}^{-1} \dot{\boldsymbol{\mu}}\right)^{-1}, \tag{12.53}$$

and this can be used to provide an estimated variance matrix by evaluating $\mathbf{V} \equiv \mathbf{V}(\boldsymbol{\theta})$ at $\widetilde{\boldsymbol{\theta}}$. Alternatively the estimate $\widehat{\mathbf{B}}$ in (12.46) can be used, which in this context is obtained as the empirical variance matrix of appropriately standardized residuals (Hardin and Hilbe 2003). However, if $\mathbf{V}$ is regarded as a working approximation to $\text{var}_{\boldsymbol{\theta}}(\boldsymbol{Y})$, then it should not be used in estimation of the variance of $\widetilde{\boldsymbol{\theta}}$, and the sandwich estimator (12.47) should then be employed.

It can be shown that the variance matrix in (12.53) is smaller (with respect to Löwner order) than (12.52) (e.g. see Exercise 12.4). That is, $\text{var}_{\boldsymbol{\theta}}(\boldsymbol{Y})$ is the optimal choice for $\mathbf{V}$. □

12.4 Wald tests and confidence regions

12.4.1 Wald test statistics

Wald test statistics are derived as an immediate consequence of the approximate normality of the MLE. However, there are several variants of the test statistic, depending on which form of the Fisher information is used to estimate the variance of the MLE. The most convenient and commonly used variant is the one that uses the observed Fisher information evaluated at $\widehat{\boldsymbol{\theta}}$. This form of information is simply the negative of the Hessian of the log-likelihood at $\widehat{\boldsymbol{\theta}}$, and is routinely provided by optimizing software (e.g. Section 5.2). However, another three variants are also presented here, and it is seen that the convenient choice is not always the best.

In the scalar parameter case, $\theta \in \Theta \subset \mathbb{R}$, squaring both sides of the convergence result in (12.17) gives (by property P3, Section 13.6.2),

$$W_1 = \mathcal{I}(\theta_0)(\widehat{\theta}_n - \theta_0)^2 \rightarrow_D N(0,1)^2 \equiv \chi_1^2, \tag{12.54}$$

where W_1 denotes a Wald test statistic for the null hypothesis $H_0 : \theta = \theta_0$.

From the definition of convergence in distribution, Equation (12.54) establishes that $P_{\theta_0}(W_1 \leq w)$ converges to $F_{\chi_1^2}(w)$, where P_{θ_0} denotes probability under $H_0 : \theta = \theta_0$, and $F_{\chi_1^2}(w)$ denotes the distribution function of a χ_1^2 random variable. In particular, letting $\chi^2_{1,1-\alpha}$ denote the $(1-\alpha)$ quantile of the χ_1^2 distribution, $P_{\theta_0}(W_1 > \chi^2_{1,1-\alpha})$ is approximately equal to α for large n if H_0 is true. Hence, the critical region

$$\{\boldsymbol{Y} : W_1 > \chi^2_{1,1-\alpha}\}$$

defines an approximate level α test of $H_0 : \theta = \theta_0$.

In the multi-dimensional case, $\boldsymbol{\theta} \in \Theta \subset \mathbb{R}^s (s \geq 1)$, the corresponding Wald test statistic for $H_0 : \boldsymbol{\theta} = \boldsymbol{\theta}_0$ is a quadratic form, and it follows from Equation (15.4) that

$$W_1 = (\widehat{\boldsymbol{\theta}}_n - \boldsymbol{\theta}_0)^T \mathcal{I}(\boldsymbol{\theta}_0)(\widehat{\boldsymbol{\theta}}_n - \boldsymbol{\theta}_0) \rightarrow_D \chi_s^2. \tag{12.55}$$

Other variants of the Wald test statistic are obtained by substituting $\mathcal{I}(\boldsymbol{\theta}_0)$ with a suitable alternative. By virtue of the consistency of $\widehat{\boldsymbol{\theta}}_n$ (i.e. $\widehat{\boldsymbol{\theta}}_n \rightarrow_P \boldsymbol{\theta}_0$) one asymptotically equivalent[2] alternative is to use the expected Fisher information evaluated at $\widehat{\boldsymbol{\theta}}_n$, $\mathcal{I}(\widehat{\boldsymbol{\theta}}_n)$. Other asymptotically equivalent alternatives include using the observed Fisher information evaluated at either $\widehat{\boldsymbol{\theta}}_n$ or $\boldsymbol{\theta}_0$. These variants will be denoted

$$W_2 = (\widehat{\boldsymbol{\theta}}_n - \boldsymbol{\theta}_0)^T \mathcal{I}(\widehat{\boldsymbol{\theta}}_n)(\widehat{\boldsymbol{\theta}}_n - \boldsymbol{\theta}_0) \tag{12.56}$$

$$W_3 = (\widehat{\boldsymbol{\theta}}_n - \boldsymbol{\theta}_0)^T \mathbf{O}(\boldsymbol{\theta}_0)(\widehat{\boldsymbol{\theta}}_n - \boldsymbol{\theta}_0) \tag{12.57}$$

$$W_4 = (\widehat{\boldsymbol{\theta}}_n - \boldsymbol{\theta}_0)^T \mathbf{O}(\widehat{\boldsymbol{\theta}}_n)(\widehat{\boldsymbol{\theta}}_n - \boldsymbol{\theta}_0), \tag{12.58}$$

where $\mathbf{O}(\boldsymbol{\theta}) = -l''(\boldsymbol{\theta})$.

Example 12.10. Binomial. Let Y be distributed Bin(n, p) and let $\widehat{p} = y/n$ denote the MLE. In Example 11.1 it was shown that

$$l''(p) = -\frac{y}{p^2} - \frac{n-y}{(1-p)^2}$$

$$\mathcal{I}(p) = \frac{n}{p(1-p)}.$$

[2] Two sequences of random variables are said to be asymptotically equivalent if the difference between them converges in probability to 0.

Hence, the four variants of the Wald test statistic of $H_0 : p = p_0$ are

$$W_1 = \mathcal{I}(p_0)(\widehat{p} - p_0)^2 = \frac{n(\widehat{p} - p_0)^2}{p_0(1 - p_0)}$$

$$W_2 = \mathcal{I}(\widehat{p})(\widehat{p} - p_0)^2 = \frac{n(\widehat{p} - p_0)^2}{\widehat{p}(1 - \widehat{p})}$$

$$W_3 = \mathbf{O}(p_0)(\widehat{p} - p_0)^2 = \left(\frac{y}{p_0^2} + \frac{n - y}{(1 - p_0)^2}\right)(\widehat{p} - p_0)^2$$

$$W_4 = \mathbf{O}(\widehat{p})(\widehat{p} - p_0)^2 = \frac{n(\widehat{p} - p_0)^2}{\widehat{p}(1 - \widehat{p})}.$$

These four Wald test statistics all have an asymptotic χ_1^2 distribution if H_0 is true. Equivalently, their square root has an asymptotic standard normal distribution. Here, W_2 and W_4 are identical, by virtue of the fact that $\mathcal{I}(\widehat{p}) = \mathbf{O}(\widehat{p})$. This equivalence of the expected and observed Fisher information does not hold in general, although in Section 11.7.5 it is seen to hold for generalized linear models that use the canonical link function.

Statistical texts differ in their statement of the Wald test for the binomial probability, with some using test statistic W_1 (e.g. Collett 1991), and others using the form of test statistics W_2 and W_4 (e.g. Wild and Seber 2000). The justification for statistics W_2 and W_4 is that they are easier to invert for purposes of obtained confidence intervals (see Section 12.4.2). However, it has been shown that test statistic W_1 has the best performance (Agresti and Coull 1998, Brown *et al.* 2001, Brown *et al.* 2002), which can be attributed to the fact that it makes use of the value of p_0 that is specified under the null hypothesis. □

12.4.2 Wald confidence intervals and regions

The $(1 - \alpha)100\,\%$ confidence region for $\boldsymbol{\theta} \in \mathbb{R}^s$ is the collection of all values $\boldsymbol{\theta}_0$ that are not rejected by the size α test of $H_0 : \boldsymbol{\theta} = \boldsymbol{\theta}_0$ (Section 13.2). The Wald statistic therefore leads to an approximate $(1 - \alpha)100\,\%$ confidence region given by all values $\boldsymbol{\theta}_0$ such that

$$W \leq \chi^2_{s,1-\alpha},$$

where W denotes any of the choices of Wald test statistic in (12.55)–(12.58).

The computation of the confidence region is generally more tractable for statistics W_2 and W_4 because these use forms of the Fisher information that are evaluated at the MLE $\widehat{\boldsymbol{\theta}}$, and the confidence region then takes the shape of an s-dimensional ellipse.

Example 12.10 continued. Although it has been argued that the Wald test statistic W_1 is preferred for testing $H_0 : p = p_0$ (Agresti and Coull 1998, Brown *et al.* 2001, Brown *et al.* 2002), it is not widely used for obtaining confidence intervals.

Constructing a $(1-\alpha)100\,\%$ CI using this test statistic requires determination of the collection of values p_0 such that

$$\left\{ p_0 : \frac{n(\widehat{p}-p_0)^2}{p_0(1-p_0)} < \chi^2_{1,1-\alpha} \right\}. \tag{12.59}$$

This can be done explicitly, by finding the roots of a quadratic equation (Exercise 12.7), and is sometimes called the Wilson (score) interval because of its early use by Wilson (1927), and because, for this example, test statistic W_1 has an alternative justification as a score test statistic (Section 12.6).

The most common interval is obtained from the expedient inversion of test statistics W_4 (or W_2). Under this choice, the desired interval is the collection of p_0 values given by

$$\left\{ p_0 : \frac{n(\widehat{p}-p_0)^2}{\widehat{p}(1-\widehat{p})} < \chi^2_{1,1-\alpha} \right\},$$

which is the familiar interval

$$\widehat{p} \pm z_{1-\alpha/2}\sqrt{\frac{\widehat{p}(1-\widehat{p})}{n}}. \tag{12.60}$$

□

The next example demonstrates the forms of the Wald test statistic for making inference about a subset of the parameter vector.

Example 12.11. Suppose that it is desired to test a hypothesis that specifies the values of the first r elements of parameter vector $\boldsymbol{\theta} \in \mathbb{R}^s$, i.e. $H_0 : (\theta_1, \ldots, \theta_r) = (\theta_{01}, \ldots, \theta_{0r})$. Denoting $g(\boldsymbol{\theta}) = (\theta_1, \ldots, \theta_r)$, the first r columns of the $r \times s$ Jacobian matrix $\mathbf{G}(\boldsymbol{\theta}_0)$ form the $r \times r$ identity matrix, and the remaining $s-r$ columns contain only zeroes. Let $\boldsymbol{\psi} = (\theta_1, \ldots, \theta_r)$, $\widehat{\boldsymbol{\psi}}_n = (\widehat{\theta}_1, \ldots, \widehat{\theta}_r)$ and $\boldsymbol{\psi}_0 = (\theta_{01}, \ldots, \theta_{0r})$ denote the first r elements of $\boldsymbol{\theta}$, $\widehat{\boldsymbol{\theta}}_n$ and $\boldsymbol{\theta}_0$, respectively. Then, from application of (12.21),

$$\sqrt{n}(\widehat{\boldsymbol{\psi}}_n - \boldsymbol{\psi}_0) \to_D N_r\left(\mathbf{0}, [\,\mathcal{I}_1(\boldsymbol{\theta}_0)^{-1}]_{[rr]}\right)$$

where $[\,\mathcal{I}_1(\boldsymbol{\theta}_0)^{-1}]_{[rr]}$ is the upper $r \times r$ submatrix of $\mathcal{I}_1(\boldsymbol{\theta}_0)^{-1}$.

Consequently, it follows that a Wald statistic for testing the hypothesis $H_0 : \boldsymbol{\psi} = \boldsymbol{\psi}_0$ is

$$W_1 = (\widehat{\boldsymbol{\psi}} - \boldsymbol{\psi}_0)^T \left([(\mathcal{I}(\boldsymbol{\theta}_0)^{-1}]_{[r,r]}\right)^{-1} (\widehat{\boldsymbol{\psi}} - \boldsymbol{\psi}_0) \to \chi^2_r,$$

or any of the asymptotically equivalent variants, such as,

$$W_4 = (\widehat{\boldsymbol{\psi}} - \boldsymbol{\psi}_0)^T \left([(\mathbf{O}(\widehat{\boldsymbol{\theta}}_n)^{-1}]_{[r,r]}\right)^{-1} (\widehat{\boldsymbol{\psi}} - \boldsymbol{\psi}_0) \to \chi^2_r. \tag{12.61}$$

□

Application of the Wald statistic in (12.61), for a function $g(\boldsymbol{\theta}) : \mathbb{R}^5 \to \mathbb{R}^3$, is given in Section 3.3.4.

12.5 Likelihood ratio test statistic

The asymptotic convergence of the likelihood ratio test statistic is stated and proved in Section 12.5.1 for the scalar parameter case. Extension to the multi-dimensional case follows naturally, and is stated in Section 12.5.2.

12.5.1 Likelihood ratio test: $\theta \in \mathbb{R}$

Theorem 12.4 Asymptotic distribution of the likelihood ratio statistic.
In the scalar parameter case, $\theta \in \Theta \subset \mathbb{R}$, if the assumptions of Theorem 12.1 hold then

$$2\big(l(\widehat{\theta}_n; \boldsymbol{Y}) - l(\theta_0; \boldsymbol{Y})\big) = 2\log\left(\frac{L(\widehat{\theta}_n; \boldsymbol{Y})}{L(\theta_0; \boldsymbol{Y})}\right) \to_D \chi^2_1. \tag{12.62}$$

Proof. A Taylor series expansion of $l(\theta_0; \boldsymbol{Y})$ about $\widehat{\theta}_n$ gives

$$l(\theta_0) = l(\widehat{\theta}_n) + l'(\widehat{\theta}_n)(\theta_0 - \widehat{\theta}_n) + \frac{1}{2} l''(\theta_n^*)(\theta_0 - \widehat{\theta}_n)^2, \tag{12.63}$$

where θ_n^* is some value between θ_0 and $\widehat{\theta}_n$. Since $l'(\widehat{\theta}_n) = 0$ this gives

$$2\big(l(\widehat{\theta}_n) - l(\theta_0)\big) = -l''(\theta_n^*)(\theta_0 - \widehat{\theta}_n)^2. \tag{12.64}$$

The result follows from applying Slutsky's theorem (Section 13.6.3) to the product of

$$\big(\sqrt{n}(\theta_0 - \widehat{\theta}_n)\big)^2 \to_D N\left(0, \frac{1}{\mathcal{I}_1(\theta_0)}\right)^2 \equiv \frac{1}{\mathcal{I}_1(\theta_0)}\chi^2_1,$$

and

$$\frac{1}{n} l''(\theta_n^*) \to_P -\mathcal{I}_1(\theta_0). \tag{12.65}$$

From (12.64) and (12.65) it follows that the Wald and likelihood-ratio test statistics are asymptotically equivalent. □

Consequently, for n sufficiently large, an approximate size α likelihood ratio hypothesis test of $H_0 : \theta = \theta_0$ is given by rejecting H_0 if

$$2\big(l(\widehat{\theta}_n; \boldsymbol{Y}) - l(\theta_0; \boldsymbol{Y})\big) > \chi^2_{1,1-\alpha}.$$

12.5.2 Likelihood ratio test for $\boldsymbol{\theta} \in \mathbb{R}^s$, and $g(\boldsymbol{\theta}) \in \mathbb{R}^p$

The convergence result in Theorem 12.4 generalizes to the multi-dimensional case, $\boldsymbol{\theta} \in \mathbb{R}^s$, with the degrees of freedom of the limiting chi-square being s. That is,

$$2\big(l(\widehat{\boldsymbol{\theta}}_n; \boldsymbol{Y}) - l(\boldsymbol{\theta}_0; \boldsymbol{Y})\big) = 2\log\left(\frac{L(\widehat{\boldsymbol{\theta}}_n; \boldsymbol{Y})}{L(\boldsymbol{\theta}_0; \boldsymbol{Y})}\right) \rightarrow_D \chi^2_s. \tag{12.66}$$

The convergence results in (12.62) and (12.66) are for the simple form of hypothesis under which $\boldsymbol{\theta}_0$ is fully specified. The convergence result for composite hypotheses is similar, but requires a sub-maximization under H_0, and the degrees of freedom is equal to the reduction in dimension under the null hypothesis. The composite form of the LR test is expressed in general form in the following theorem, where it is stated for the purpose of testing the hull hypothesis $H_0 : g(\boldsymbol{\theta}) = \boldsymbol{\zeta}_0 \in \mathbb{R}^r$, where $1 \leq r \leq s$.

Theorem 12.5 Given $g(\boldsymbol{\theta}) : \Theta \rightarrow \mathbb{R}^r$, and subject to appropriate regularity conditions,

$$2\big(l(\widehat{\boldsymbol{\theta}}_n; \boldsymbol{Y}) - \max_{\boldsymbol{\theta}: g(\boldsymbol{\theta}) = \boldsymbol{\zeta}_0} l(\boldsymbol{\theta}; \boldsymbol{Y})\big) \rightarrow_D \chi^2_r, \tag{12.67}$$

under $H_0 : g(\boldsymbol{\theta}) = \boldsymbol{\zeta}_0$.

The proof of Theorem 12.5 follows from results found in Wilks (1938) for testing a subset of the parameter vector, by assuming that the model $\{f(\boldsymbol{y}; \boldsymbol{\theta}) : \boldsymbol{\theta} \in \Theta \subset \mathbb{R}^s\}$ can be re-parameterized using $\boldsymbol{\zeta} = (g(\boldsymbol{\theta}), \lambda(\boldsymbol{\theta}))$ for some suitable transformation $\lambda(\boldsymbol{\theta}) \in \mathbb{R}^{s-r}$.

12.6 Rao-score test statistic

The Rao-score test (Rao 1948) is not so widely used in practice as the Wald and LR tests, perhaps due to its less intuitive motivation, and only brief presentation is given here.

For $\theta \in \mathbb{R}$, the Rao-score statistic can be obtained from Equation (12.7), by noting that $\sqrt{n}(\widehat{\theta}_n - \theta_0)$ is asymptotically equivalent to $l'(\theta_0)/(\sqrt{n}\,\mathcal{I}_1(\theta_0))$. The asymptotic convergence of the square of this statistic is therefore

$$\frac{l'(\theta_0)^2}{n\,\mathcal{I}_1(\theta_0)^2} \rightarrow_D N\left(0, \frac{1}{\mathcal{I}_1(\theta_0)}\right)^2 \equiv \frac{1}{\mathcal{I}_1(\theta_0)}\chi^2_1,$$

which leads to the Rao-score statistic

$$\frac{l'(\theta_0)^2}{\mathcal{I}(\theta_0)} \rightarrow_D \chi^2_1. \tag{12.68}$$

In practice, $\mathcal{I}(\theta_0)$ is usually replaced with the observed information $\mathbf{O}(\theta_0)$ or $\mathbf{O}(\widehat{\theta})$.

The convergence result in (12.68) can be obtained more directly from Equation (12.8). However, the approach taken above establishes the asymptotic equivalence of the Wald and Rao-score statistics, and hence also the LR and Rao-score statistics.

Using the Rao-score statistic to test $H_0 : \theta = \theta_0$ is the Rao (or score) test. One advantage of the Rao test is that it does not require calculation of the MLE $\widehat{\theta}$ (unless the Fisher information evaluated at $\widehat{\theta}$ is used in the denominator).

In the vector parameter case, $\boldsymbol{\theta} \in \mathbb{R}^s$,

$$l'(\boldsymbol{\theta}_0)^T \mathcal{I}(\boldsymbol{\theta}_0)^{-1} l'(\boldsymbol{\theta}_0) \rightarrow_D \chi_s^2$$

is the Rao-score test statistic of the simple hypothesis $H_0 : \boldsymbol{\theta} = \boldsymbol{\theta}_0$. In the composite hypothesis case, with $\widehat{\boldsymbol{\theta}}_{0n}$ denoting the MLE under the r-dimensional restriction $H_0 : \boldsymbol{\theta} \in \Theta_0 \subset \Theta$, the Rao-score statistic (also known as the Lagrange multiplier statistic) is

$$l'(\widehat{\boldsymbol{\theta}}_{0n})^T \mathcal{I}(\widehat{\boldsymbol{\theta}}_{0n})^{-1} l'(\widehat{\boldsymbol{\theta}}_{0n}) \rightarrow_D \chi_r^2.$$

Again, note that these test statistics do not require maximization of the likelihood with respect to the full model, and that the expected information can be replaced with the observed information.

Example 12.10 continued. Let Y be distributed Bin(n, p). Now,

$$l'(p) = \frac{y - np}{p(1-p)}$$
$$\mathcal{I}(p) = \frac{n}{p(1-p)},$$

and hence the Rao-score test statistic is

$$\begin{aligned} \frac{l'(p_0)^2}{\mathcal{I}(p_0)} &= \frac{(y - np_0)^2}{p_0^2(1-p_0)^2} \frac{p_0(1-p_0)}{n} \\ &= \frac{(y-np_0)^2}{np_0(1-p_0)} \\ &= \frac{n(y/n - p_0)^2}{p_0(1-p_0)}. \end{aligned} \tag{12.69}$$

In this particular case, the Rao-score test is identical to the Wald test statistic W_1. Recall that W_1 is the form of the Wald statistic that is used to obtain the Wilson confidence interval in Equation (12.59), and hence this interval is also often called the Wilson score interval. □

12.7 Exercises

12.1 Let $A_1, ..., A_n$ and $B_1, ..., B_n$ denote sequences of events such that $P(A_n) \to 1$ and $P(B_n) \to 1$ as $n \to \infty$. Show that $P(A_n \cap B_n) \to 1$ as $n \to \infty$.

12.2 Let $Y_1, \ldots, Y_n$ be iid Pois(λ) and consider the problem of estimating $\zeta = P(Y = 0) = \exp(-\lambda)$. The MLE of λ is the sample mean $\overline{Y}_n$ and so the ML estimator of ζ is $\widehat{\zeta}_n = \exp(-\overline{Y}_n)$. In Exercise 11.2 is was seen that $\mathcal{I}_1(\lambda) = 1/\lambda$.

1. The asymptotic convergence result (12.5) gives

$$\sqrt{n}(\overline{Y}_n - \lambda) \to_D N(0, \lambda).$$

Use the delta theorem to determine the convergence (in distribution) of $\sqrt{n}(\widehat{\zeta}_n - \zeta)$.

2. Instead of using the ML estimator of ζ one could simply use the proportion of Y_i which take the value zero, which will be denoted $T_n(\boldsymbol{Y})$. This is a binomial experiment and so by the central limit theorem

$$\sqrt{n}(T_n(\boldsymbol{Y}) - \zeta) \to_D N(0, \zeta(1 - \zeta)).$$

Show that the MLE $\widehat{\zeta}_n$ is asymptotically more efficient than $T_n(\boldsymbol{Y})$. That is, show that $\zeta(1 - \zeta)$ is greater than the limiting variance of $\sqrt{n}(\widehat{\zeta}_n - \zeta)$ (calculated in Part 1 above).

12.3 Let U be an estimating function satisfying the zero-mean property (12.27), and with $\mathbf{A}$ and $\mathbf{B}$ defined in (12.43) and (12.44), respectively. Then, the estimating function $U^{(s)} = \mathbf{B}^{-1}\mathbf{A}U$ has the same solution as the estimating function U.

1. Show that $U^{(s)}$ satisfies the zero-mean property.

2. Show that $U^{(s)}$ is standardized, that is, $-E_{\boldsymbol{\theta}_0}[U^{(s)\prime}(\boldsymbol{\theta}_0, \boldsymbol{Y})] = \text{var}_{\boldsymbol{\theta}_0}(U^{(s)}(\boldsymbol{\theta}_0, \boldsymbol{Y}))$.

12.4 Consider the (weighted) linear regression model

$$\boldsymbol{Y} \sim N_n(\mathbf{X}\boldsymbol{\beta}, \mathbf{V}),$$

where $\mathbf{V}$ is a diagonal matrix with diagonal elements $v_{ii} = w_i^{-1}\sigma^2$ for known weights w_i, and $\mathbf{X}$ is an $n \times p$ design matrix of rank p. For convenience, attention here is restricted to estimation of $\boldsymbol{\beta} \in \mathbb{R}^p$, and without loss of generality it can be assumed that σ^2 is known because of the parameter orthogonality of $\boldsymbol{\beta}$ and σ^2.

If $\boldsymbol{\beta}$ is estimated using an ordinary (unweighted) linear regression then the least-squares estimator $\widetilde{\boldsymbol{\beta}}$ is the MLE from a misspecified model in which all w_i are taken to be unity.

1. Show that the variance of $\widetilde{\boldsymbol{\beta}}$ is

$$\text{var}(\widetilde{\boldsymbol{\beta}}) = (\mathbf{X}^T\mathbf{X})^{-1}\mathbf{X}^T\mathbf{V}\mathbf{X}(\mathbf{X}^T\mathbf{X})^{-1}.$$

2. Verify that $\text{var}(\widetilde{\boldsymbol{\beta}})$ is of the sandwich form, $\mathbf{A}^{-1}\mathbf{B}\mathbf{A}^{-T}$, where

$$\mathbf{A} = -E_{\boldsymbol{\beta}}[\widetilde{l}''(\boldsymbol{\beta}, \boldsymbol{Y})], \quad \text{and} \quad \mathbf{B} = \text{var}_{\boldsymbol{\beta}}(\widetilde{l}'(\boldsymbol{\beta}, \boldsymbol{Y})),$$

where $\widetilde{l}$ denotes log-likelihood under the (mis-specified) ordinary least-squares model.

3. The MLE $\widehat{\boldsymbol{\beta}}$ is the weighted least-squares estimator, and $\text{var}(\widehat{\boldsymbol{\beta}}) = (\mathbf{X}\mathbf{V}^{-1}\mathbf{X})^{-1}$. Show that $\text{var}(\widetilde{\boldsymbol{\beta}}) \geq \text{var}(\widehat{\boldsymbol{\beta}})$, where $\geq$ is defined using the Löwner ordering. That is, show that

$$\boldsymbol{b}^T\text{var}(\widetilde{\boldsymbol{\beta}})\boldsymbol{b} \geq \boldsymbol{b}^T\text{var}(\widehat{\boldsymbol{\beta}})\boldsymbol{b},$$

for any $\boldsymbol{b} \in \mathbb{R}^p$.

12.5 For independent Y_i and $\theta \in \mathbb{R}$, let $\widetilde{\theta}$ be the QLE obtained from Wedderburn's quasi-likelihood equation. For sufficiently large n, it can be shown (see Section 12.3.2) that $\text{var}(\widetilde{\theta}) \approx A/B^2$ where

$$A = \sum \frac{\dot{\mu}_i^2 \text{var}_\theta(Y_i)}{v_i^2}, \quad \text{and} \quad B = \sum \frac{\dot{\mu}_i^2}{v_i},$$

where $\dot{\mu}_i = \partial\mu_i/\partial\theta$.

Show that this approximate variance is minimized when $v_i = \text{var}_\theta(Y_i)$. *Hint:* Denote $x_i = \dot{\mu}_i/\sqrt{\text{var}_\theta(Y_i)}$ and $y_i = \dot{\mu}_i\sqrt{\text{var}_\theta(Y_i)}/v_i$, and apply the form of the Cauchy-Schwarz inequality given in (13.6).

12.6 The method-of-moments estimator for the parameters of the zero-inflated Poisson was derived using the estimating equation in Exercise 8.2. For the ZIP data of Exercise 3.9, calculate the approximate variance of $(\widetilde{\lambda}, \widetilde{p})$ given in (12.42), and estimate this variance using (12.45) and (12.46).

12.7 Determine the closed form of the Wilson confidence interval (Equation 12.59) for the binomial probability, for y successes from n trials. Calculate this interval with $y = 10$ successes observed from $n = 100$ trials, and compare to the more expedient interval obtained from (12.60).

12.8 Consider the nonlinear regression model $Y_i \sim N(g_i(\beta), \sigma^2)$, $i = 1, \ldots, n$, where $\beta \in \mathbb{R}$ and each g_i is a differentiable function of β (and may depend on covariates associated with observation i). For simplicity, it may be assumed that σ^2 is known. Let $\widehat{\beta}_n$ be the MLE of β. Determine the form of the Wald statistics W_1 and W_3.

12.9 Let A_i, $i = 1, \ldots, n$ be independent and identically distributed Bernoulli random variables with $P(A_i = 0) = P(A_i = 1) = 0.5$. If $A_i = 0$ then

Y_i is observed from a Normal(μ, σ_0^2) distribution, otherwise Y_i is observed from a Normal(μ, σ_1^2) distribution. The values of σ_0 and σ_1 are known and are unequal. Both A_i and Y_i are observed, so the data are of the form $(a_1, y_1), \ldots, (a_n, y_n)$, or equivalently $(\boldsymbol{y}, \boldsymbol{a})$ where $\boldsymbol{y} = (y_1, \ldots, y_n)$ and $\boldsymbol{a} = (a_1, \ldots, a_n)$.

1. Write down the likelihood function $L(\mu) = f(\boldsymbol{y}, \boldsymbol{a}; \mu)$. *Hint:* It may simplify notation to denote the distribution of $Y_i|A_i = a_i$ as Normal($\mu, \sigma_{a_i}^2$).)
2. Show that the maximum likelihood estimator of μ is the weighted mean
$$\widehat{\mu} = \frac{\sum_{i=1}^n y_i/\sigma_{a_i}^2}{\sum_{i=1}^n 1/\sigma_{a_i}^2}.$$
3. Calculate the observed Fisher information $\mathbf{O}(\mu)$ for the data.
4. Calculate the expected Fisher information $\mathcal{I}(\mu)$ (for a sample of size n, $(A_1, Y_1), \ldots, (A_n, Y_n)$).
5. $\mathcal{I}(\mu)$ and $\mathbf{O}(\mu)$ calculated above are unequal except when n is even and $\sum_{i=1}^n a_i = n/2$. Which do you prefer as an estimator of the variance of $\widehat{\mu}$, $1/\mathcal{I}(\widehat{\mu})$ or $1/\mathbf{O}(\widehat{\mu})$? Justify your choice.

12.10 The representation of expected Fisher information as the variance matrix of the score function gives rise to an alternative empirical estimate of information that can be used when Y_i are independent. This is

$$\mathbf{B}(\boldsymbol{\theta}) = \sum_{i=1}^n \frac{\partial l(\boldsymbol{\theta}; y_i)}{\partial \boldsymbol{\theta}} \frac{\partial l(\boldsymbol{\theta}; y_i)}{\partial \boldsymbol{\theta}}^T.$$

In practice, this form of information is not used for estimation of the variance of $\widehat{\boldsymbol{\theta}}$, although in SAS it is available using the `COV=5` procedure option in `PROC NLP`. (Note that the empirical estimate $\widehat{\mathbf{B}}$ in (12.46) is given by replacing the score function with $\mathcal{U}_i$.)

1. If $Y_i, i = 1, \ldots n$ are iid Bernoulli(p), determine the forms of $\mathbf{O}(\widehat{p})$ and $\mathbf{B}(\widehat{p})$, and show that they are identical.
2. Suppose that Y_i are iid from a mixture of two known distributions (i.e. from a distribution with density $f(y_i; p) = pf_1(y_i) + (1-p)f_2(y_i)$, leaving only p to estimate.) For any p, $0 < p < 1$, determine the forms of $\mathbf{O}(p)$ and $\mathbf{B}(p)$, and show that they are identical.

13

Tools of the trade

Even for practical purposes, theory generally turns out the most important thing in the end – Oliver Wendell Holmes

13.1 Introduction

Sections 13.2 and 13.3 present some fundamental statistical properties that are utilized throughout Parts I and II of this text. For example, it is noted in Box 13.1 that ML inference is invariant to fixed transformations of the data, because the model likelihood changes by a fixed constant that does not depend on the parameter. It is likely that these properties will be well known to the reader, but are included for completeness.

The results in Sections 13.4–13.6 are primarily used in this part of the text. For example, the identities in Section 13.4 can be used to determine the mean and variance of mixture distributions (e.g. Exercise 14.6). The inequalities in Section 13.5 were utilized in the previous two chapters. Indeed, the Cramér-Rao inequality (Section 11.2) reduces to an application of the Cauchy-Schwarz inequality. Section 13.6 contains the asymptotic probability theory that is required to fully comprehend the theoretical concepts used throughout Chapter 12.

13.2 Equivalence of tests and confidence intervals

Test statistics can be used to construct confidence intervals, and vice-versa. Indeed, the following theorem formalizes the equivalency between hypothesis tests and confidence intervals.

Maximum Likelihood Estimation and Inference: With Examples in R, SAS and ADMB, First Edition. Russell B. Millar.

Theorem 13.1

I A $(1-\alpha)100\,\%$ confidence region for parameter $\boldsymbol{\theta}$ is given by the collection of all values $\boldsymbol{\theta}_0$ for which the size α hypothesis test of $H_0 : \boldsymbol{\theta} = \boldsymbol{\theta}_0$ is not rejected.

II A size α hypothesis test is given by rejecting $H_0 : \boldsymbol{\theta} = \boldsymbol{\theta}_0$ if and only if the $(1-\alpha)100\,\%$ confidence region does not include $\boldsymbol{\theta}_0$.

Proof.

I The confidence region contains $\boldsymbol{\theta}_0$ if and only if $H_0 : \boldsymbol{\theta} = \boldsymbol{\theta}_0$ is not rejected. The latter has probability $(1-\alpha)$ when $\boldsymbol{\theta}_0$ is the true parameter value, as required.

II The hypothesis test $H_0 : \boldsymbol{\theta} = \boldsymbol{\theta}_0$ is rejected if and only if the confidence region fails to contain $\boldsymbol{\theta}_0$. The latter has probability α when $\boldsymbol{\theta}_0$ is the true parameter value, as required. □

13.3 Transformation of variables

Theorem 13.2 Let Y be a continuous random variable with density function f_Y and sample space S_Y. Let $Z = u(Y)$ where the function u is a monotone (increasing or decreasing) function from S_Y to S_Z (the sample space of Z) with inverse $v(z) = y$ having derivative $v'(z)$. Then, Z has density function

$$f_Z(z) = \begin{cases} f_Y(v(z))|v'(z)|, & z \in S_Z \\ 0, & \text{otherwise.} \end{cases}$$

Proof. Since they are invertible, u and v are either both monotone increasing or both monotone decreasing. If u and v are monotone increasing, then, for any $y \in \mathbb{R}$,

$$\begin{aligned} F_Z(z) = P(Z \le z) &= P(Z \in (-\infty, z]) \\ &= P(Y \in (-\infty, v(z)]) \\ &= \int_{-\infty}^{v(z)} f_Y(t)dt. \end{aligned}$$

Making the substitution $s = u(t)$, with inverse $t = v(s)$, the change-of-variables formula gives

$$F_Z(z) = \int_{(-\infty,z]\cap S_Z} f_Y(v(s))v'(s)ds = \int_{-\infty}^{z} f_Z(s)ds,$$

where f_Z is defined in (13.1) . Hence, f_Z is the density function of Z. The proof is similar in the monotone decreasing case. □

The above theorem extends in a natural way to the multi-dimensional case where $\boldsymbol{Y}$ and $\boldsymbol{Z} = u(\boldsymbol{Y})$ are both in $\mathbb{R}^n$. In this case, $u : \mathbb{R}^n \to \mathbb{R}^n$ is assumed to be one-to-one with differentiable inverse $v(z) = \mathbf{y}$. Then,

$$f_Z(z) = \begin{cases} f_Y(v(z))|\det\left(v'(z)\right)|, & z \in S_Z \\ 0, & \text{otherwise.} \end{cases} \tag{13.1}$$

Here, $|\det(v'(z))|$ is the absolute value of the determinant of the $n \times n$-dimensional derivative matrix (i.e. Jacobian) of v. This has (i, j) element $\partial v_i(z)/\partial z_j$, where v_i is the ith element of v.

It follows from (13.1) that maximum likelihood inference is invariant to transformation of the data $\mathbf{y}$. This assumes that the transformation $z = u(\mathbf{y})$ does not depend on paramater $\boldsymbol{\theta}$. Then, $\det(v'(z))$ is a constant (with respect to $\boldsymbol{\theta}$), and hence $f_Z(z; \boldsymbol{\theta}) \propto f_Y(v(z); \boldsymbol{\theta}) = f_Y(\mathbf{y}; \boldsymbol{\theta})$.

Box 13.1

13.4 Mean and variance conditional identities

The mean and variance identities given below can be very useful if it is difficult to obtain these moments directly from the distribution of Y, but the distribution of X, and of Y given X, are of convenient form (e.g. Exercise 13.10).

Assuming existence of the means and variances, then for random variables X and Y,

$$E[Y] = E_X\left[E[Y|X]\right], \tag{13.2}$$

and

$$\text{var}(Y) = E_X\left[\text{var}(Y|X)\right] + \text{var}_X\left(E[Y|X]\right), \tag{13.3}$$

where E_X and var_X denote expectation and variance with respect to X, respectively.

The proof of (13.2) is relatively painless, and is given below. See Casella and Berger (1990, Section 4.4 therein) for proof of (13.3).

Proof:

$$\begin{aligned} E[Y] &= \int\int yf(x, y)dxdy \\ &= \int\int yf(y|x)f(x)dxdy \\ &= \int\left(\int yf(y|x)dy\right) f(x)dx \\ &= \int E[Y|X]f(x)dx \\ &= E_X\left[E[Y|X]\right]. \end{aligned}$$

Example 13.1. Let $N \sim \text{Pois}(\lambda)$ and suppose that $Y|N \sim \text{Bin}(N, p)$. From (13.2) and (13.3), the mean and variance of Y are

$$\begin{aligned} E[Y] &= E_N\big[E[Y|N]\big] \\ &= E_N\big[Np\big] = pE_N[N] = p\lambda, \end{aligned}$$

and

$$\begin{aligned} \text{var}(Y) &= E_N\big[\text{var}(Y|N)\big] + \text{var}_N\big(E[Y|N]\big) \\ &= E_N\big[Np(1-p)\big] + \text{var}_N(Np) \\ &= p(1-p)\lambda + p^2\lambda = p\lambda. \end{aligned}$$

In fact, it can be shown that $Y \sim \text{Pois}(p\lambda)$ (Exercise 13.13). □

13.5 Relevant inequalities

13.5.1 Jensen's inequality for convex functions

Let D be an interval in $\mathbb{R}$. A function $\phi : D \to \mathbb{R}$ is convex if the graph of $(y, \phi(y))$ is such that the chord between any two points is completely on or above the curve described by $\phi(y)$ (Figure 13.1). That is,

$$\phi(ay_1 + (1-a)y_2) \leq a\phi(y_1) + (1-a)\phi(y_2),$$

for all $y_1, y_2 \in D$ and $0 \leq a \leq 1$.

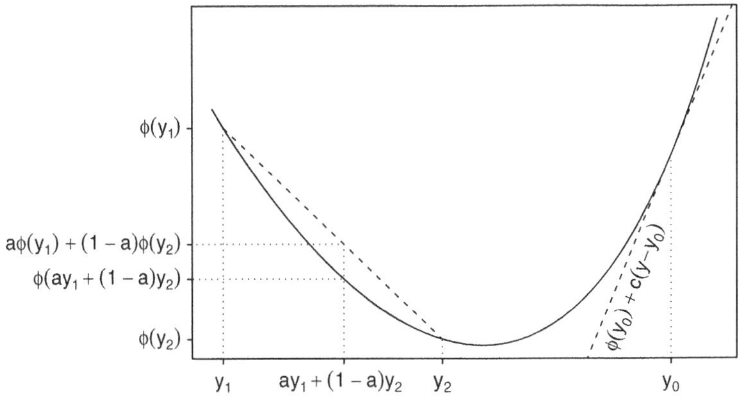

Figure 13.1 Graph of a convex function. The chord between points $(y_1, \phi(y_1))$ and $(y_2, \phi(y_2))$ lies completely on or above the curve $(y, \phi(y))$. The tangent to the point $(y_0, \phi(y_0))$ lies completely on or below the curve.

Jensen's inequality states that if ϕ is convex, and Y is any random variable on D with finite expectation, then

$$\phi(E[Y]) \leq E[\phi(Y)].$$

Proof. An equivalent definition of convexity is that, for any fixed $y_0 \in D$, there exists a value c such that

$$\phi(y_0) + c(y - y_0) \leq \phi(y) \text{ for all } y \in D. \tag{13.4}$$

If ϕ is differentiable, then $c = \phi'(y_0)$, and (13.4) states that the tangent at $(y_0, \phi(y_0))$ lies completely on or below $\phi(y)$ (Figure 13.1).

Jensen's inequality results from setting $y_0 = E[Y]$, and then replacing y by the random variable Y, and taking expectations on both sides of (13.4). □

It can be shown that if $\phi(y)$ is twice differentiable then it is convex if and only if the second derivative is non-negative on D. This can be used to quickly establish that $\phi(y) = \exp(y)$ and $\phi(y) = y^2$ are convex functions on $\mathbb{R}$. The functions $\phi(y) = -\log(y)$ and $\phi(y) = -\sqrt{y}$ are convex on $\mathbb{R}^+$.

Example 13.2. Let Y be distributed $N(\mu, \sigma)$. Then $X = \exp(Y)$ has a lognormal distribution. The exponential function is convex, and application of Jensen's inequality gives

$$E[X] = E[\exp(Y)] \geq \exp(E[Y]) = \exp(\mu).$$

In fact, $E[X] = \exp(\mu + \sigma^2/2)$. □

Example 13.3. Negative bias of the sample standard deviation. Let $Y_1, ..., Y_n$ be iid from some distribution with finite mean μ and variance σ^2. Consider the random variable given by the sample standard deviation

$$S(\boldsymbol{Y}) = \left(\frac{1}{n-1} \sum_{i=1}^{n} (Y_i - \overline{Y})^2 \right)^{1/2}.$$

Using Jensen's inequality with $\phi(S) = S^2$, it follows that

$$\begin{aligned} \phi(E[S]) &= (E[S])^2 \\ &\leq E[\phi(S)] \\ &= E[S^2] \\ &= \sigma^2. \end{aligned}$$

This shows that $E[S] \leq \sigma$. This inequality will be strict provided that the distribution of Y_i is not degenerate, because $\phi(S)$ is strictly convex. □

13.5.2 Cauchy-Schwarz inequality

For any two random variables X and Y such that $E[X^2] < \infty$ and $E[Y^2] < \infty$,

$$(E[XY])^2 \leq E[X^2]E[Y^2].$$

Proof. For any $t \in \mathbb{R}$, it is the case that

$$E[(X - tY)^2] \geq 0.$$

That is,

$$E[X^2] - 2tE[XY] + t^2E[Y^2] \geq 0. \tag{13.5}$$

Equation (13.5) is a quadratic in t of the form $at^2 + bt + c$, where $a = E[Y^2]$, $b = -2E[XY]$, and $c = E[X^2]$. This quadratic takes only non-negative values and so it is necessarily the case that $b^2 - 4ac \leq 0$ (Stewart 1999). That is,

$$4(E[XY])^2 - 4E[Y^2]E[X^2] \leq 0,$$

from which the Cauchy-Schwarz inequality follows.

Corollary. Let (X, Y) be a discrete bivariate random variable having point mass $1/n$ at values (x_i, y_i), $i = 1, ..., n$. The pairs (x_i, y_i) are not required to be distinct. From application of the Cauchy-Schwarz inequality, it follows that

$$\left(\sum_{i=1}^{n} x_i y_i\right)^2 \leq \sum_{i=1}^{n} x_i^2 \sum_{i=1}^{n} y_i^2. \tag{13.6}$$

Example 13.4. Covariance inequality. Let X and Y be random variables with finite means and variances μ_X, σ_X^2, and μ_Y, σ_Y^2, respectively. Then

$$\begin{aligned}\text{cov}(X, Y)^2 &= \big(E[(X - \mu_X)(Y - \mu_Y)]\big)^2 \\ &\leq E\big[(X - \mu_X)^2\big]E\big[(Y - \mu_Y)^2\big] = \sigma_X^2\sigma_Y^2.\end{aligned}$$

□

13.6 Asymptotic probability theory

This section focuses on the concept of convergence for sequences of random variables. The context in which this is relevant to maximum likelihood estimation is that the ML estimator is a random variable under repetition of the experiment, with distribution that depends on the sample size n. This section provides the tools that are used in Chapter 12 to derive the behaviour of this sequence of random variables (after suitable standardization) as n tends to infinity.

13.6.1 Convergence in distribution and probability

Definition 13.1 Convergence in distribution. Let $X_1, X_2, \ldots$ be a sequence of random variables with distribution functions denoted $F_n(x) = P(X_n \le x), n = 1, 2, \ldots$, and let X be a random variable with distribution function $F(x)$. Then the sequence X_n is said to *converge in distribution* (or *in law*), $X_n \to_D X$, if

$$F_n(x) \to F(x) \tag{13.7}$$

for all points x at which F is continuous.

In practice, notation such as $X_n \to_D N(0, 1)$, say, is used to denote $X_n \to_D X$ where X is distributed according to a standard normal distribution. Some of the more theoretical texts use the terminology that F_n converges *weakly* to F, and for this reason convergence in distribution is often called *weak convergence*.

Convergence in distribution extends naturally to the random vector case, $\boldsymbol{X} \in \mathbb{R}^s$, with (13.7) being replaced by the requirement that $F_n(\boldsymbol{x}) \to F(\boldsymbol{x})$ for all points $\boldsymbol{x} \in \mathbb{R}^s$ at which F is continuous. In this case, $F_n(\boldsymbol{x}) = P(X_{n1} \le x_1, \ldots, X_{ns} \le x_s)$ where $\boldsymbol{X}_n = (X_{n1}, \ldots, X_{ns})$ and $\boldsymbol{x} = (x_1, \ldots, x_s)$.

Example 13.5. Let X be distributed $N(0, 1)$, and define X_n as

$$X_n = (-1)^n X + Y_n,$$

where Y_n are independently distributed $N(0, \frac{1}{n})$ random variables. Note that

$$X_n \sim N(0, 1 + 1/n),$$

and hence the distribution function of X_n is

$$\begin{aligned} F_n(x) &= P(X_n \le x) \\ &= P\left(\frac{X_n}{\sqrt{1+1/n}} \le \frac{x}{\sqrt{1+1/n}}\right) \\ &= \Phi\left(\frac{x}{\sqrt{1+1/n}}\right), \end{aligned}$$

where Φ is the distribution function of the standard normal. From the continuity of Φ it follows that $F_n(x)$ converges to $\Phi(x)$ as $n \to \infty$, for all $x \in \mathbb{R}$. This establishes that $X_n \to_D N(0, 1)$, that is, $X_n \to_D X$. □

Convergence in distribution is often used to express the asymptotic behaviour of some suitably standardized estimator, as sample size becomes large. This is demonstrated below by the central limit theorem for the sample mean (see Billingsley 1979, for proof). This theorem is extended to ML estimators by Theorem 12.1 in Chapter 12.

Example 13.6. Central Limit Theorem (CLT). If $Y_1, Y_2, \ldots$ are iid random variables with finite mean and variance, μ and σ^2, and $\overline{Y}_n$ is the mean of $Y_1, \ldots, Y_n$, then

$$X_n = \sqrt{n}(\overline{Y}_n - \mu) \to_D N\left(0, \sigma^2\right). \tag{13.8}$$

□

Example 13.7. Normal approximation to the binomial. A binomial random variable $Y \sim \text{Bin}(n, p)$ is the sum of n iid Bernoulli(p) random variables, each with mean $\mu = P(Y_i = 1) = p$ and variance $\sigma^2 = p(1-p)$. The sample mean of these n Bernoulli random variables is the MLE $\widehat{p}_n = Y/n$, where the subscript n is used to denote that this is viewed as a sequence of estimators, indexed by the number of Bernoulli trials n. From the central limit theorem, it follows that

$$\sqrt{n}(\widehat{p}_n - p) \to_D N\big(0, p(1-p)\big).$$

Figure 13.2 shows the increasing accuracy of the $N(0, p(1-p))$ distribution function as an approximation to the distribution function of $\sqrt{n}(\widehat{p}_n - p)$ as n increases, for the parameter value $p = 0.5$. □

Definition 13.2 Convergence in probability. Let $X_1, X_2, \ldots$ and X be random variables on some probability space. Then the sequence X_n is said to converge in probability, $X_n \to_P X$, if for every $\epsilon > 0$,

$$P(|X_n - X| > \epsilon) \to 0 \text{ as } n \to \infty. \tag{13.9}$$

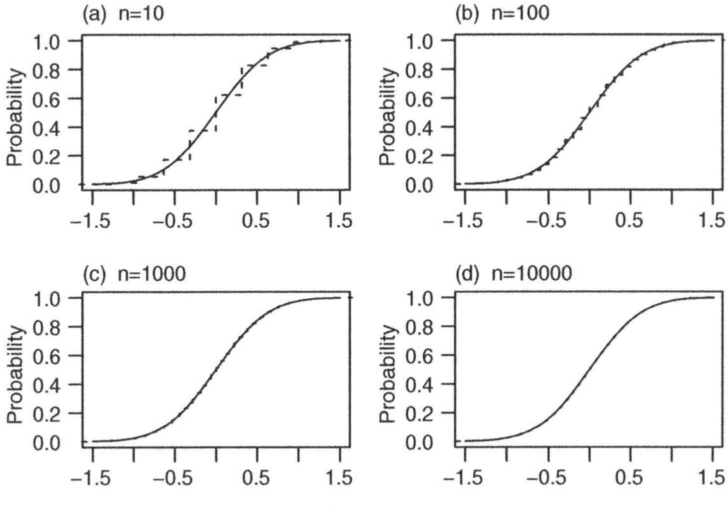

Figure 13.2 Distribution function (dashed stepwise line) of $\sqrt{n}(\widehat{p}_n - 0.5)$ where the estimator $\widehat{p}_n$ is the observed proportion from a Bin(n,0.5) experiment, for n = 10,100,1000,10000. The limiting N (0,0.25) distribution is overlaid (solid line).

Convergence in distribution simply requires the distribution function of X_n to converge to the distribution function of X, but does not make any other assumptions. In particular, it does not assume that X_n takes similar values to X. In contrast, convergence in probability requires that X_n and X be defined on the same probability space (so that the random variable $X_n - X$ is defined) and that X_n takes values close to X with high probability.

Example 13.8. Weak law of large numbers. Let $Y_1, Y_2, \ldots$ be iid with distribution function F, and let $\overline{Y}_n$ denote the mean of $Y_1, \ldots, Y_n$. Then,

$$\overline{Y}_n \to_P \mu$$

for some constant μ if and only if

$$nP(|Y_1| > n) \to 0 \quad \text{and} \quad \int_{[-n,n]} y dF(y) \to \mu \quad \text{as} \quad n \to \infty.$$

The above condition is weaker than existence of $E[Y_1]$. If $E[Y_1]$ exists then convergence is almost sure (Billingsley 1979), and $\mu = E[Y_1]$ (this is the strong law of large numbers). □

13.6.2 Properties

P1: $X_n \to_P X$ implies $X_n \to_D X$. The reverse does not hold in general (but see P2 below).

P2: $X_n \to_D c$, where c is a constant, implies $X_n \to_P c$. That is, convergence in probability and convergence in distribution are equivalent when convergence is to a constant (Exercise 13.5).

P3: If g is a continuous function, and $X_n \to_P X$, then $g(X_n) \to_P g(X)$. The same is true for convergence in distribution, i.e. $X_n \to_D X$ implies $g(X_n) \to_D g(X)$. If X is a constant, c, then g need only be continuous at c to conclude that $g(X_n) \to_D g(c)$.

The following continuation of Example 13.6.2 shows that the reverse of property P1 does not hold in general.

Example 13.5 continued. It was seen earlier that X_n converges in distribution to X, where X has a standard normal distribution. However, X_n does not converge in probability to X because when n is odd then $X_n = -X + Y_n$, and so $X_n - X = -2X + Y_n$ which has a $N(0, 4 + \frac{1}{n})$ distribution. □

Example 13.9. If X_n is the random variable defined in the statement of the central limit theorem in Example 13.6, then applying property P3 with the function

$g(x) = x/\sigma$ gives

$$\frac{\sqrt{n}(\overline{Y}_n - \mu)}{\sigma} \rightarrow_D N(0, 1),$$

which is another commonly used and equivalent statement of the CLT. Moreover, using property P3 once again, but with $g(x) = x^2$, it follows that

$$\frac{n(\overline{Y}_n - \mu)^2}{\sigma^2} \rightarrow_D \chi_1^2,$$

since the square of a standard normal random variable has a χ_1^2 distribution. □

13.6.3 Slutsky's theorem

Slutsky's theorem.
If $X_n \rightarrow_D X$, $A_n \rightarrow_P a$, and $B_n \rightarrow_P b$ (where a and b are constants), then

$$A_n + B_n X_n \rightarrow_D a + bX.$$

Example 13.10. Asymptotic normality of the *t*-statistic. Let $Y_1, ..., Y_n$ be iid from a distribution with mean μ and variance σ^2 and let $S_n^2 = \sum_{i=1}^n (Y_i - \overline{Y}_n)^2/(n-1)$ denote the sample variance. It can be shown (Exercise 13.6) that $S_n \rightarrow_P \sigma$, and hence that $B_n = \sigma/S_n \rightarrow_P 1$ by property P3 above. From the central limit theorem (Example 13.6), $X_n = \sqrt{n}(\overline{Y}_n - \mu)/\sigma \rightarrow_D N(0, 1)$, and so application of Slutsky's theorem (with A_n omitted) gives

$$\begin{aligned} T &= \frac{\sqrt{n}(\overline{Y}_n - \mu)}{S_n} \\ &= \frac{\sigma}{S_n} \frac{\sqrt{n}(\overline{Y}_n - \mu)}{\sigma} \\ &\rightarrow_D 1 \times N(0, 1) = N(0, 1). \end{aligned}$$

□

13.6.4 Delta theorem

The delta theorem is of great practical importance because, given convergence in distribution of (suitably standardized) X_n, it gives the asymptotic behaviour of (suitably standardized) functions of X_n. The practical version of this, the delta method, is implemented in R function `deltamethod` and SAS macro `DeltaMethod` (Section 15.5.5) and in several SAS procedures. See Section 4.2 for examples of its use.

Delta theorem, $X_n \in \mathbb{R}$.
Suppose that $\sqrt{n}(X_n - b) \to_D X$. If $g : \mathbb{R} \to \mathbb{R}$ is differentiable and its derivative g' is continuous at b, then

$$\sqrt{n}\big(g(X_n) - g(b)\big) \to_D g'(b)X.$$

Proof. A Taylor series expansion of $g(X_n)$ about b, with Lagrange's form of remainder (i.e, the mean-value theorem, Apostol 1967), gives

$$g(X_n) = g(b) + g'(X_n^*)(X_n - b),$$

where X_n^* lies between X_n and b. Thus,

$$\begin{aligned}\sqrt{n}\big(g(X_n) - g(b)\big) &= \sqrt{n}\big(g'(X_n^*)(X_n - b)\big)\\ &= g'(X_n^*)\big(\sqrt{n}(X_n - b)\big).\end{aligned} \tag{13.10}$$

By Slutsky's theorem

$$X_n - b = \frac{1}{\sqrt{n}}\big(\sqrt{n}(X_n - b)\big) \to_D 0 \times X = 0,$$

since $1/\sqrt{n} \to_p 0$. Hence $X_n - b \to_p 0$ (by Property P2). Since $X_n - b \to_p 0$ then $X_n^* - b \to_p 0$ also, since $|X_n^* - b| \leq |X_n - b|$. This is equivalent to $X_n^* \to_p b$. Now, g' is continuous at b, so $g'(X_n^*) \to_p g'(b)$ by Property P3. Applying Slutsky's theorem to (13.10) gives the required result. □

Example 13.11. If $\sqrt{n}(\overline{Y}_n - \mu) \to_D N(0, \sigma^2)$ and $g(x) = x^2$ then

$$\sqrt{n}(\overline{Y}_n^2 - \mu^2) \to_D 2\mu N(0, \sigma^2) = N(0, 4\mu^2\sigma^2).$$

□

Delta theorem, $X_n \in \mathbb{R}^s$.
The delta theorem extends naturally to the random vector case. Suppose that

$$\sqrt{n}(\boldsymbol{X}_n - \boldsymbol{b}) \to_D \boldsymbol{X}.$$

Let $g : \mathbb{R}^s \to \mathbb{R}^p$ where

$$g(\boldsymbol{x}) = \begin{pmatrix} g_1(\boldsymbol{x}) \\ \cdot \\ \cdot \\ \cdot \\ g_p(\boldsymbol{x}) \end{pmatrix},$$

and each co-ordinate $g_i : \mathbb{R}^s \to \mathbb{R}$ has derivative $g_i' = (\frac{\partial g_i}{\partial x_1}, ..., \frac{\partial g_i}{\partial x_s})^T$ that is continuous at $\boldsymbol{b}$. Then,

$$\sqrt{n}(g(\boldsymbol{X}_n) - g(\boldsymbol{b})) \to_D \mathbf{G}(\boldsymbol{b})\boldsymbol{X},$$

where $\mathbf{G}(\boldsymbol{b})$ is the $p \times s$ Jacobian matrix of partial derivatives

$$\mathbf{G}(\boldsymbol{b}) = \begin{pmatrix} g_1'(\boldsymbol{b})^T \\ \cdot \\ \cdot \\ \cdot \\ g_p'(\boldsymbol{b})^T \end{pmatrix}.$$

13.7 Exercises

13.1 Provide a counter example to show that $X_n \to_D X$ and $Y_n \to_D Y$ does *not* imply that $X_n + Y_n \to_D X + Y$.

13.2 The definition of convergence in probability (Definition 13.2) used a strict inequality. Show that this inequality does not need to be strict, that is, show that the condition

$$P(|X_n - X| > \epsilon) \to 0 \text{ as } n \to \infty \qquad \text{for all } \epsilon > 0, \tag{13.11}$$

is equivalent to the condition that

$$P(|X_n - X| \geq \varepsilon) \to 0 \text{ as } n \to \infty \qquad \text{for all } \varepsilon > 0. \tag{13.12}$$

13.3 Let $k < 0$ be a constant and let $X_1, X_2, ...$ be a sequence of random variables. Prove that $X_n \to_P k$ implies $P(X_n < 0) \to 1$.

13.4 Suppose that $X_n \to_P 0$ and that B_n is bounded in probability, that is, there exists a $b > 0$ such that $P(|B_n| > b) \to 0$. Show that $X_n B_n \to_P 0$.

13.5 Let c be a constant and let $X_1, X_2, ...$ be a sequence of random variables. Prove that

1. $X_n \to_P c$ implies $X_n \to_D c$.
2. $X_n \to_D c$ implies $X_n \to_P c$.

13.6 Let $Y_1, ..., Y_n$ be iid with mean μ and variance σ^2. Show that the sample variance S_n^2 converges in probability to σ^2.
Hint: Show that the sample variance can be expressed as

$$S_n^2 = \sum (Y_i - \overline{Y}_n)^2/(n-1) = \frac{n}{n-1}\left(\sum (Y_i - \mu)^2/n + K_n\right),$$

where $K_n \to_P 0$.

13.7 Let $X_n \sim \text{Pois}(n)$. Using the fact that X_n has distribution equal to that of the sum of n iid Pois(1) random variables, use the CLT to show that $\sqrt{n}\left(\frac{X_n}{n} - 1\right) \to_D N(0, 1)$. (For large n, this result can be used to state that a Pois(n) random variable is approximately normally distributed, $X_n \overset{\cdot}{\sim} N(n, n)$.)

13.8 Let $X_n \sim \chi^2_n$. Using the fact that X_n has distribution equal to that of the sum of n iid χ^2_1 random variables,

1. Show that $\sqrt{n}\left(\frac{X_n}{n} - 1\right) \to_D N(0, 2)$,
2. Hence, show that $\sqrt{n}\sigma^2\left(\frac{X_{n-1}}{n} - 1\right) \to_D N(0, 2\sigma^4)$ (this result is noted in Example 12.4).

13.9 The following counter examples show that convergence in probability (and hence also convergence in distribution) do not imply convergence of moments.

1. Construct an example in which $X_n \to_P X$ where $E[X_n]$ exist for all n, and $\lim_{n\to\infty} E[X_n] = \mu$, but $E[X] = \mu_X \neq \mu$.
2. Construct an example in which $X_n \to_P X$ where $E[X]$ exists and $E[X_n]$ exist for all n, but $\lim_{n\to\infty} E[X_n]$ does not exist.

13.10 Let Y be distributed according to the five-parameter binormal mixture model of Example 2.9. By conditioning on the unobserved Bernoulli random variable B, use (13.2) and (13.3) to find $E[Y]$ and var(Y).

13.11 For Y distributed according to the zero-inflated Poisson (see Section 7.6.2), use (13.2) and (13.3) to show that $E[Y] = (1 - p)\lambda$ and $\text{var}(Y) = (1 + p\lambda)E[Y]$.

13.12 Let $\lambda \sim \text{Gamma}(m, \mu/m)$ with density function given by (15.5), and let $Y|\lambda \sim \text{Pois}(\lambda)$. Using (13.2) and (13.3), show that Y has mean μ and variance $\mu(1 + \mu/m)$. (Note: It can be shown that Y has a negative binomial distribution, NB(m, μ), with density given by (15.2).)

13.13 For Example 13.1, show that $Y \sim \text{Pois}(p\lambda)$.
Hint: Use the so-called law of total probability, which in this case can be written as

$$f(y) = \sum_{n \geq y} f(y|N = n)P(N = n).$$

14

Fundamental paradigms and principles of inference

The statistician cannot excuse himself from the duty of getting his head clear on the principles of scientific inference, but equally no other thinking man can avoid a like obligation – Sir Ronald Fisher

14.1 Introduction

Not all statisticians think alike. In fact, there are several schools of thought regarding the appropriate framework for statistical inference. These include frequentist, fiducial, and Bayesian.[1]

The frequentist approach is also known as the 'classical' or 'traditional' approach. It is (at the time of writing) the form of statistical inference that is most widely taught at schools and universities, and with the widespread use of familiar frequentist-based methods (e.g. linear regression, ANOVA, chi-squared tests, generalized linear models) it continues to be the most widely used of the statistical paradigms. This could change in the future, due to the recent computational and methodological advances in Bayesian inference. Bayesians can also lay claim to a more axiomatic and complete approach to inference. Of course, Bayesian inference requires the specification of priors, and much of its current popularity is a consequence of the willingness to utilize so-called non-informative or reference priors

[1] See Section 15.3 for a quick self-test of frequentist versus Bayesian thinking.

Maximum Likelihood Estimation and Inference: With Examples in R, SAS and ADMB, First Edition. Russell B. Millar.

(e.g. Gelman, Carlin, Stern and Rubin 2003). This could be viewed as a pact among Bayesians to seek validity by all agreeing to do the same thing.[2]

The frequentist approach is built around the idea of the (hypothetical) repeatability of experiments. Indeed, frequentists define the probability of an event to be the proportion of times it occurs out of a large number of independent repeat experiments, and the tools of frequentist inference are based on this notion. For example, the method used to construct a 95 % confidence interval is such that, if the experiment is repeated a large number of times, the calculated 95 % confidence intervals will contain the unknown parameter about 95 % of the time.

The fiducial approach (Fisher 1933) essentially advocates that inference should be based on the likelihood function alone, without recourse to prior knowledge or the concept of repeated sampling. However, this approach appears to have clouded rather than clarified the debate, and has now gone into statistical obscurity.

The Bayesian approach also advocates using the likelihood function, but in conjunction with a prior distribution on the parameters. Moveover, the Bayesian approach departs markedly from the frequentist and fiducial approaches, by viewing probability as an expression of belief.

This chapter scratches the surface on some of the considerations behind the justifications of an appropriate paradigm for statistic inference. The presentation takes a look at three principles, namely, the sufficiency, conditionality and likelihood principles. Birnbaum (1962, with discussion) caused a statistical furor between frequentist and Bayesian statisticians with his ground-breaking work on the interplay between these relationships. It is the flavour of this controversy that the presentation here attempts to capture.

Conceptually, it may help the reader to (temporarily) unlearn their previous statistical training. So, for example, when conducting an experiment to make inference about the probability that a passenger car contains more than one occupant, suspend your knowledge that an appropriately conducted experiment will yield a binomial random variable. Instead, consider what it means to repeat the experiment (e.g. see Example 14.4), and what it is that the observer actually measures (e.g. see Example 14.2). In the following sections, notation of the form $(E, \boldsymbol{y})$ is used to denote the observation of $\boldsymbol{y}$ from an experiment E.

14.2 Sufficiency principle

Speaking very crudely, the notion behind the sufficiency principle is that the data $\boldsymbol{y} = (y_1, ..., y_n)$ can be partitioned into a useful part and an irrelevant part. It is only the useful part that should be utilized in making inference, and the irrelevant part should be discarded because it will simply add noise (i.e. increase statistical variability). This is seen in the case of iid $N(\mu, \sigma^2)$ data where it is

[2] The author enjoys modelling under both the frequentist and Bayesian paradigms. One of the first questions in any analysis is to determine the most appropriate paradigm, or mix of paradigms (e.g. Gelman and Hill 2007), to use in order to best answer the research question and to meet the needs of scientific colleagues or clients.

only the sample mean, $\overline{y}$, and the sample variance $s^2 = \sum_{i=1}^{n}(y_i - \overline{y})^2/(n-1)$ that are used for inference about μ or σ^2. The data, $\boldsymbol{y}$, can be ignored once $\overline{y}$ and s^2 have been calculated.[3]

Somewhat more formally, the concept of sufficiency is used to reduce the data in such a way that no relevant information about the parameters is lost. In the definition of sufficiency below, the statistic $T(\boldsymbol{Y}) : \mathbb{R}^n \to \mathbb{R}^m$, where $m \leq n$, can be considered a 'reduction' of the data $\boldsymbol{Y}$ if $m < n$. The essence of sufficiency is that the data $\boldsymbol{Y}$ can, loosely speaking, be partitioned into the pieces $T(\boldsymbol{Y})$ and $\boldsymbol{Y}|T(\boldsymbol{Y})$ ($\boldsymbol{Y}$ given $T(\boldsymbol{Y})$), where the distribution of the latter does not depend on parameter (vector) $\boldsymbol{\theta}$.

Definition 14.1 Sufficient statistic. $T(\boldsymbol{Y}) : \mathbb{R}^n \to \mathbb{R}^m$ is a *sufficient statistic* for $\boldsymbol{\theta}$ if the distribution of $\boldsymbol{Y}$ given $T(\boldsymbol{Y})$ does not depend on $\boldsymbol{\theta}$. □

The argument can be made that if $\boldsymbol{Y}|T(\boldsymbol{Y})$ does not depend on $\boldsymbol{\theta}$ then it can not provide information about $\boldsymbol{\theta}$. Hence, all of the information in the data about $\boldsymbol{\theta}$ is contained solely in the sufficient statistic $T(\boldsymbol{Y})$. This logic is formally embodied in the sufficiency principle, which is stated below.

The sufficiency principle:
Let $\boldsymbol{y}$ be observed from experiment E_1, and let $T(\boldsymbol{Y})$ be any sufficient statistic for parameter $\boldsymbol{\theta}$. Let E_2 be the corresponding experiment in which only the value of $T(\boldsymbol{y})$ is observed (but not $\boldsymbol{y}$ itself). Then, $(E_1, \boldsymbol{y})$ and $(E_2, T(\boldsymbol{y}))$ have the same information content about $\boldsymbol{\theta}$. □

14.2.1 Finding sufficient statistics

In simple cases, sufficiency of a statistic $T(\boldsymbol{Y})$ can be established directly from the above definition, as demonstrated below.

Example 14.1. Let Y_1 and Y_2 be iid Pois(λ). It will be established that the statistic $T(\boldsymbol{Y}) = Y_1 + Y_2$ is sufficient for λ, by showing that the distribution of (Y_1, Y_2) given T does not depend on λ.

The conditional density is

$$\begin{aligned} f(\boldsymbol{y}|T(\boldsymbol{Y}) = t; \lambda) &= f(y_1, y_2|t; \lambda) \\ &= \frac{f(y_1, y_2, t; \lambda)}{f_T(t; \lambda)} \\ &= \begin{cases} \frac{f(y_1, y_2; \lambda)}{f_T(t; \lambda)}, & y_1 + y_2 = t \\ 0, & y_1 + y_2 \neq t\,. \end{cases} \end{aligned}$$

[3] This assumes that the specified model is correct – the raw data $\boldsymbol{y}$ would be needed for model assessment.

Now, since Y_1 and Y_2 are independent, $f(y_1, y_2; \lambda)$ is the product of their Pois(λ) density functions. Also, using moment generating functions, or from direct calculation, it can be shown that T has a Pois(2λ) distribution, and so

$$\frac{f(y_1, y_2; \lambda)}{f_T(t; \lambda)} = \frac{\frac{e^{-\lambda}\lambda^{y_1}}{y_1!}\frac{e^{-\lambda}\lambda^{y_2}}{y_2!}}{\frac{e^{-2\lambda}(2\lambda)^t}{t!}}$$
$$= \frac{t!}{y_1!y_2!}\left(\frac{1}{2}\right)^t . \tag{14.1}$$

This does not depend on λ, and hence T is sufficient for λ. In fact, since $y_2 = t - y_1$ (with probability one), it follows that (14.1) can be written

$$f(y_1|t) = \frac{t!}{y_1!(t-y_1)!}\left(\frac{1}{2}\right)^t ,$$

which establishes that, conditional on t, Y_1 is distributed Bin(t, $\frac{1}{2}$). □

In general, determining the conditional distribution of $\boldsymbol{Y}|T(\boldsymbol{Y})$ can be cumbersome or intractable. The factorization theorem (also known as the Fisher-Neyman criterion) provides a much more convenient method to identity sufficient statistics.

Theorem 14.1 Factorization theorem.
$T(\boldsymbol{Y})$ is sufficient for $\boldsymbol{\theta}$ if and only if there exist non-negative functions $g(T(\boldsymbol{y}); \boldsymbol{\theta})$ and $h(\boldsymbol{y})$, such that the density function of $\boldsymbol{y}$, $f(\boldsymbol{y}; \boldsymbol{\theta})$, can be written in the form

$$f(\boldsymbol{y}; \boldsymbol{\theta}) = g(T(\boldsymbol{y}); \boldsymbol{\theta})h(\boldsymbol{y}) , \tag{14.2}$$

where $h(\boldsymbol{y})$ does not depend on $\boldsymbol{\theta}$.

Proof. To show that sufficiency implies the above factorization of $f(\boldsymbol{y}; \boldsymbol{\theta})$, assume that $T(\boldsymbol{Y})$ is sufficient for $\boldsymbol{\theta}$. Then,

$$f(\boldsymbol{y}; \boldsymbol{\theta}) = f(\boldsymbol{y}|T(\boldsymbol{y}); \boldsymbol{\theta})f_T(T(\boldsymbol{y}); \boldsymbol{\theta})$$
$$= f(\boldsymbol{y}|T(\boldsymbol{y}))f_T(T(\boldsymbol{y})|\boldsymbol{\theta}) \equiv h(\boldsymbol{y})g(T(\boldsymbol{y}); \boldsymbol{\theta}) ,$$

where $h(\boldsymbol{y}) = f(\boldsymbol{y}|T(\boldsymbol{y}))$ does not depend on $\boldsymbol{\theta}$ because $T(\boldsymbol{Y})$ is sufficient for $\boldsymbol{\theta}$.

To show that the factorization of $f(\boldsymbol{y}; \boldsymbol{\theta})$ implies sufficiency, assume that (14.2) holds, and note that the density of $T(\boldsymbol{Y})$ can then be simplified to

$$f_T(t; \boldsymbol{\theta}) = \int_{\boldsymbol{y}:T(\boldsymbol{y})=t} f(\boldsymbol{y}; \boldsymbol{\theta})d\boldsymbol{y}$$
$$= \int_{\boldsymbol{y}:T(\boldsymbol{y})=t} g(t; \boldsymbol{\theta})h(\boldsymbol{y})d\boldsymbol{y}$$
$$= g(t; \boldsymbol{\theta})\int_{\boldsymbol{y}:T(\boldsymbol{y})=t} h(\boldsymbol{y})d\boldsymbol{y} = k_t g(t; \boldsymbol{\theta}) , \tag{14.3}$$

where k_t depends only on t. Thus, the conditional density of $\boldsymbol{Y}$ given $T(\boldsymbol{y}) = t$ is

$$\begin{aligned} f(\boldsymbol{y}|T(\boldsymbol{y}) = t) &= \frac{f(\boldsymbol{y};\boldsymbol{\theta})}{f_T(t;\boldsymbol{\theta})} \\ &= \frac{g(t;\boldsymbol{\theta})h(\boldsymbol{y})}{k_t g(t;\boldsymbol{\theta})} = \frac{h(\boldsymbol{y})}{k_t}, \end{aligned}$$

which does not depend on $\boldsymbol{\theta}$. Hence, $T(\boldsymbol{Y})$ is sufficient for $\boldsymbol{\theta}$. □

It is immediate from the factorization theorem that maximum likelihood estimators are functions of sufficient statistics, because the maximization of $L(\boldsymbol{\theta}) = f(\boldsymbol{y};\boldsymbol{\theta})$ is equivalent to maximization of $g(T(\boldsymbol{y});\boldsymbol{\theta})$, and depends on $\boldsymbol{y}$ only through $T(\boldsymbol{y})$.

14.2.2 Examples of the sufficiency principle

Example 14.1 continued. The density function of $\boldsymbol{y} = (y_1, y_2)$ is

$$f(y_1, y_2; \lambda) = \frac{e^{-2\lambda}\lambda^{y_1+y_2}}{y_1!y_2!},$$

which can be written in the form of (14.2) with $T(\boldsymbol{y}) = y_1 + y_2$, $g(T(\boldsymbol{y}); \lambda) = e^{-2\lambda}\lambda^{T(\boldsymbol{y})}$, and $h(\boldsymbol{y}) = 1/(y_1!y_2!)$. Hence, $T(\boldsymbol{Y}) = Y_1 + Y_2$ is sufficient for λ by the factorization theorem. That is, the information content about λ from observing iid Y_1 and Y_2 distributed Pois(λ) is the same as that from observing $T = Y_1 + Y_2$ distributed Pois(2λ). □

Example 14.2. Bernoulli trials. Suppose that n iid Bernoulli(p) random variables, $Y_1, \ldots, Y_n$, are observed. Then, by the factorization theorem, $T = \sum_{i=1}^n Y_i$ is sufficient for p because $f(y_1, \ldots, y_n; p) = p^t(1-p)^{n-t}$. The sufficiency principle says that $T \sim \text{Bin}(n, p)$ has the same information content about p as the experiment in which the Bernoulli sequence $Y_1, \ldots, Y_n$ is observed. □

Example 14.3. Normal. For iid $Y_i \sim N(\mu, \sigma^2)$, the factorization theorem quickly establishes that $(\overline{Y}, \sum Y_i^2)$ is sufficient for $\boldsymbol{\theta} = (\mu, \sigma^2)$. The sufficiency principle can therefore be used to say that experiment E_2 in which only the bivariate outcome $(\overline{y}, \sum y_i^2)$ is observed has the same information content about μ and σ^2 as the experiment E_1 in which the entire data vector $\boldsymbol{y} = (y_1, \ldots, y_n)$ is observed. Note that observing $(\overline{y}, \sum y_i^2)$ is equivalent to observing $(\overline{y}, \sum(y_i - \overline{y})^2)$ or $(\overline{y}, s^2)$, because each of these pairs of values can be calculated from any other pair. □

In frequentist statistics there is a standard body of theory based on the use of sufficient statistics for the construction of unbiased estimators with minimum variance (MVUE's). At the core of this theory is the Rao-Blackwell theorem which states that if a MVUE exists then it must be a function of only a minimal[4] sufficient statistic.

14.3 Conditionality principle

The idea behind the conditionality principle is to condition upon (i.e. treat as fixed) aspects of the experiment which contain no information about $\boldsymbol{\theta}$. Some form of conditioning is vital in frequentist statistics, because it is needed to determine what exactly is meant by 'repetition of the experiment'.

The statement of the conditionality principle that is given below uses a mixture of just two experiments (see Example 14.4), but it applies more generally to a mixture of an arbitrary number of experiments (see Example 14.4). Here, E_1 and E_2 are two experiments with the same parameter space Θ. The observations from these two experiments have densities $f_1(\boldsymbol{y}_1; \boldsymbol{\theta})$ and $f_2(\boldsymbol{y}_2; \boldsymbol{\theta})$ respectively, where the unknown $\boldsymbol{\theta} \in \Theta$ is the same in both experiments.

The conditionality principle:
Let the mixture experiment, E, consist of $(E_1, \boldsymbol{y}_1)$ (i.e. performing experiment E_1, from which $\boldsymbol{y}_1$ is observed) with probability p, or $(E_2, \boldsymbol{y}_2)$ with probability $1 - p$. If $(E_i, \boldsymbol{y}_i)$, $i = 1, 2$, is the experiment actually performed, then the information content about $\boldsymbol{\theta}$ from the mixture experiment $(E, (E_i, \boldsymbol{y}_i))$ is equal to that from $(E_i, \boldsymbol{y}_i)$. □

In the above statement of the conditionality principle, probability p does not depend on $\boldsymbol{\theta}$, and hence there is no information about $\boldsymbol{\theta}$ from the choice of whether it is experiment E_1 or E_2 that is observed. The implication is that it is appropriate to condition on the experiment actually performed.

Example 14.4. Random experiment. Suppose that the experiment consists of a laboratory making four iid measurements of the concentration of a chemical in a supplied sample. The laboratory has two spectrometers, an older machine, and a new one that is an order of magnitude more precise than its predecessor. For ease of exposition, it will be assumed that the measurements (under repeated use) from the two machines are normally distributed about the true concentration, μ, with standard deviations of 1 and 0.1 micro-moles per gram, respectively. Unfortunately, on the day of measuring, the new machine was not available and it was necessary to use the old machine.

Having taken a sample of four measurements using the old machine, inference would proceed by calculating the mean of these measurements and using knowledge about the distribution of this sample mean under replication of the experiment. Specifically, since individual measurements from the old machine are distributed $N(\mu, 1)$, it follows that their sample mean is normally distributed with mean μ and

[4] A sufficient statistic $T(Y)$ is said to be minimal sufficient if for any other sufficient statistic $S(Y)$ there exists a function h_S such that $T(Y) = h_S(S(Y))$.

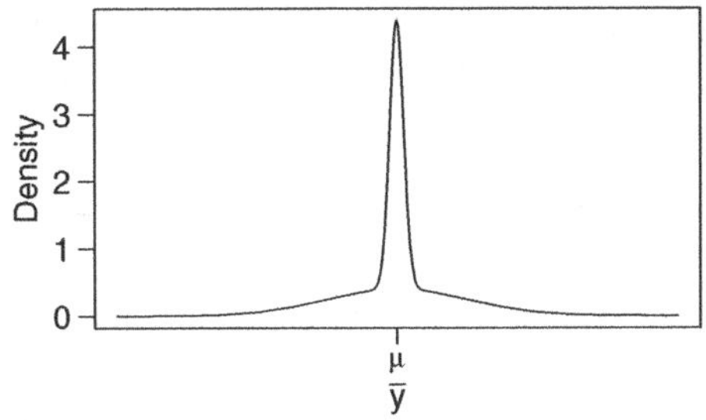

Figure 14.1 Binormal density function for an equal mixture of $N(\mu, 0.5^2)$ and $N(\mu, 0.05^2)$ distributions, corresponding to the distribution of $\overline{Y}$ under the mixture experiment.

standard deviation of $1/\sqrt{4} = 0.5$ micro-moles g^{-1}. But, is this *really* the sampling distribution under repetition of the experiment?

The experiment described above is a mixture experiment, in the sense that the measured concentrations are obtained from either using the new spectrometer (experiment E_1, say) or the old one (experiment E_2), depending on the availability of the new spectrometer. In the above scenario the old machine was used, and it is totally irrelevant that there was a more precise machine that could have been used on another occasion. By calculating a standard deviation of 0.5 micro-moles g^{-1} for the sample mean of the four measurements, the experiment that is being hypothetically repeated is the observation of measurements $\boldsymbol{Y}_2$ from experiment E_2. That is, four iid measurements from the old machine.

If the conditionality principle was not used here, then the distribution of the sample mean under repetition of the experiment would be binormal (see Example 2.9). Figure 14.1 shows the shape of this binormal density assuming that the more accurate spectrometer is available 50 % of the time (at random). The mean and standard deviation of this binormal distribution are μ and $\sqrt{0.12625} = 0.3553$, respectively (see Exercise 14.6). □

Example 14.5. Random number of trials. In practice, the number of trials in a binomial experiment is often random. (This is equivalent to a random sample size in a sequence of Bernoulli trials.) For example, to estimate the probability that a passenger car has more than one occupant, one might record the number of occupants in all passenger cars passing through an intersection over the period of an hour. The sample size is therefore random, because if this hour-long experiment were to be repeated on another occasion then it would most likely result in a different sample size. Thus, this experiment is actually a mixture experiment that consists of conducting one of many different individual binomial experiments, each having a different number of trials.

To make inference about the probability of multi-occupancy in passenger cars, it would not be appropriate to use the mixture experiment in which the sample size is random. Rather, one would condition on the binomial experiment actually performed by using a binomial model with the total number of trials fixed and equal to the number of passenger cars that were observed during the hour of observation. □

> Chapter 9 presents further uses of conditioning, but in the context of coping with incidental parameters. There, it is seen that conditioning may be appropriate even when the probabilities of performing the component experiments (e.g. E_1 and E_2 in the statement of the conditionality principle) do depend on $\boldsymbol{\theta}$. This is because the knowledge of which experiment was performed may be confounded by lack of knowledge about the incidental parameters, and therefore be unable to contribute information about $\boldsymbol{\theta}$.

Box 14.1

14.4 The likelihood principle

The likelihood principle is the third and final principle that is presented here. In essence, it states that if two experiments result in the same likelihood function, then those two experiments have the same information content about $\boldsymbol{\theta}$.

The likelihood principle:
If $\boldsymbol{y}_1$ observed from experiment E_1, and $\boldsymbol{y}_2$ observed from experiment E_2, have the same likelihood functions (to within a constant), i.e. $f_1(\boldsymbol{y}_1;\boldsymbol{\theta}) = cf_2(\boldsymbol{y}_2;\boldsymbol{\theta})$, for all $\boldsymbol{\theta} \in \Theta$, then the information content about $\boldsymbol{\theta}$ is the same from both $(E_1, \boldsymbol{y}_1)$ and $(E_2, \boldsymbol{y}_2)$. □

The likelihood principle provides a rationale for ignoring measurements that have a distribution that does not depend on $\boldsymbol{\theta}$. The density function of such measurements is a constant (with respect to $\boldsymbol{\theta}$) and, as seen in the continuation of Example 14.5 below, the likelihood principle asserts that they contain no relevant information about $\boldsymbol{\theta}$.

Example 14.5 continued. In the context of Example 14.5 let E_1 be the binomial experiment that observes y_1 multi-occupant passenger cars out of a total of n passing cars.[5] A second experiment, E_2, is given by recording additional variables, such as the colour of the car, gender of the driver, etc, and these additional observations will be denoted z, so that $\boldsymbol{y}_2 = (y_1, z)$. If these additional observations have distributions that do not depend on the probability of multiple occupancy (but could be dependent on the occupancy status of the cars), then

$$\begin{aligned} f_2(\boldsymbol{y}_2; p) &\equiv f_2(y_1, z; p) \\ &= f_Z(z|y_1) f_1(y_1; p) \equiv cf_1(y_1; p) , \end{aligned}$$

[5] The conditionality principle has been applied here, and n is taken as fixed.

where p denotes the probability of multiple occupancy. That is, the two experiments have the same information content about p because the likelihood corresponding to these extra observations is incorporated into the constant c. □

However, it has been argued that the likelihood principle is contradictory to some aspects of frequentist inference (e.g. Berger and Wolpert 1988), and this is demonstrated by the following example. Some discussion of this potential conflict is given in Section 14.4.1.

Example 14.6. Negative binomial vs binomial. A market researcher is interested in households that have gross income within a certain range. She needs to question four such households, and so contacts randomly chosen households until four such households are obtained. This required contacting 12 households to get the required four. What proportion, p, of households are in the desired income range?

This experiment, E_1, is negative binomial NB(4, p), where the random variable Y_1 is the number of 'failures' (households *not* in the income range of interest) that occur prior to observing the fourth 'success' (Section 15.4.1). The observation was $y_1 = 8$, and from (15.1), the likelihood function from this negative binomial experiment is

$$L_1(p, 8) = \binom{11}{3} p^4(1 - p)^8.$$

To within a constant, this is the same as the likelihood for a binomial experiment, E_2, where $y_2 = 4$ successes are observed from 12 trials,

$$L_2(p, 4) = \binom{12}{4} p^4(1 - p)^8.$$

The likelihood principle states that the negative binomial experiment $(E_1, 8)$ has the same information content as the binomial experiment $(E_2, 4)$. The MLE of p is 1/3 for both of these experiments.

However, rather than using the MLE, some frequentists might prefer to use the minimum variance unbiased estimator (MVUE) of p, should it exist. For a negative binomial requiring $m \geq 2$ successes, each with probability p, it can be shown that the MVUE of p is $\widetilde{p} = (m - 1)/(m + Y - 1)$ (Lehmann 1983, p. 134). Here, $\widetilde{p} = 3/11$ for the negative binomial experiment $(E_1, 8)$. In contrast, for the binomial experiment $(E_2, 4)$, the minimum variance unbiased estimator (and also MLE in this case) is $\widehat{p} = 1/3$. It could be argued that this conflicts with the assertion of the likelihood principle, that these two experiments have the same information content about p, and this issue is briefly examined in Section 14.4.1. □

14.4.1 Relationship with sufficiency and conditionality

It is relatively easy to show that the likelihood principle implies both the sufficiency and conditionality principles, and the proof of this is given in Theorem 14.2. Birnbaum (1962) showed that the converse also holds, that is, the sufficiency and conditionality principles jointly imply the likelihood principle. (Also of note, Evans, Fraser and Monette (1986) presented a modified version of the conditionality principle that by itself implies the likelihood principle.) Birnbaum's result caused a lot of disquiet to the frequentist community, because the sufficiency and conditionality principles are generally considered necessary tenets of frequentist inference, but the likelihood principle is considered to be unpalatable (due to the type of conflict demonstrated in Example 14.6). This is because the likelihood principle is based solely on the likelihood function that results from the data that were actually observed, and makes no use of the concept of repeatable experiments.

Theorem 14.2 The likelihood principle implies both the sufficiency principle and the conditionality principle.

Proof: To show that the likelihood principle implies the sufficiency principle, let $T(\boldsymbol{Y})$ be a sufficient statistic. Then, from the factorization theorem and Equation (14.3), the density function of $\boldsymbol{y}$ can be written

$$\begin{aligned} f(\boldsymbol{y};\boldsymbol{\theta}) &= g(T(\boldsymbol{y});\boldsymbol{\theta})h(\boldsymbol{y}) \\ &= \frac{f_T(T(\boldsymbol{y});\boldsymbol{\theta})}{k_t}h(\boldsymbol{y}) \\ &= c(\boldsymbol{y})f_T(T(\boldsymbol{y});\boldsymbol{\theta})\,, \end{aligned}$$

where $f_T(T(\boldsymbol{y});\boldsymbol{\theta})$ is the density function of $T(\boldsymbol{y})$, and $c(\boldsymbol{y})$ does not depend on $\boldsymbol{\theta}$. Applying the likelihood principle, with E_1 being the experiment where $\boldsymbol{y}$ is observed from the distribution with density function $f_1 = f(\boldsymbol{y};\boldsymbol{\theta})$, and E_2 being the experiment where $T(\boldsymbol{y})$ is observed from the distribution with density function $f_2 = f_T(T(\boldsymbol{y});\boldsymbol{\theta})$, it is concluded that the information content of $(E_1, \boldsymbol{y})$ and $(E_2, T(\boldsymbol{y}))$ is the same. This is the sufficiency principle.

To show that the likelihood principle implies the conditionality principle, let the mixture experiment assign probabilities p and $1-p$ to experiments 1 and 2, respectively. Hence, under the mixture experiment, the density function for observing $\boldsymbol{y}_i$ from experiment $i = 1, 2$ is given by the probability that experiment i is performed, multiplied by the density function for $\boldsymbol{y}_i$ under experiment i. That is, the observation $(E_i, \boldsymbol{y}_i)$ has probability density function

$$p_i f_i(\boldsymbol{y}_i;\boldsymbol{\theta}) \quad , i = 1, 2,$$

where $p_1 = p$ and $p_2 = 1-p$ do not depend on $\boldsymbol{\theta}$. This density function is proportional to the density function for observing $\boldsymbol{y}_i$ from experiment E_i, $f_i(\boldsymbol{y}_i;\boldsymbol{\theta})$. The likelihood principle asserts that the information content of $(E, (E_i, \boldsymbol{y}_i))$ is the same as that of $(E_i, \boldsymbol{y}_i)$. This is the conditionality principle. □

Birnbaum (1962) and others (e.g. Pawitan 2001) comment that it is not clear that the likelihood principle really is contrary to frequentist inference. The likelihood principle is saying that the information content is the same from two experiments if the likelihoods are equal (to within a constant). However, this does not necessarily imply that these experiments should result in the same *inference*. It may be that the manner in which the experiment could be repeated is relevant to inference, but this knowledge is not captured in the likelihood. In Example 14.6, the observed proportion is the parameter value that is most supported by the likelihood under both the negative binomial and binomial models. However, under the negative binomial model this estimator is biased, but it is unbiased under the binomial model.

The monograph of Berger and Wolpert (1988) argues in favour of the likelihood principle, and moreover, that the Bayesian paradigm is compatible with this principle. Of course, it must be noted that the Bayesian approach does not lead to identical inference from identical likelihoods if different priors are used.

14.5 Statistical significance versus statistical evidence

To conclude this chapter, Example 14.7 below is a well-known contrived example that is often used to question the common interpretation that p-values provide measures of statistical evidence. However, this is rather moot, because the real lesson from this example is that hypothesis testing should not have been performed in the first place, since there is no motivation for preferring to falsify one model (Popper 1959) rather than the other. Exercise 14.9 provides a more meaningful approach to choosing between two simple hypotheses. The second example reiterates this theme, and provides the details that were omitted from Example 2.4.

Royall (1997) provides a more in-depth coverage of the relationship between statistical significance and statistical evidence, and Lehmann (2006) considers other strategies for circumventing some of the shortcomings of Neyman-Pearson tests of hypotheses.

Neyman-Pearson theorem for simple hypotheses.
Consider testing the null hypothesis $H_0 : \boldsymbol{\theta} = \boldsymbol{\theta}_0$ versus the alternative $H_a : \boldsymbol{\theta} = \boldsymbol{\theta}_a$, and let $L(\boldsymbol{\theta}; \boldsymbol{y}) = f(\boldsymbol{y}; \boldsymbol{\theta})$ denote the likelihood function. The hypothesis test with critical region

$$C = \left\{ \boldsymbol{y} : \frac{L(\boldsymbol{\theta}_a; \boldsymbol{y})}{L(\boldsymbol{\theta}_0; \boldsymbol{y})} > k \right\} , \tag{14.4}$$

is the most powerful size α test of $H_0 : \boldsymbol{\theta} = \boldsymbol{\theta}_0$ versus $H_a : \boldsymbol{\theta} = \boldsymbol{\theta}_a$, where k is such that $P_{\boldsymbol{\theta}_0}(\boldsymbol{Y} \in C) = \alpha$. □

Example 14.7. Suppose that $Y \in \{1, 2, 3\}$, and it is required to test $H_0 : \theta = 0$ versus $H_a : \theta = 1$, where $P_\theta(Y = y), \theta = 0, 1$, is given by

	y = 1	2	3
P_0	.009	.001	.99
P_1	.001	.989	.01

To test the null hypothesis H_0, from (14.4) it can be seen that the critical region $\{1, 2\}$ corresponds to the most powerful test of size 0.01. Thus, if $y \in \{1, 2\}$ then the p-value can be no bigger than 0.01. However, the likelihood ratio is 9 to 1 in favour of H_0 when $y = 1$ is observed! □

Example 2.4 continued. Recall that the observation Y was either from a $N(0, 1)$ distribution or $N(0, 100^2)$ distribution, but it is unknown which. To test $H_0 : Y \sim N(0, 1)$ versus $H_a : Y \sim N(0, 100^2)$, the critical region in (14.4) reduces to the form

$$C = \{y : |y| > k\} .$$

Therefore, the most powerful 5 %-level test of H_0 rejects for all y outside of the interval $(z_{0.025}, z_{0.975}) \approx (-1.96, 1.96)$. Conversely, the most powerful test of $H_0 : Y \sim N(0, 100^2)$ versus $H_a : Y \sim N(0, 1)$ rejects H_0 at the 5 % level for all y inside the interval $(-6.2707, 6.2707)$ (Exercise 14.8). Thus, the value of $y = 2$ would be rejected by both tests.

Using the likelihood ratio as a measure of support (Figure 14.2), it can be shown that the $N(0, 1)$ model has greater support than the $N(0, 100^2)$ model for all y in the interval $(-3.0351, 3.0351)$.

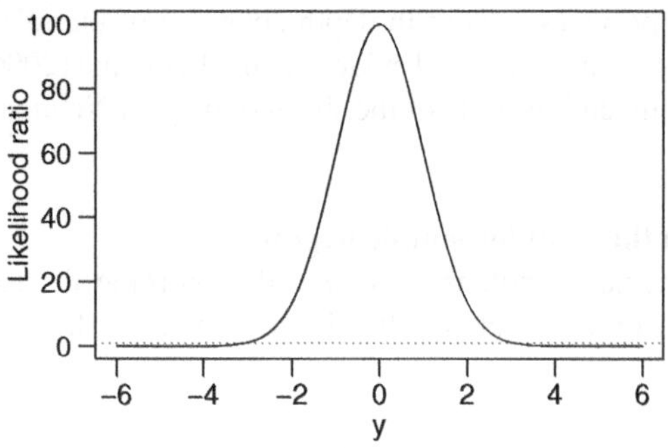

Figure 14.2 Likelihood ratio of the N(0,1) and N(0,100²) models, with a dashed horizontal line at unity. Note: Over the range of y values plotted, the likelihood of the N(0,100²) model is so flat that the shape of the likelihood ratio is very much like that of the bell-shaped curve of the N(0,1) model. The likelihood ratio exceeds unity on the interval (−3.0351,3.0351).

14.6 Exercises

14.1 Let $Y_i, i = 1, ..., n$ be iid from a chi-square distribution with ν degrees of freedom. This distribution has density

$$f(y; \nu) = \frac{y^{\nu/2-1} e^{-y/2}}{\Gamma(\nu/2) 2^{\nu/2}}, \ y > 0 .$$

Use the factorization theorem to obtain a univariate sufficient statistic for ν.

14.2 Let $Y_i, i = 1, ..., n$ be iid from a Pareto(α, m) distribution, for some $\alpha > 0, m > 0$ (see Exercise 2.8 for definition of the Pareto density function). Find a bivariate sufficient statistic for (α, m).

14.3 Let Y (possibly vector valued) be iid with exponential family density of the form

$$f(y; \boldsymbol{\theta}) = c(y) k(\boldsymbol{\theta}) \exp \left\{ \sum_{j=1}^{s} \theta_j t_j(y) \right\} . \tag{14.5}$$

For an iid sample, $Y_i, i = 1, ..., n$, from this distribution, use the factorization theorem to establish that $(T_1(\boldsymbol{y}), ..., T_s(\boldsymbol{y}))$ is a sufficient statistic for $\boldsymbol{\theta} = (\theta_1, ..., \theta_s)$ where $T_j(\boldsymbol{y}) = \sum_{i=1}^{n} t_j(y_i)$.

14.4 Under Hardy-Weinberg equilibrium, alleles A and B occur independently and thus the probabilities of genotypes AA, AB and BB are $p^2, 2p(1-p)$ and $(1-p)^2$, respectively, where p is the probability of allele type A. Let n be the number of genotypes measured and let (N_1, N_2, N_3) (where $N_3 = n - N_1 - N_2$) be the observed number of AA, AB and BB alleles, respectively. Use the factorization theorem to show that the number of A alleles observed, $T = 2N_1 + N_2$, is sufficient for p.

14.5 In Example 14.2 the factorization theorem was used to show that $T = \sum_{i=1}^{n} Y_i$ is a sufficient statistic for p when Y_i are iid Bernoulli(p) random variables. What is the distribution of $(Y_1, Y_2, ..., Y_n)|T$?

14.6 In Example 14.4 the mixture experiment corresponded to a binormal distribution consisting of an equal mixture of $N(\mu, 0.5^2)$ and $N(\mu, 0.05^2)$ distributions. Using (13.2) and (13.3), verify that this binormal distribution has mean μ and variance 0.12625.

14.7 Let Y_1 and Y_2 be independent Poisson random variables, with means parameterized as $p\lambda$ and $(1-p)\lambda$ respectively, for $0 < p < 1$ and $\lambda \in \mathbb{R}^+$. Show that $Y_1 + Y_2 \sim \text{Pois}(\lambda)$, and hence that $Y_1|(Y_1 + Y_2 = n) \sim \text{Bin}(n, p)$. If interest lies only in parameter p, argue that the information content from observing Y_1, Y_2 is the same as that from observing $Y_1|(Y_1 + Y_2)$.

14.8 In the continuation of Example 2.4 in Section 14.5, show that the most powerful test of $H_0 : Y \sim N(0, 100^2)$ versus $H_a : Y \sim N(0, 1)$ has critical region $C = (-100z_{0.525}, 100z_{0.525}) = (-6.2707, 6.2707)$.

14.9 Given the observation of $\mathbf{y}$, it is required to choose between the two hypotheses $H_0 : \mathbf{Y} \sim f(\mathbf{Y}; \theta_0)$ and $H_1 : \mathbf{Y} \sim f(\mathbf{Y}; \theta_1)$, by specifying a region, C, such that H_0 is chosen if $\mathbf{y} \in C$, else H_1 is chosen. Let p_i, $i = 0, 1$ denote the probability of falsely choosing H_i when H_{1-i} is true.

1. Show that the region

$$C = \{\mathbf{y} : f(\mathbf{y}; \theta_0) > f(\mathbf{y}; \theta_1)\} \tag{14.6}$$

minimizes $p_0 + p_1$. That is, the sum of error probabilities is minimized by choosing the hypothesis that has greater support from the likelihood.

2. Show that the region obtained by replacing the strict inequality in (14.6) by a nonstrict inequality also minimizes $p_0 + p_1$.

15

Miscellanea

It's the little details that are vital. Little things make big things happen. – John Wooden

15.1 Notation

Y	Generic notation for a random variable.
$\boldsymbol{Y}$	Generic notation for a random vector.
y	Observed value of Y.
$\boldsymbol{y}$	Observed value of vector $\boldsymbol{Y}$.
θ	Generic notation for a scalar parameter.
$\boldsymbol{\theta}$	Generic notation for a vector-valued parameter.
$\boldsymbol{\theta}_0$	A specific value of $\boldsymbol{\theta} \in \Theta$.
Θ	Parameter space.
Θ_0	Reduced parameter space under a null hypothesis.
$f(\boldsymbol{y};\boldsymbol{\theta})$	Density (or probability mass function) of $\boldsymbol{Y}$. In some places this notation is used generically, e.g. in the expression $f(\boldsymbol{u}, \boldsymbol{v};\boldsymbol{\theta}) = f(\boldsymbol{u};\boldsymbol{\theta})f(\boldsymbol{v}\|\boldsymbol{u};\boldsymbol{\theta})$ it is understood (from the arguments) that the density functions are different.
$L(\boldsymbol{\theta}; \boldsymbol{y})$	Likelihood function. Sometimes abbreviated to $L(\boldsymbol{\theta})$.

Maximum Likelihood Estimation and Inference: With Examples in R, SAS and ADMB, First Edition. Russell B. Millar.

$l(\boldsymbol{\theta}; \boldsymbol{y})$	Log-likelihood function. Sometimes abbreviated to $l(\boldsymbol{\theta})$.
$\mathcal{I}_1(\boldsymbol{\theta})$	Expected Fisher information from a single observation Y.
$\mathcal{I}(\boldsymbol{\theta})$	Expected Fisher information from observation of $Y_1, ..., Y_n$.
$\mathbf{O}(\boldsymbol{\theta})$	Observed Fisher information.
$\mathbb{R}$	Real numbers.
$\mathbb{R}^+$	Positive reals, i.e. $(0, \infty)$.
$\vert$	Conditional on, e.g. $Y\vert B$ denotes Y given B.
$\sim$	Distributed as.
$\dot{\sim}$	Approximately distributed as.
$\approx_D$	Approximately equal in distribution.
$\rightarrow_D$	Converges in distribution.
$\rightarrow_P$	Converges in probability.
$E_{\boldsymbol{\theta}}$	Expectation under the distribution with density $f(\boldsymbol{y};\boldsymbol{\theta})$. e.g. $E_{\boldsymbol{\theta}}[h(\boldsymbol{Y})] = \int h(\boldsymbol{y})f(\boldsymbol{y};\boldsymbol{\theta})d\boldsymbol{y}$.
$P_{\boldsymbol{\theta}}(A)$	Probability of event A, under the distribution with density $f(\boldsymbol{y};\boldsymbol{\theta})$.
$\text{var}_{\boldsymbol{\theta}}$	Variance (matrix) under $f(\boldsymbol{y};\theta)$.
P_{χ^2}	Pearson chi-square statistic.
$\chi^2_{r,q}$	The q quantile of a χ^2_r distribution, i.e. if $X \sim \chi^2_r$ then $P(X \leq \chi^2_{r,q}) = q$.
z_q	The q quantile of a $N(0, 1)$ distribution, e.g. $z_{0.975} \approx 1.96$.
$\Gamma(\alpha)$	Gamma function (see Figure 6.5).
$\vert x\vert$	Absolute value of $x \in \mathbb{R}$.
$\det(\mathbf{X})$	Determinant of square matrix $\mathbf{X}$.
$\mathbf{X}^{-T}$	$(\mathbf{X}^T)^{-1}$, or equivalently, $(\mathbf{X}^{-1})^T$.

Table 15.1 The Greek symbols used within this text.

α	alpha	β	beta	γ	gamma	δ	delta
ϵ	epsilon	ε	epsilon (variation)	ζ	zeta	η	eta
θ	theta	λ	lambda	μ	mu	ν	nu
π	pi	ρ	rho	σ	sigma	τ	tau
ϕ	phi	χ	chi	ψ	psi	ω	omega
Θ	Theta	Φ	Phi	Ψ	Psi	Ω	Omega

15.2 Acronyms

AIC	Akaike's information criterion
ADMB	Automatic Differentiation Model Builder
CI	Confidence interval
CR	Cramér-Rao
CV	Coefficient of variation
d.o.f.	Degrees of freedom
GEE	Generalized estimating equation
GLM	Generalized linear model
GLMM	Generalized linear mixed model
LR	Likelihood ratio
LRCI	Likelihood ratio confidence interval
ML	Maximum likelihood
MLE	Maximum likelihood estimate (or estimator)
NLMM	Nonlinear mixed model
QLE	Quasi-likelihood estimate (or estimator)
ODS	The SAS Output Delivery System
R	The R software environment
REML	Restricted (or residual) maximum likelihood
SAS	Statistical Analysis System (deprecated usage)
TLA	Three letter acronym

15.3 Do you think like a frequentist or a Bayesian?

The customary way to toss a coin is to flip it into the air, catch it with one hand, and to place that hand over the top of the other hand so that the coin remains concealed. Assume that the coin is balanced and tossed in a fair way, so that neither heads or tails is favoured.

Question: The coin has been tossed and caught, and is being held in the hidden position between the two hands. What is the probability that the coin shows heads when the top hand is lifted? □

See the Appendix for self diagnosis.

15.4 Some useful distributions

Most of the distributions used in this text are listed below. The description of each distribution includes the specification of its density function, its mean and variance, and a very brief explanation of its common usage. An authoritative reference on a vast assortment of distributions is provided by Johnson, Kotz and Kemp (1992), Johnson, Kotz and Balakrishnan (1994) and Johnson, Kotz and Balakrishnan (1995).

15.4.1 Discrete distributions

Bernoulli. Bernoulli(p), $0 \le p \le 1$:

A Bernoulli(p) random variable takes the value 1 or 0, with probabilities p and $(1-p)$, respectively. It is typically an indicator variable for the occurrence of a particular event, and this event is often generically called 'success'. That is, $y = 1$ if a 'success' occurs, otherwise $y = 0$ if a 'failure' occurs.

Density: $f(y; p) = p^y(1-p)^{1-y}, \quad y = 0, 1$

Mean: $E[Y] = p$

Variance: $\text{var}(Y) = p(1-p)$.

Binomial. Bin(n, p), $0 \le p \le 1$:

A Bin(n, p) random variable takes an integer value between 0 and n, corresponding to the number of 1's observed from n iid Bernoulli(p) trials.

Density: $f(y; p) = \binom{n}{y} p^y(1-p)^{n-y}, \quad y = 0, 1, ..., n$

Mean: $E[Y] = np$

Variance: $\text{var}(Y) = np(1-p)$.

Multinomial process. MultProc($p_1, ..., p_s$), $0 \le p_i \le 1$, $\sum_{i=1}^{s} p_i = 1$:

This random variable is a generalization of the Bernoulli to $s \ge 3$ possible outcomes. It takes an integer value between 1 and s with probabilities $p_1, ..., p_s$, respectively, and is used to model a single trial from an experiment with s possible outcomes (e.g. Yes, No, or Undecided).

Density: $f(y; p_1, ..., p_s) = p_y, \quad y = 1, ..., s$

Mean: $E[Y] = \sum_{i=1}^{s} i p_i$

Variance: $\text{var}(Y) = \sum_{i=2}^{s} i(i-1)p_i$.

In many situations, the multinomial-process random variable will be a factor variable. That is, the values 1, ..., s are simply labels for the possible outcomes (e.g. $1 \equiv$ Yes, $2 \equiv$ No, $3 \equiv$ Undecided), in which case the mean and variance are not meaningful.

Multinomial. Mult$(n, p_1, ..., p_s)$, $0 \le p_i \le 1$, $\sum_{i=1}^{s} p_i = 1$:
A multinomial random vector, $\boldsymbol{Y} = (Y_1, ..., Y_s)$ of length s, is generated from n iid trials of a multinomial process. The value of Y_i is the number of MultProc$(p_1, ..., p_s)$ trials that resulted in outcome i.

Density: $f(y_1, ..., y_s; p_1, ..., p_s) = \frac{n!}{y_1!...y_s!} p_1^{y_1} ... p_s^{y_s}$,

$y_i = 0, ..., n$, $\sum_{i=1}^{s} y_i = n$

Mean: $E[Y_i] = np_i$

Variance: $\text{var}(Y_i) = np_i(1 - p_i)$, $\text{cov}(Y_i, Y_j) = -np_i p_j$, $i \ne j$.

Poisson. Pois(λ), $\lambda > 0$:
A Poisson random variable takes non-negative integer values, and is commonly used to model count data. However, it is often the case that the Poisson model will be inadequate, in which case a more flexible alternative (e.g. negative binomial) may be required.

Density: $f(y; \lambda) = \frac{e^{-\lambda}\lambda^y}{y!}$, $y = 0, 1, ...$

Mean: $E[Y] = \lambda$

Variance: $\text{var}(Y) = \lambda$.

Negative binomial. NB(m, p), $m > 0$, $0 \le p \le 1$:
A negative binomial random variable takes non-negative integer values, and is often used to model count data that are over-dispersed relative to the Poisson, that is, count data showing extra-Poisson variation.

Density: $$f(y; m, p) = \frac{\Gamma(y + m)}{\Gamma(m)y!} p^m (1 - p)^y, \quad y = 0, 1, ... \qquad (15.1)$$

Mean: $E[Y] = \frac{m(1 - p)}{p}$

Variance: $\text{var}(Y) = \frac{m(1 - p)}{p^2} = \frac{E[Y]}{p} = E[Y]\left(1 + \frac{E[Y]}{m}\right)$.

The Poisson is obtained in the limit as $m \to \infty$ and $p \to 0$, such that $\mu = E[Y] = m(1 - p)/p$ converges to a fixed limit.

When m is a positive integer, then Y is distributed as the number of Bernoulli(p) trials that result in a 0 (i.e. 'failure') before m 1's (i.e. 'successes') are observed. The geometric distribution is the special case of the negative binomial with $m = 1$.

The negative binomial density is sometimes parameterized using m and $\mu = E[Y]$. Since $p = m/(\mu + m)$, this gives

$$f(y; m, \mu) = \frac{\Gamma(y + m)}{\Gamma(m)y!} \left(\frac{m}{\mu + m}\right)^m \left(\frac{\mu}{\mu + m}\right)^y . \qquad (15.2)$$

For a constant value of m, negative binomial data can be modelled using generalized linear models (see Section 7.7). With constant m, note that var(Y) is a quadratic function of μ. An alternative is to allow m to vary for each i, such that $p = m_i/(\mu_i + m_i)$ is constant. This parameterization results in a linear mean-variance relationship, var(Y) $= \mu/p$. However, this is a more challenging model to fit.

Hypergeometric. H(r, n, m):
A H(r, n, m) random variable is distributed as the number of white balls removed from an urn containing n white and m black balls, when r ($\leq n + m$) balls are removed by random sampling *without* replacement. Note that Y can not exceed the smaller of r and n. Also, if $r > m$ then at least $r - m$ balls must be white.

$$\text{Density:} \qquad f(y; r, n, m) = \frac{\binom{n}{y}\binom{m}{r-y}}{\binom{n+m}{r}} \tag{15.3}$$

$$\text{Mean:} \qquad E[Y] = rp$$

$$\text{Variance:} \qquad \text{var}(Y) = rp(1-p)(n+m-r)/(n+m-1),$$

where $\max(0, r - m) \leq y \leq \min(r, n)$, and $p = n/(n + m)$ is the proportion of white balls in the urn.

In Section 6.4.1 the hypergeometric distribution was used to model recaptures of marked animals.

15.4.2 Continuous distributions

Uniform. Unif(a, b), $b > a$:
The Unif(a, b) density function is constant between a and b, and zero otherwise.

$$\text{Density:} \qquad f(y; a, b) = \frac{1}{b-a}, \quad a \leq y \leq b$$

$$\text{Mean:} \qquad E[Y] = (a + b)/2$$

$$\text{Variance:} \qquad \text{var}(Y) = (b - a)^2/12\,.$$

Normal. $N(\mu, \sigma^2)$, $\mu \in \mathbb{R}$, $\sigma > 0$:
The normal distribution is also known as the Gaussian distribution. It is often the 'default' distribution for continuous data because of the well developed and exact statistical methods of statistical inference for normally distributed data (e.g. regression, ANOVA, etc.). Moreover, many types of estimators (including MLEs)

have a distribution that is approximately normal (under appropriate regularity conditions) when the sample size is large (Section 12.3).

Density: $f(y;\mu,\sigma^2) = \dfrac{e^{-(y-\mu)^2/(2\sigma^2)}}{\sqrt{2\pi}\sigma}, \quad y \in \mathbb{R}$

Mean: $E[Y] = \mu$

Variance: $\text{var}(Y) = \sigma^2$.

Lognormal. $\text{LN}(\mu,\sigma^2),\ \mu \in \mathbb{R},\ \sigma > 0$:
A lognormal random variable is so named because its log is normal. That is, $Y \sim \text{LN}(\mu,\sigma^2)$ takes positive values, and $\log(Y)$ is distributed $N(\mu,\sigma^2)$. The lognormal distribution is often used for the modelling of right-skew data, and this is equivalent to modelling the data as normally distibuted on the log scale.

Density: $f(y;\mu,\sigma^2) = \dfrac{e^{-(\log(y)-\mu)^2/(2\sigma^2)}}{\sqrt{2\pi}\sigma y}, \quad 0 < y$

Mean: $E[Y] = e^{\mu+\sigma^2/2}$

Variance: $\text{var}(Y) = (e^{\sigma^2} - 1)E[Y]^2$.

Exponential. $\text{Exp}(\mu),\ \mu > 0$:
The exponential distribution is commonly used to model the duration of time until an event occurs. For example, it is often used in survival analysis to model the time until failure (Example 6.1) .

Density: $f(y;\mu) = \dfrac{e^{-y/\mu}}{\mu}, \quad 0 < y$

Mean: $E[Y] = \mu$

Variance: $\text{var}(Y) = \mu^2$.

The exponential density is often re-parameterized using the rate parameter $\lambda = \mu^{-1}$.

Chi-square. $\chi^2_r,\ r > 0$:
The chi-square distribution arises from quadratic functions of normally distributed random variables, such as residual sum-of-squares. This is the property that results in many familiar test statistics having a (possibly approximate) chi-square distribution. Parameter r is the degrees of freedom.

Density: $f(y;r) = \dfrac{y^{r/2-1}e^{-y/2}}{2^{r/2}\Gamma(r/2)}, \quad 0 < y$

Mean: $E[Y] = r$

Variance: $\text{var}(Y) = 2r$.

The square of a standard normal random variable is distributed χ_1^2, and so if $Z \sim N(\mu, \sigma^2)$, then

$$\frac{(Z-\mu)^2}{\sigma^2} \sim \chi_1^2 \,.$$

In the multi-dimensional case, if $\mathbf{Z} \sim N_r(\boldsymbol{\mu}, \boldsymbol{\Sigma})$ is an r-dimensional multivariate normal random vector, then (Seber and Lee 2003, p. 30)

$$(\mathbf{Z}-\boldsymbol{\mu})^T \boldsymbol{\Sigma}^{-1}(\mathbf{Z}-\boldsymbol{\mu}) \sim \chi_r^2 \,. \tag{15.4}$$

Gamma. Gamma(α, β), $\alpha > 0, \beta > 0$:

The gamma family of distributions provides a lesser-used alternative to the log-normal for the modelling of right-skewed data on $\mathbb{R}^+$, although it can be fitted using most GLM software since it is a member of the exponential family of distributions (see Section 7.2.1 and Exercise 7.2).

In the parameterization of the Gamma(α, β) density used below, α is the shape parameter and β is the scale parameter (i.e. the mean and standard deviation are proportional to β). The gamma is also commonly parameterized using α and $\beta^* = \beta^{-1}$, where β^* is called the rate parameter.

Density: $$f(y; \alpha, \beta) = \frac{y^{\alpha-1}e^{-y/\beta}}{\beta^\alpha \Gamma(\alpha)}, \quad 0 < y \tag{15.5}$$

Mean: $E[Y] = \alpha\beta$

Variance: $\text{var}(Y) = \alpha\beta^2$.

An Exp(μ) distribution corresponds to a Gamma$(1, \mu)$ distribution, and a χ_r^2 distribution corresponds to a Gamma$(r/2, 2)$ distribution.

***F*-distribution.** F_{r_1,r_2}, $r_1 > 0$, $r_2 > 0$:

The F-distribution arises as the ratio of two independent chi-square random variables that have each been divided by their degrees of freedom. Specifically, if Y_1 and Y_2 are independently distributed as $\chi_{r_1}^2$ and $\chi_{r_2}^2$, respectively, then

$$F = \frac{Y_1/r_1}{Y_2/r_2} \sim F_{r_1,r_2} \,.$$

As r_2 tends to infinity, F_{r_1,r_2} converges to the distribution of a $\chi_{r_1}^2/r_1$ random variable. The F_{1,r_2}-distribution is equivalent to the square of a t_{r_2}-distribution.

Density: $$f(y; r_1, r_2) = \frac{(r_1/r_2)^{r_1/2}}{B(r_1/2, r_2/2)} \frac{y^{r_1/2-1}}{(1+r_1 y/r_2)^{(r_1+r_2)/2}}, \quad 0 < y$$

Mean: $E[Y] = r_2/(r_2-2)\,, \quad r_2 > 2$

Variance: $\text{var}(Y) = \dfrac{2r_2^2(r_1 + r_2 - 2)}{r_1(r_2 - 2)^2(r_2 - 4)}\,, r_2 > 4\,,$

where $B()$ denotes the beta function.

Multivariate normal. $N_s(\boldsymbol{\mu}, \boldsymbol{\Sigma})$, $\boldsymbol{\mu} \in \mathbb{R}^s$, $\boldsymbol{\Sigma} \in \mathbb{R}^{s \times s}$ is a positive definite matrix: The multivariate normal is the approximating distribution (subject to regularity conditions) of vector-valued MLEs $\widehat{\boldsymbol{\theta}} \in \mathbb{R}^s$, $s \geq 2$ (Section 12.3).

Density: $f(\mathbf{y}; \boldsymbol{\mu}, \boldsymbol{\Sigma}) = \dfrac{1}{(2\pi)^{s/2}\det(\boldsymbol{\Sigma})^{1/2}} e^{-(\mathbf{y}-\boldsymbol{\mu})^T \boldsymbol{\Sigma}^{-1}(\mathbf{y}-\boldsymbol{\mu})}, \quad \mathbf{y} \in \mathbb{R}^s$

Mean: $E[Y_i] = \mu_i$

Variance: $\text{var}(Y_i) = \sigma_{ii}^2, \quad \text{cov}(Y_i, Y_j) = \sigma_{ij}^2\,,$

where $\boldsymbol{\mu} = (\mu_1, ..., \mu_s)^T$, and σ_{ij}^2 is the (i, j) element of the variance matrix $\boldsymbol{\Sigma}$.

15.5 Software extras

One of the recurring themes in this text is that it is preferable, where practicably possible, to work directly with the likelihood function for the purpose of conducting hypothesis tests and calculating confidence intervals. The utilities described in Sections 15.5.1–15.5.4 are provided for this purpose.

15.5.1 R function `Plkhci` for likelihood ratio confidence intervals

At the time of writing, the current version of the `plkhci` function in library `Bhat` does not accept additional optional arguments. This text makes use of a modified version of `plkhci`, whereby the data $\mathbf{y}$ can be included as an argument for the purpose of being passed to the negative log-likelihood function – see the R code in Section 3.4.1.

The modified function is called `Plkhci` (with a capital ‘P’) and is defined in the text file `Plkhci.R`, which can be downloaded from `http://www.stat.auckland.ac.nz/~millar`. It is obtained from a very minor modification to the `plkhci` function. Specifically, the first two lines of `plkhci` are

```
function (x,nlogf,label,prob=0.95,eps=0.001,nmax=10,nfcn=0)
{
```

and in `Plkhci` these are replaced by

```
function (x,nllhood,label,prob=0.95,eps=0.001,nmax=10,nfcn=0,...)
{
  nlogf=function(x) nllhood(x,...)
```

15.5.2 R function `Profile` for calculation of profile likelihoods

The `Profile` function calculates the profile log-likelihood

$$l^*(\boldsymbol{\psi}; \boldsymbol{y}) = \max_{\boldsymbol{\lambda}} l(\boldsymbol{\psi}, \boldsymbol{\lambda}; \boldsymbol{y}) . \tag{15.6}$$

Parameter $\boldsymbol{\psi}$ may be vector valued, and can be any subset of $\boldsymbol{\theta}$.

By way of example, in the binormal mixture model with parameter vector $\boldsymbol{\theta} = (p, \mu, \sigma, \nu, \tau)$ labelled as `parnames = c("p","mu","sigma","nu","tau")`, and negative log-likelihood function `nllhood` (Section 3.3.4), suppose that $\boldsymbol{\psi} = (\sigma, \tau)$. The profile log-likelihood evaluated at $\boldsymbol{\psi} = (5, 7)$, denoted $l^*(5, 7; \boldsymbol{y})$, is obtained by the following code.

```
> Profile(parnames,nllhood,label=c("sigma","tau"),psi=c(5,7),
+            lambda=c(0.5,55,80),y=waiting)$value
[1] -1037.363
```

Argument `label` specifies the subset of parameter names in `parnames` that correspond to $\boldsymbol{\psi}$, and argument `lambda` gives the start values for the maximization over $\boldsymbol{\lambda}$ in (15.6), in the order in which they appear in `parnames`. Here, $\boldsymbol{\lambda} = (p, \mu, \nu)$.

The code

```
> Profile(parnames,nllhood,label=c("p","mu","sigma"),
+            psi=c(1/3,55,5),lambda=c(80,5),y=waiting)$value
[1] -1036.829
```

provides an easier implementation of the likelihood ratio test of $H_0 : (p, \mu, \sigma) = (1/3, 55, 5)$ than that performed in Section 3.4.1.

15.5.3 SAS macro `Plkhci` for likelihood ratio confidence intervals

The `Plkhci` macro is used for calculation of likelihood-ratio confidence intervals, and may be useful for SAS users who do not have access to `PROC NLP`, and/or require the functionality of `PROC NLMIXED`. This macro repeatedly calls a user-specified macro in which the model is fitted in `PROC NLMIXED` with the parameter of interest passed as an argument to that macro. Macro `Plkhci` simply uses the bisection method to calculate the likelihood ratio CI at the desired level.

Examples of the use of the `Plkhci` macro are provided in Sections 1.4.1 and 3.4.1. The first argument to `Plkhci` is the name of the user-specified macro. It searches for the CI bound within the range specified by its next two arguments. The fourth argument is the maximal value of the log-likelihood, $l(\widehat{\boldsymbol{\theta}})$. The fifth argument is used to specify whether it the left side (`side="L"`) or right side (`side="R"`) bound that is desired. Optional argument alpha specifies that a $(1-\alpha)100\,\%$ confidence interval is desired, and optional argument tol specifies the desired convergence tolerance of the computed confidence interval bound.

15.5.4 SAS macro **Profile** for calculation of profile likelihoods

The Profile macro can be used to calculate the profile log-likelihood $l^*(\psi; \mathbf{y})$ for individual model parameters, $\psi = \theta_k$. It uses the same user-defined macro mentioned in the description of the Plkhci macro, and an example of its use is provided in Section 3.6.1.

15.5.5 SAS macro **DeltaMethod** for application of the delta method

The DeltaMethod macro is for application of the delta method to general transformations of the parameters, $g(\boldsymbol{\theta}) : \mathbb{R}^s \to \mathbb{R}^p$ for $p \leq s \leq 2$. The macro constructs a call to PROC NLMIXED in which the ESTIMATE statement is used to obtain the variance of $g(\widehat{\boldsymbol{\theta}})$, and is demonstrated in Section 4.2.3 and Example 4.4.

15.6 Automatic differentiation

The ADMB template file contains the program code for calculation of (the negative of) the log-likelihood $l(\boldsymbol{\theta}) = \log f(\mathbf{y}; \boldsymbol{\theta})$. Automatic differentiation facilitates the exact (to within machine precision) algebraic calculation of the partial derivatives of $l(\boldsymbol{\theta})$. The computation is highly efficient, and the vector of derivatives is typically obtained using less than three times the number of numerical operations that are required to evaluate $l(\boldsymbol{\theta})$ (Griewank 2003).

In simplified form, automatic differentiation is essentially the automatic application of the chain rule of differentiation. Variables that are defined in an ADMB template file are such that the corresponding object also contains derivative information with respect to $\boldsymbol{\theta}$. Numerical operators are overloaded to apply the chain rule to these derivatives.

By way of example, if the data are iid $N(\mu, \sigma^2)$, then each y_i contributes a term of the form

$$-\log \sigma - \frac{(y_i - \mu)^2}{2\sigma^2}\,, \qquad (15.7)$$

to $l(\boldsymbol{\theta})$. To match the template files used in Chapter 10, this model will be parameterized using $\boldsymbol{\theta} = (\theta_1, \theta_2) = (\mu, \log \sigma)$. So, if $\sigma^2(\boldsymbol{\theta}) = \exp(2\theta_2)$ is a variable defined in the ADMB template file, then this object is stored in ADMB as

$$\begin{aligned}
\sigma^2 &\equiv \left(\sigma^2, \left[\frac{\partial\sigma^2}{\partial\theta_1}, \frac{\partial\sigma^2}{\partial\theta_2}\right]\right) \\
&= \left(\exp(2\theta_2), \left[\frac{\partial\exp(2\theta_2)}{\partial\theta_1}, \frac{\partial\exp(2\theta_2)}{\partial\theta_2}\right]\right) \\
&= \left(\exp(2\theta_2), \left[0, 2\exp(2\theta_2)\right]\right) . \qquad (15.8)
\end{aligned}$$

Similarly, if the second term in (15.7) is denoted by ζ, then

$$\begin{aligned}
\zeta &= \frac{(y_i - \mu)^2}{2\sigma^2} \\
&= \frac{(y_i - \theta_1)^2}{2\exp(2\theta_2)} ,
\end{aligned}$$

is stored in ADMB as

$$\zeta \equiv \left(\zeta, \left[\frac{\partial\zeta}{\partial\theta_1}, \frac{\partial\zeta}{\partial\theta_2}\right]\right) ,$$

where, for example,

$$\begin{aligned}
\frac{\partial\zeta}{\partial\theta_2} &= \frac{\partial\frac{(y_i-\theta_1)^2}{2\sigma^2}}{\partial\sigma^2} \times \frac{\partial\sigma^2}{\partial\theta_2} \qquad (15.9) \\
&= -\frac{(y_i - \theta_1)^2}{2\sigma^4} \times 2\exp(2\theta_2) \\
&= -\frac{(y_i - \theta_1)^2}{\exp(2\theta_2)} . \qquad (15.10)
\end{aligned}$$

On the right-hand side of (15.9), the first partial derivative is obtained from ADMB's application of the chain rule of differentiation. The second partial derivative is a component of the ADMB object σ^2, as shown in (15.8).

Griewank (2003) comments that obtaining the derivative is far from 'automatic', and recommends 'algorithmic' differentiation as a more apt name for this methodology.

Appendix: Partial solutions to selected exercises

Chapter 1

1.1 The 95 % Wald CI for λ is $(-0.395, 6.395)$ and the 95 % likelihood ratio CI is $(0.746, 7.779)$. The Wald CI is clearly inappropriate because λ can not be negative – this interval suffers from the nonquadratic shape of the log-likelihood.

Chapter 2

2.3 The density function $f(t)$ of $Y_{\max}$ is

$$f(t) = \frac{\partial F(t)}{\partial t} = \begin{cases} 0, & t < 0 \\ \frac{n}{M}\left(\frac{t}{M}\right)^{n-1}, & 0 < t < M \\ 0, & M < t\,, \end{cases}$$

and the result follows from evaluating

$$E[Y_{\max}] = \int_0^M tf(t)dt.$$

2.9 Note: This example has similarities with Example 2.7 because the sample space depends on the parameter value.

The log-likelihood function is

$$l(\alpha, \lambda) = -n\log(\lambda) - \frac{\sum_{i=1}^n (y_i - \alpha)}{\lambda}, \quad y_{\min} \geq \alpha,\ \lambda > 0,$$

where $y_{min} = \min\{y_1, ..., y_n\}$. Note that the log-likelihood is a monotone increasing function of α, and hence is maximized by $\widehat{\alpha} = y_{\min}$.

The MLE of λ can be obtained from the likelihood equation for α,

$$\frac{\partial l(\alpha, \lambda)}{\partial \lambda} = -\frac{n}{\lambda} + \frac{\sum_{i=1}^{n}(y_i - \alpha)}{\lambda^2} = 0,$$

from which it follows that $\widehat{\lambda} = \overline{y} - y_{\min}$.

2.10 The log-likelihood function is

$$l(\alpha) = -n \log(2) - \sum_{i=1}^{n} |y_i - \alpha|.$$

Now,

$$-|y_i - \alpha| = \begin{cases} y_i - \alpha, & \alpha > y_i \\ \alpha - y_i, & \alpha < y_i, \end{cases}$$

and so

$$-\frac{\partial |y_i - \alpha|}{\partial \alpha} = \text{sign}(y_i - \alpha).$$

The derivative of the log-likelihood, $l'(\alpha)$, is therefore the number of observations that are greater than α minus the number of observations that are less than α. (It is not defined if $\alpha = y_i$ for any y_i.) This derivative is a non-increasing function of α, and is uniformly zero in the interval $(y_{[m]}, y_{[m+1]})$ where $y_{[i]}$ denotes the ith smallest of the observed values. This is enough to conclude that any value in $(y_{[m]}, y_{[m+1]})$ is an MLE. By continuity of the likelihood, the end points of this interval, $y_{[m]}$ and $y_{[m+1]}$ are also MLEs.

Figure 15.1 gives an example of the log-likelihood function for a random sample of size ten from the Laplace distribution with $\alpha = 1$. For these data the MLEs are the values in the interval $[y_{[5]}, y_{[6]}] = [1.313, 1.653]$.

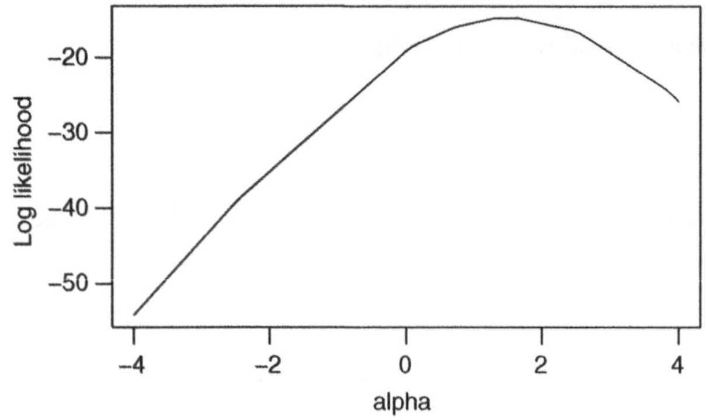

Figure 15.1 Log-likelihood for a random sample of size 10 from the Laplace distribution with $\alpha = 1$.

Chapter 3

3.5 Note that the likelihood under the independence assumption reduces to

$$(r_1c_1)^{y_{11}}(r_1(1-c1))^{y_{12}}((1-r_1)c_1)^{y_{21}}((1-r_1)(1-c_1))^{y_{22}}$$
$$= r_1^{y_{11}+y_{12}}(1-r_1)^{y_{21}+y_{22}}c_1^{y_{11}+y_{21}}(1-c_1)^{y_{12}+y_{22}},$$

from which the MLEs of r_1 and c_1 are obtained by analogy with the binomial likelihood.

3.6 To make the solution more general, let n_1, n_2 and n_3 denote the number of AA, AB and BB genotypes, respectively, and let $n_+ = n_1 + n_2 + n_3$. The likelihood is given by (15.11). This likelihood is, to within a multiplicative constant, the same as that of a binomial experiment where $t = 2n_1 + n_2$ successes are observed from $2n_+$ trials. Hence the MLE is simply the proportion of the $2n_+$ alleles that were of type A,

$$\widehat{p} = \frac{2n_1 + n_2}{2n_+},$$

and so, under the assumption of Hardy-Weinberg equilibrium,

$$(\widehat{n}_1, \widehat{n}_2, \widehat{n}_3) = n_+(\widehat{p}^2, 2\widehat{p}(1-\widehat{p}), (1-\widehat{p})^2).$$

Using the stated data values $(n_1, n_2, n_3) = (14, 21, 25)$, $\widehat{p} = 49/120$, $(\widehat{n}_1, \widehat{n}_2, \widehat{n}_3) = (10.004, 28.992, 21.004)$ and the G-test statistic is

$$G = 2\sum_{i=1}^{3} n_i \log(n_i/\widehat{n}_i) \approx 4.57.$$

The unrestricted multinomial model has two parameters, but there is only one under the Hardy-Weinberg restriction, and hence G has degrees of freedom $r = 1$ and the p-value is about 0.0325.

3.8 The approximate 95 % likelihood ratio CIs for b and $l_{50\,\%}$ are found to be (0.2605, 0.3535) and (34.12, 35.79), respectively.

Chapter 4

4.1 Letting $x(p)$ denote the $\sin^{-1}\sqrt{p}$ transformation, it is required to show that $\text{var}(X) = \left(\frac{\partial x(p)}{\partial p}\right)^2 \text{var}(\widehat{p})$ does not depend on p. Now, $\text{var}(\widehat{p}) = p(1-p)/n$, and $\frac{\partial x}{\partial p} = \left(\frac{\partial p}{\partial x}\right)^{-1}$ where

$$p = \sin(x)^2.$$

Using the identity $\sin^2(x) + \cos^2(x) = 1$,

$$\frac{\partial p}{\partial x} = 2\sin(x)\cos(x) = 2\sin(x)\sqrt{1-\sin(x)^2} = 2\sqrt{p}\sqrt{1-p},$$

which gives the required result.

4.3 The delta method approximation of the variance of $\widehat{l}_{50\,\%} = -\widehat{\beta}_1/\widehat{\beta}_2$ is given by the formula of Exercise 4.4. R and SAS code to answer this question are given in the case study in section 7.5.

4.4 The derivative vector of $g(\widehat{\boldsymbol{\theta}}) = \widehat{\theta}_1/\widehat{\theta}_2$ is $G(\widehat{\boldsymbol{\theta}}) = (\widehat{\theta}_2^{-1}, -\widehat{\theta}_1\widehat{\theta}_2^{-2})$ and so

$$\widehat{\text{var}}(g(\widehat{\boldsymbol{\theta}})) = \frac{1}{\widehat{\theta}_2^2}\left(\widehat{\text{var}}(\widehat{\theta}_1) + \frac{\widehat{\theta}_1^2}{\widehat{\theta}_2^2}\widehat{\text{var}}(\widehat{\theta}_2) - 2\frac{\widehat{\theta}_1}{\widehat{\theta}_2}\widehat{\text{cov}}(\widehat{\theta}_1, \widehat{\theta}_2)\right).$$

4.7 Note that the probability that toxin is present in a batch of 10 samples is

$$\theta = 1 - P(\text{Not present in the batch}) = 1 - (1-p)^{10}.$$

1. The testing of the 100 batches is a binomial experiment with $y = 12$ observed to have toxin. Hence $\widehat{\theta} = 0.12$ and $\widehat{p} = 1 - (1-\widehat{\theta})^{0.1} = 0.0127$.

2. The MLE $\widehat{\theta}$ has approximate variance $\frac{\widehat{\theta}(1-\widehat{\theta})}{100}$ and so from the delta method for MLEs (Equation 4.5)

$$\widehat{p} \dot{\sim} N\left(p_0, \left(\frac{\partial p(\widehat{\theta})}{\partial\theta}\right)^2 \frac{\widehat{\theta}(1-\widehat{\theta})}{100}\right) = N(p_0, 0.0000133).$$

The approximate 95 % Wald CI is therefore (0.0056, 0.0198).

3. By using the SAS or R code of the motivating Example 1.1, it is found that the approximate 95 % likelihood ratio CI for θ is (0.0662, 0.1930). Inverting this gives the CI (0.0068, 0.0212) for p.

4.8 If the log-likelihood is quadratic, then a Taylor series expansion around $\widehat{\theta}$, and evaluated at θ_0, yields

$$2\big(l(\widehat{\theta}; \boldsymbol{Y}) - l(\theta_0; \boldsymbol{Y})\big) = -l''(\widehat{\theta})(\theta_0 - \widehat{\theta})^2.$$

Note that $\widehat{v} = -1/l''(\widehat{\theta})$ is the standard estimator of the variance of $\widehat{\theta}$.

Chapter 5

5.5 Evaluating the maximal multinomial log-likelihood (but ignoring the constant terms) under H_0 (i.e. using $\widehat{p} = \widehat{r}_1\widehat{c}_1$, etc.) gives -143.795. The maximal unrestricted log-likelihood (i.e. using $\widehat{p}_i = y_i/100$) is -143.056.

Twice the difference is 1.478, which has a p-value of about 0.22 when compared to the χ_1^2 distribution.

5.7 The solution proceeds along the same lines as Examples 5.1 and 5.2, with $f_1(y)$ replaced by the density (i.e. probability) function with all its point mass at zero, and $f_0(y)$ the Pois(λ) density. In particular, when $y \geq 1$ then $f_1(y) = 0$ (i.e. the observation can not have come from the zero-distribution) and so

$$b_i^* = E[B_i|y_i > 0] = 0, \quad y_i \geq 1.$$

When $y = 0$ then

$$b_i^* = \frac{p^{(k)} f_1(y)}{f^{(k)}(y)} = \frac{p^{(k)}}{p^{(k)} + (1 - p^{(k)})e^{-\lambda}}.$$

Maximizing the complete data likelihood $\prod_{i=1}^n f(y_i, b_i^*; p^{(k)}, \lambda)$ gives

$$p^{(k+1)} = \frac{1}{n}\sum_{i=1}^{n} b_i^* = \frac{1}{n}\sum_{i:y_i=0} \frac{p^{(k)}}{p^{(k)} + (1 - p^{(k)})e^{-\lambda}}$$

$$\lambda^{(k+1)} = \frac{\sum_{i=1}^n (1 - b_i^*)y_i}{n(1 - p^{(k+1)})} = \frac{\sum_{i=1}^n y_i}{n(1 - p^{(k+1)})},$$

since b_i^* is equal to zero whenever y_i is nonzero.

Chapter 6

6.6 The result follows from showing

$$\Delta l(N; n_1, n_2) \equiv l(N; n_1, n_2) - l(N - 1; n_1, n_2) = \log\left(\frac{N(1 - p)^2}{N - n_1 - n_2}\right).$$

Chapter 7

7.2 The Gamma(α, β) density function can be written

$$\exp\left\{\frac{-y/(\alpha\beta) - \log(\alpha\beta)}{1/\alpha} + c(y, \alpha)\right\},$$

which is exponential family form with $\psi = -1/(\alpha\beta)$, $b(\psi) = -\log(-\psi)$ and $\phi = 1/\alpha$.

7.3 The negative binomial with known m can be written as exponential family with $\psi = \log(1 - p)$, $b(\psi) = m \log(1 - e^{\psi})$ and $\phi = 1$.

Chapter 8

8.2

$$\widetilde{\lambda} = \frac{\overline{y^2}}{\overline{y}} - 1 = 4.837 \quad \text{and} \quad \widetilde{p} = 1 - \frac{\overline{y}}{\widetilde{\lambda}} = 0.4935.$$

Chapter 11

11.6 Calculating the expected Fisher information as the negative of the expected value of the second derivative of the log-likelihood (formula (11.14)) results in the integral

$$\mathcal{I}_1(\theta) = 2\int_{-\infty}^{\infty} \frac{e^{2(y-\theta)}}{\left(1 + e^{(y-\theta)}\right)^4} dy,$$

and the required result can be obtained after making the substitution $v = 1 + e^{(y-\theta)}$.

11.8 The required moments are obtained by applying the identities to

$$\frac{\partial l}{\partial \psi} = \frac{y - b'(\psi)}{a(\phi)}, \quad \text{and} \quad \frac{\partial^2 l}{\partial \psi^2} = -\frac{b''(\psi)}{a(\phi)}.$$

11.10.b Because $\mathcal{I}(\boldsymbol{\theta})$ is positive definite, then so too are the submatrices $\mathcal{I}_{\psi\psi}$ and $\mathcal{I}_{\lambda\lambda}$, and their inverses. Hence, $\mathcal{I}_{\psi\lambda}\mathcal{I}_{\lambda\lambda}^{-1}\mathcal{I}_{\lambda\psi}$ is also positive definite since

$$\boldsymbol{h}^T \mathcal{I}_{\psi\lambda}\mathcal{I}_{\lambda\lambda}^{-1}\mathcal{I}_{\lambda\psi}\boldsymbol{h} = (\mathcal{I}_{\lambda\psi}\boldsymbol{h})^T \mathcal{I}_{\lambda\lambda}^{-1}(\mathcal{I}_{\lambda\psi}\boldsymbol{h}) \geq 0,$$

because $\mathcal{I}_{\lambda\lambda}^{-1}$ is positive definite. Thus,

$$\begin{aligned} \boldsymbol{h}^T[\mathcal{I}^{\psi\psi}]^{-1}\boldsymbol{h} &= \boldsymbol{h}^T(\mathcal{I}_{\psi\psi} - \mathcal{I}_{\psi\lambda}\mathcal{I}_{\lambda\lambda}^{-1}\mathcal{I}_{\lambda\psi})\boldsymbol{h} \\ &\leq \boldsymbol{h}^T I_{\psi\psi}\boldsymbol{h}, \end{aligned}$$

from which the result follows by using the given hint.

Chapter 12

12.1

$$\begin{aligned} P(A_n \cap B_n) = 1 - P(\overline{A_n \cap B_n}) &= 1 - P(\overline{A_n} \cup \overline{B_n}) \\ &\geq 1 - P(\overline{A_n}) - P(\overline{B_n}) \\ &= 1 - (1 - P(A_n)) - (1 - P(B_n)) \to 1. \end{aligned}$$

12.2

1. From the delta theorem,

$$\sqrt{n}(\widehat{p} - p_0) \to g'(p_0)N(0, \lambda_0) = -e^{-\lambda_0}N(0, \lambda_0) = N(0, \lambda_0 e^{-2\lambda_0}).$$

2. It is required to show that $\lambda e^{-2\lambda} < e^{-\lambda}(1 - e^{-\lambda})$ for all $\lambda > 0$. After a bit or re-arranging, it can be seen that this is equivalent to showing $1 + \lambda < e^{\lambda}$ which is immediate from the identity $e^{\lambda} = \sum_{i=0}^{\infty} \lambda^i / i!$.

12.6 The estimates of **A** and **B** are

$$\mathbf{A} = \begin{pmatrix} 1 - \widetilde{p} & -\widetilde{\lambda} \\ (1 - \widetilde{p})(1 + 2\widetilde{\lambda}) & -\widetilde{\lambda}(1 + \widetilde{\lambda}) \end{pmatrix} = \begin{pmatrix} 0.50654 & -4.83674 \\ 5.40654 & -28.23074 \end{pmatrix}.$$

$$\mathbf{B} = n^{-1}\widehat{\text{var}}\begin{pmatrix} Y \\ Y^2 \end{pmatrix} = \frac{1}{40}\begin{pmatrix} 8.5103 & 61.3487 \\ 61.3487 & 489.2410 \end{pmatrix}.$$

12.7 In the general case where y is observed from a Bin(n, p) experiment, the confidence interval endpoints are given by the p_0 values satisfying

$$\frac{n(\widehat{p} - p_0)^2}{p_0(1 - p_0)} = \chi^2_{1,1-\alpha}.$$

Re-arranging leads to the quadratic

$$(n + \chi^2_{1,1-\alpha})p_0^2 - (2n\widehat{p} + \chi^2_{1,1-\alpha})p_0 + n\widehat{p}^2 = 0.$$

This is a quadratic in p_0, and the roots are therefore

$$\frac{-b \pm \sqrt{b^2 - 4ac}}{2a},$$

where $a = n + \chi^2_{1,1-\alpha}$, $b = -(2n\widehat{p} + \chi^2_{1,1-\alpha})$ and $c = n\widehat{p}^2$.

With $n = 100$ and $y = 10$, the coefficients of the quadratic are $a = 103.841$, $b = -23.8415$, $c = 1$, and the 95 % Wilson interval is therefore (0.0552, 0.1744).

12.8 For testing $H_0 : \beta = \beta_0$, Wald statistics W_1 and W_3 are of the form $W_i = c_i(\widehat{\beta} - \beta_0)^2$ where

$$c_3 = -l''(\beta_0) = \frac{1}{\sigma^2}\sum_{i=1}^{n}\left(g_i'(\beta_0)^2 - g''(\beta_0)\left(y_i - g_i(\beta_0)\right)\right),$$

and

$$c_1 = E[c_3] = \frac{\sum_{i=1}^{n} g_i'(\beta_0)^2}{\sigma^2}.$$

12.9 The distribution of the A_i doesn't depend on μ, and so the likelihood function can be written

$$l(\mu; \boldsymbol{y}, \boldsymbol{a}) = f(\boldsymbol{y}, \boldsymbol{a}; \mu) = f(\boldsymbol{y}|\boldsymbol{a}; \mu) f(\boldsymbol{a})$$
$$= \frac{1}{(2\pi)^{n/2} \prod_i \sigma_{a_i}} \exp\left(-\frac{\sum (y_i - \mu)^2}{2\sigma_{a_i}^2}\right) \times \left(\frac{1}{2}\right)^n .$$

If follows that the observed Fisher information is $\mathbf{O}(\mu) = \sum_i 1/\sigma_{a_i}^2$. To calculate the expected information, note that $E(1/\sigma_{A_i}^2) = \frac{1}{2}(1/\sigma_0^2 + 1/\sigma_1^2)$ and so $\mathcal{I}(\mu) = \frac{n}{2}(1/\sigma_0^2 + 1/\sigma_1^2)$.

The distribution of each observation is a mixture of a $N(\mu, \sigma_0^2)$ and a $N(\mu, \sigma_1^2)$ distribution, and the conditionality principle states that inference should be based on the distribution actually used. That is, it is stating that a_i should not be regarded as random realizations of A_i, and that they should be treated as though fixed in advance. In this case $\mathbf{O}(\mu)$ is preferred because its inverse is the variance of μ for fixed values of a_i.

12.10

1. For Bernoulli(p) observations y_i, $l_i'(p) = (y_i - p)/p(1-p)$. That is,

$$l_i'(p) = \begin{cases} \frac{-1}{(1-p)}, & y_i = 0 \\ \frac{1}{p}, & y_i = 1. \end{cases}$$

   Hence, if an un-subscripted y is used to denote the binomial observation $\sum_i y_i$, then

$$\mathbf{B}(p) = \frac{y}{p^2} + \frac{n-y}{(1-p)^2},$$

   and this is equivalent to $\mathbf{O}(p)$ (see the continuation of Example 11.1 in Section 11.4). In this particular case, this equivalence holds for all p, not just $\widehat{p}$.

2. The result follows from the fact that, for each y_i,

$$l' = \frac{f_1(y_i) - f_2(y_i)}{f(y_i; p)} \quad \text{and} \quad l'' = \frac{-\left(f_1(y_i) - f_2(y_i)\right)^2}{f(y_i; p)^2}.$$

Chapter 13

13.1 Here is one possible counter-example. Let X, Y and X_n, $n = 1, 2, \ldots$ all be independent $N(0, 1)$ random variables and define $Y_n = -X_n$. Then every

X_n and Y_n is exactly distributed $N(0, 1)$ and hence (trivially) $X_n \rightarrow_D X$ and $Y_N \rightarrow_D Y$. However, $X_n + Y_n = 0$ for all n, whereas $X + Y \sim N(0, 2)$.

13.2 It is immediate that (13.12) implies (13.11) because

$$P(|X_n - X| \geq \epsilon) \geq P(|X_n - X| > \epsilon),$$

and the quantity on the left-hand side converges to zero as $n \rightarrow \infty$. To show that (13.11) implies (13.12), let ε be given, and take $\epsilon = \varepsilon/2$. Then

$$P(|X_n - X| > \epsilon) = P(|X_n - X| > \varepsilon/2) \geq P(|X_n - X| \geq \varepsilon),$$

and hence the quantity on the right-hand side converges to zero as $n \rightarrow \infty$.

13.3 Let $\epsilon = -k/2 > 0$. Now,

$$\begin{aligned} P(|X_n - k| < \epsilon) &= P(k - \epsilon < X_n < k + \epsilon) \\ &\leq P(X_n < k + \epsilon) \\ &\leq P(X_n < 0), \end{aligned}$$

since $k + \epsilon < 0$. By the convergence in probability of X_n to k, $P(|X_n - k| < \epsilon) \rightarrow 1$, and hence $P(X_n < 0) \rightarrow 1$.

13.4 For any $\epsilon > 0$,

$$P(|B_n X_n| < \epsilon) \geq P(|B_n| < b \cap |X_n| < \epsilon/b) \rightarrow 1,$$

because $P(|B_n| < b) \rightarrow 1$ and $P|X_n| < \epsilon/b) \rightarrow 1$.

13.5

1. To prove that $X_n \rightarrow_P c$ implies $X_n \rightarrow_D c$, it is required to show that $F_n(x) \rightarrow F(x)$ (at all points of continuity of $F(x)$) where $F(x)$ is the distribution function of the random variable that takes the value c with probability one. That is

$$F(x) = \begin{cases} 0, & x < c \\ 1, & x \geq c \end{cases}.$$

The distribution function $F(x)$ is continuous at all values other than c.

Consider any value $x < c$ (the $x > c$ case is similar) and let $\epsilon = c - x > 0$. Now,

$$\begin{aligned} F_n(x) &= P(X_n \leq x) \\ &= 1 - P(x < X_n) \\ &\leq 1 - P(x < X_n < c + \epsilon) \\ &= 1 - P(c - \epsilon < X_n < c + \epsilon) \\ &= 1 - P(|X_n - c| < \epsilon) = P(|X_n - c| \geq \epsilon). \end{aligned}$$

The right-hand side converges to zero unity as $n \to \infty$ (see Exercise 13.2), and hence so to does the left-hand side, giving the required result that $F_n(x) \to 0$.

2. To show $X_n \to_D c$ implies $X_n \to_P c$, let $\epsilon > 0$ be given. Then,

$$\begin{aligned} P(|X_n - c| \leq \epsilon) &= P(c - \epsilon \leq X_n \leq c + \epsilon) \\ &= P(X_n \leq c + \epsilon) - P(X_n < c - \epsilon) \\ &\geq P(X_n \leq c + \epsilon) - P(X_n \leq c - \epsilon) \\ &= F_n(c + \epsilon) - F_n(c - \epsilon) \\ &\quad \to F(c + \epsilon) - F(c - \epsilon) = 1 - 0 = 1, \end{aligned}$$

as required.

13.6 Since Y_i are iid, it follows that so too are $(Y_i - \mu)^2$. These have expected value σ^2, and hence

$$\sum_{i=1}^{n} \frac{(Y_i - \mu)^2}{n} \to_P \sigma^2,$$

by the weak law of large numbers. After showing that $K_n \to_p 0$, the result $S_n^2 \to_P \sigma^2$ can then be obtained from application of Slutsky's theorem.

13.10 Now, $B \sim \text{Bernoulli}(p)$ and Y is distributed $N(\mu, \sigma^2)$ if $B = 1$, or distributed $N(\nu, \tau^2)$ if $B = 0$. Thus

$$E[Y|B] = \begin{cases} \mu, & B = 1 \\ \nu, & B = 0, \end{cases}$$

and

$$\text{var}(Y|B) = \begin{cases} \sigma^2, & B = 1 \\ \tau^2, & B = 0. \end{cases}$$

Hence,

$$\begin{aligned} E[Y] &= E_B[E[Y|B]] \\ &= p\mu + (1 - p)\nu, \end{aligned}$$

and

$$\begin{aligned} \text{var}(Y) &= \text{var}_B(E[Y|B]) + E_B[\text{var}(Y|B)] \\ &= (\mu - \nu)^2 p(1 - p) + p\sigma^2 + (1 - p)\tau^2. \end{aligned}$$

Chapter 14

14.1 In the joint density function $f(\mathbf{y};\nu)$ the only term involving both $\mathbf{y}$ and ν is of the form $(\prod_i y_i)^{(\nu/2-1)}$, and hence $\prod_i y_i$ (or equivalently, $\sum_i \log y_i$) is a sufficient statistic for ν.

14.3 In the joint density function of $\mathbf{y}$, the only term involving both $\mathbf{y}$ and $\boldsymbol{\theta}$ has the form

$$\exp\left(\sum_{j=1}^{s}\theta_j\sum_{i=1}^{n}t_j(y_i)\right),$$

and hence $(T_1(\mathbf{y}), ..., T_s(\mathbf{y}))$ is an s-dimensional sufficient statistic for $\boldsymbol{\theta} = (\theta_1, ..., \theta_s)$.

14.4 The data (N_1, N_2, N_3) are multinomial with cell probabilities of $p^2, 2p(1-p)$ and $(1-p)^2$, respectively. Let (n_1, n_2, n_3) denote the observed genotypes frequencies and let $t = 2n_1 + n_2$ be the observed frequency of allele A. Ignoring constants, the likelihood is

$$\begin{aligned}L(p;n_1,n_2,n_3) &= (p^2)^{n_1}(2p(1-p))^{n_2}((1-p)^2)^{n_3}\\ &= 2^{n_2}p^{2n_1+n_2}(1-p)^{2n_3+n_2}\\ &= 2^{n_2}p^t(1-p)^{2n-t}. \end{aligned}\tag{15.1}$$

14.7 Note that the joint density function can be written

$$f(y_1, y_2; p, \lambda) = \frac{p^{y_1}(1-p)^{n-y_1}}{y_1!(n-y_1)!}e^{-\lambda}\lambda^n,$$

where $n = y_1 + y_2$. Using the identity $\sum_{y_1=0}^{n}\binom{n}{y_1}p^{y_1}(1-p)^{n-y_1} = 1$, it follows that the density of $N = Y_1 + Y_2$ is

$$f(n;\lambda) = \frac{e^{-\lambda}\lambda^n}{n!},$$

and hence

$$f(y_1|N=n;p) = \frac{f(y_1, y_2; p, \lambda)}{f(n;\lambda)} = \binom{n}{y_1}p^{y_1}(1-p)^{(n-y_1)},\tag{15.2}$$

as required.

The experiment in which (Y_1, Y_2) is observed is equivalent to a Bin(n, p) experiment with random N (not depending on p), and hence the information content about p is the same as that of the Bin(n, p) experiment actually performed. (This can also be argued using the likelihood principle and the fact that $f(y_1, y_2; p, \lambda) = f(y_1|y_1 + y_2; p)f(y_1 + y_2; \lambda)$. For the more general case, see Equation (9.5).)

14.9 The result follows from showing that

$$p_0 + p_1 = 1 - \int_C \big(f(y; \theta_0) - f(y; \theta_1)\big)dy.$$

Chapter 15

15.1 Section 15.3: Do you think like a frequentist or a Bayesian?

A Bayesian would answer that the probability of heads is 0.5. Prior to the toss, heads and tails are equally favoured, corresponding to a prior probability of 0.5, and since the coin remains hidden, there is no reason to update that probability.

The frequentist would also like to answer that the probability is 0.5, but the question is contrived to present this. The question says 'The coin has been tossed...' so the frequentist can not use a concept of probability based on repetition of the experiment. So, the frequentist is forced to say that this particular coin is either heads, or it is not, since there is no longer a random experiment to be considered.[1] However, the frequentist could say that, for this particular coin, heads and tails are equally likely outcomes (and the Bayesian would concur).

[1] Note that this is the same interpretation that is made of confidence intervals, for example. Once a CI has been calculated from the data, no probabilistic statements can be made about that particular interval. Of course, one has the knowledge that, under repetition of the experiment, approximately $(1 - \alpha)100\,\%$ of the CIs will contain the true unknown parameter value.

Bibliography

ADMB-project (2008a) *An Introduction to AD Model Builder Version 9.0.0 for use in Nonlinear Modeling and Statistics*, ADMB Project, admb-project.org/documentation.

ADMB-project (2008b) *Random Effects in AD Model Builder: ADMB-RE User Guide*, ADMB Project, admb-project.org/documentation.

Agresti, A. and Coull, B. A. (1998) Approximate is better than 'exact' for interval estimation of binomial proportions, *Am. Stat.* **52**: 119–126.

Aitken, A. C. (1926) On Bernoulli's numerical solution of algebraic equations, *Proc. Roy. Soc. Edin.* **46**: 289–305.

Akaike, H. (1974) A new look at the statistical model identification, *IEEE Trans. Automat. Contr.* **15**: 716–723.

Aldrich, J. (1997) R. A. Fisher and the making of maximum likelihood 1912–1922, *Stat. Sci.* **12**: 162–176.

Apostol, T. M. (1967) *Calculus. Volume 1.*, 2nd edn, John Wiley & Sons, Inc., New York.

Bartlett, M. S. (1936) Statistical information and properties of sufficiency, *Proc. Roy. Soc. Lond. A* **154**: 124–137.

Beitler, P. J. and Landis, J. R. (1985) A mixed-effects model for categorical data, *Biometrics* **41**: 991–1000.

Berger, J. O. (1985) *Statistical Decision Theory and Bayesian Analysis*, 2nd edn, Springer-Verlag, New York.

Berger, J. O., Liseo, B. and Wolpert, R. L. (1999) Integrated likelihood methods for eliminating nuisance parameters, *Stat. Sci.* **14**: 1–28.

Berger, J. O. and Wolpert, R. L. (1988) *The Likelihood Principle, 2nd edn*, Vol. 6, IMS Lecture Notes – Monograph Series.

Billingsley, P. (1979) *Probability and Measure*, John Wiley & Sons, Inc., New York.

Birnbaum, A. (1962) On the foundations of statististical inference (with discussion), *J. Am. Stat. Assoc.* **57**: 269–326.

Bjørnstad, J. F. (1996) On the generalization of the likelihood function and the likelihood principle, *J. Am. Stat. Assoc.* **91**: 791–806.

Maximum Likelihood Estimation and Inference: With Examples in R, SAS and ADMB, First Edition. Russell B. Millar.

Bøhning, D., Dietz, E. and Schlattmann, P. (1999) The zero-inflated poisson model and the decayed, missing and filled teeth index in dental epidemiology, *Am. Stat.* **46**: 327–333.

Borchers, D. L., Buckland, S. T. and Zucchini, W. (2002) *Esimating Animal Abundance: Closed Populations*, Springer-Verlag, London.

Box, G. E. P. and Cox, D. R. (1964) An analysis of transformations (with discussion), *J. Roy. Stat. Soc. B* **26**: 211–252.

Breslow, N. E. and Clayton, D. G. (1993) Approximate inference in generalized linear mixed models, *J. Am. Stat. Assoc.* **88**: 9–25.

Breslow, N. E. and Lin, X. (1995) Bias correction in generalised linear mixed models with a single component of dispersion, *Biometrika* **82**: 81–91.

Brown, L. D., Cai, T. T. and DasGupta, A. (2001) Interval estimation for a binomial proportion (with discussion), *Stat. Sci.* **2**: 101–133.

Brown, L. D., Cai, T. T. and DasGupta, A. (2002) Confidence intervals for a binomial proportion and asymptotic expansions, *Ann. Stat.* **30**: 160–201.

Burnham, K. P. and Anderson, D. R. (2002) *Model Selection and Multimodel Inference – A Practical Information-Theoretic Approach*, 2nd edn, Springer-Verlag, New York.

Carey, V., Zeger, S. L. and Diggle, P. (1993) Modelling multivariate binary data with alternating logistic regressions, *Biometrika* **80**: 517–526.

Casella, G. and Berger, R. L. (1990) *Statistical Inference*, Duxbury Press, California.

Chambers, E. A. and Cox, D. R. (1967) Discrimination between alternative binary response models, *Biometrika* **54**: 573–578.

Chernick, M. R. (2008) *Bootstrap Methods: A Guide for Practitioners and Researchers*, 2nd edn, John Wiley & Sons, Inc., Hoboken, New Jersey.

Clark, J. R. (1957) Effect of length of haul on cod end escapement, *Technical Report Paper S25*, ICNAF/ICES/FAO Workshop on Selectivity, Lisbon.

Collett, D. (1991) *Modelling Binary Data*, Chapman & Hall, London.

Cormack, R. M. (1992) Interval estimation for mark–recapture studies of closed populations, *Biometrics* **48**: 567–576.

Cox, D. R. (1972) Regression models and life tables (with discussion), *J. Roy. Stat. Soc. B* **34**: 187–220.

Cox, D. R. (1975) Partial likelihood, *Biometrika* **62**: 269–276.

Cox, D. R. and Reid, N. (1987) Parameter orthogonality and approximate conditional inference, *J. R. Statist. Soc. B.* **49**: 1–39.

Crainiceanu, C. M. and Ruppert, D. (2004) Likelihood ratio tests in linear mixed models with one variance component, *J. Roy. Stat. Soc. B* **66**: 165–185.

Davison, A. C. and Hinkley, D. V. (1997) *Bootstrap Methods and their Applications*, Cambridge University Press, New York.

Delwiche, L. D. and Slaughter, S. J. (2003) *The Little SAS Book: A Primer*, 3rd edn, SAS Institute Inc., Cary, N.C.

Dempster, A. P., Laird, N. M. and Rubin, D. B. (1977) Maximum likelihood from incomplete data via the EM algorithm, *J. Roy. Stat. Soc. B* **39**: 1–38.

Der, G. and Everitt, B. S. (2009) *A Handbook of Statistical Analyses using SAS*, 3rd edn, Chapman & Hall/CRC, Boca Raton, FL.

Draper, N. R. and Smith, H. (1981) *Applied Regression Analysis*, 2nd edn, John Wiley & Sons, Inc., New York.

Edwards, A. W. F. (1972) *Likelihood*, Cambridge University Press, Cambridge.

Efron, B. (1979) Bootstrap methods: another look at the jackknife, *Ann. Stat.* **7**: 1–26.

Efron, B. (1987) Better bootstrap confidence intervals (with discussion), *J. Am. Stat. Assoc.* **82**: 171–200.

Efron, B. and Hinkley, D. V. (1978) Assessing the accuracy of the maximum likelihood estimator: observed versus expected Fisher information, *Biometrika* **65**: 457–487.

Efron, B. and Tibshirani, R. J. (1993) *An Introduction to the Bootstrap*, Chapman & Hall, New York.

Evans, M. A., Kim, H. and O'Brien, T. E. (1996) An application of profile-likelihood based confidence interval to capture-recapture estimators, *J. Agric. Biol. Envir. Stat.* **1**: 131–140.

Evans, M. J., Fraser, D. A. S. and Monette, G. (1986) On principles and arguments to likelihood, *Can. J. Stat.* **14**: 181–199.

Everitt, B. S. and Hothorn, T. (2006) *A Handbook of Statistical Analyses using R*, Chapman & Hall/CRC, Boca Raton, FL.

Faraggi, D., Izikson, P. and Reiser, B. (2003) Confidence intervals for the 50 per cent response dose, *Stat. Med.* **22**: 1977–1988.

Fears, T. R., Benichou, J. and Gail, M. H. (1996) A reminder of the fallibility of the Wald statistic, *Am. Stat.* **50**: 226–227.

Felsenstein, J. (1981) Evolutionary trees from DNA sequences: a maximum likelihood approach, *J. Mol. Evol.* **17**: 368–376.

Fieller, E. C. (1954) Some problems in interval estimation, *J. Roy. Stat. Soc. B* **16**: 175–185.

Finney, D. J. (1971) *Probit Analysis*, 3rd edn, Cambridge University Press, London.

Fisher, R. A. (1912) On an absolute criterion for fitting frequency curves, *Messeng. Math.* **41**: 155–160.

Fisher, R. A. (1921) On the 'probable error' of a coefficient of correlation deduced from a small sample, *Metron* **1**: 3–32.

Fisher, R. A. (1933) The concepts of inverse probability and fiducial probability referring to unknown parameters, *Proc. Roy. Soc. A* **139**: 343–348.

Freireich, E. J. *et al.* (1963) The effect of 6-mercaptopurine on the duration of steroid-induced remissions of acute leukemia: A model for evaluation of other potentially useful therapy, *Blood* **21**: 699–716.

Fushiki, T., Komaki, F. and Aihara, K. (2004) On parametric bootstrapping and Bayesian prediction, *Scand. J. Stat.* **31**: 403–416.

Gail, M. H. *et al.* (1996) Reproducibility studies and interlaboratory concordance for assays of serum hormone levels: estrone, estradiol, estrone sulfate, and progesterone, *Cancer Epidem. Biomar.* **5**: 835–844.

Gelman, A., Carlin, J. B., Stern, H. S. and Rubin, D. B. (2003) *Bayesian Data Analysis*, 2nd edn, Chapman & Hall, New York.

Gelman, A. and Hill, J. (2007) *Data Analysis using Regression and Multilevel/Hierarchical Models*, Cambridge University Press, New York.

Ghosh, M. (1995) Inconsistent maximum likelihood estimators for the Rasch model, *Stat. Probabil. Lett.* **23**: 165–170.

Golub, G. H. and Pereyra, V. (1973) The differentiation of pseudo-inverses and nonlinear least squares problems whose variables separate, *SIAM J. Numer. Anal.* **10**: 413–432.

Goodman, L. A. (1960) On the exact variance of products, *J. Am. Stat. Assoc.* **55**: 708–713.

Gribben, P. E., Helson, J. and Millar, R. B. (2004) Population abundance estimates of the New Zealand geoduck clam, *Panopea zelandica*, using North American methodology: is the technology transferable, *J. Shellfish Res.* **23**: 683–692.

Griewank, A. (2003) A mathematical view of automatic differentiation, *Acta Numerica* **12**: 321–398.

Hall, P., Peng, L. and Tajvidi, N. (1999) On prediction intervals based on predictive likelihood or bootstrap methods, *Biometrika* **86**: 871–880.

Hardin, J. W. and Hilbe, J. M. (2003) *Generalized Estimating Equations*, Chapman & Hall, New York.

Harrell, F. E. (2001) *Regression Modeling Strategies with Applications to Linear Models Logistic Regression, and Survival Analysis*, Springer-Verlag, New York.

Harris, I. R. (1989) Predictive fit for natural exponential families, *Biometrika* **76**: 675–684.

Harville, D. A. (1974) Bayesian inference for variance components using only error contrasts, *Biometrika* **61**: 383–385.

Harville, D. A. (1977) Maximum likelihood approaches to variance component estimation and to related problems, *J. Am. Stat. Assoc.* **72**: 320–340.

Hathaway, R. J. (1985) A constrained formulation of maximum-likelihood estimation for normal mixture distributions, *Ann. Stat.* **13**: 795–800.

Heyde, C. C. (1997) *Quasi-likelihood and its application: a general approach to optimal parameter estimation*, Springer-Verlag, New York.

Hoadley, B. (1971) Asymptotic properties of maximum likelihood estimators for the independent not indentically distributed case, *Ann. Math. Stat.* **42**: 1977–1991.

Huber, P. J. (1964) Robust estimation of a location parameter, *Ann. Math. Stat.* **35**: 73–101.

Hurlbert, S. H. (1984) Pseudoreplication and the design of ecological field experiments, *Ecol. Monogr.* **54**: 187–211.

Hutton, J. L. and Monaghan, P. F. (2002) Choice of parametric accelerated life and proportional hazards models for survival data: asymptotic results, *Lifetime Data Anal.* **8**: 375–393.

ICES (1979) *Reports of the ICES Advisory Committee on Fisheries Management, 1978. ICES Co-operative Research Report 85*, International Council for the Exploration of the Sea, Charlottenlund, Denmark.

Ihaka, R. and Gentleman, R. (1996) R: a language for data analysis and graphics, *J. Comput. Graph. Stat.* **5**: 299–314.

Joe, H. and Xu, J. J. (1996) The estimation method of inference functions for margins for multivariate models, *Technical Report 166*, Department of Statistics, University of British Columbia. 21 pp.

Johnson, N. L., Kotz, S. and Balakrishnan, N. (1994) *Continuous Univariate Distributions*, Vol. 1, 2nd edn, John Wiley & Sons, Inc., New York.

Johnson, N. L., Kotz, S. and Balakrishnan, N. (1995) *Continuous Univariate Distributions*, Vol. 2, 2nd edn, John Wiley & Sons, Inc., New York.

Johnson, N. L., Kotz, S. and Kemp, A. W. (1992) *Univariate Discrete Distributions*, 2nd edn, John Wiley & Sons, Inc., New York.

Jørgensen, B. (1993) A review of conditional inference: is there a universal definition of nonformation?, *Bull. Int. Stat. Inst.* **55**: 323–340.

Kashiwagi, N. and Yanagimoto, T. (1992) Smoothing serial count data through a state-space model, *Biometrics* **48**: 1187–1194.

Kauermann, G. and Carroll, R. J. (2001) A note on the efficiency of sandwich covariance matrix estimation, *J. Am. Stat. Assoc.* **96**: 1387–1396.

Kleinbaum, D. G. and Klein, M. (2005) *Survival Analysis: A Self-learning Text*, 2nd edn, Springer, New York.

Kullback, S. (1959) *Information Theory and Statistics*, John Wiley & Sons, Inc., New York.

Laird, N. M. and Ware, J. H. (1982) Random-effects models for longitudinal data, *Biometrics,* **38**: 963–974.

LaMotte, L. R. (2007) A direct derivation of the REML likelihood function, *Stat. Pap.* **48**: 321–327.

Lange, K. (2002) *Mathematical and Statistical Methods for Genetic Analysis*, 2nd edn, Springer, New York.

Lee, Y., Nelder, J. A. and Pawitan, Y. (2006) *Generalized Linear Models with Random Effects: Unified Analysis via H-likelihood,* Chapman & Hall, New York.

Lehmann, E. L. (1983) *Theory of Point Estimation,* John Wiley & Sons, Inc., New York.

Lehmann, E. L. (2006) On likelihood ratio tests, *IMS Lecture Notes* **49**: 1–8.

Liang, K. Y. and Zeger, S. L. (1986) Longitudinal data analysis using generalized linear models, *Biometrika.* **73**: 13–22.

Lin, X. and Breslow, N. E. (1996) Bias correction in generalized linear mixed models with multiple components of dispersion, *J. Am. Stat. Assoc.* **91**: 1007–1016.

Lindstrom, M. J. and Bates, D. M. (1990) Nonlinear mixed effects models for repeated measures data, *Biometrics* **46**: 673–687.

Littell, R. C., Milliken, G. A., Stroup, W. W. and Wolfinger, R. D. (1996) *SAS System for Mixed Models*, SAS Institute Inc., Cary, NC.

Lo, Y. (2005) Likelihood ratio tests of the number of components in a normal mixture with unequal variances, *Stat. Probabil. Lett.* **71**: 225–235.

Louis, T. A. (1982) Finding the observed information matrix using the *EM* algorithm, *J. Roy. Stat. Soc. B* **44**: 226–233.

Ludwig, D. (1996) Uncertainty and the assessment of extinction probabilities, *Ecol. Appl.* **6**: 1067–1076.

Manly, B. F. J. (1997) *Randomization, Bootstrap and Monte Carlo methods in Biology*, 2nd edn, Chapman & Hall, London.

Marin, J. A., Jones, O. P. and Hadlow, W. C. C. (1993) Micropropagation of columnar apple trees, *J. Hortic. Sci.* **68**: 289–297.

McCullagh, P. and Nelder, J. A. (1989) *Generalized Linear Models*, 2nd edn, Chapman & Hall, New York.

McCulloch, C. E., Searle, S. R. and Neuhaus, J. M. (2008) *Generalized, Linear, and Mixed Models*, 2nd edn, John Wiley & Sons, Inc., New York.

McDonald, J. H., Verrelli, B. C. and Geyer, L. B. (1996) Lack of geographic variation in anonymous nuclear polymorphisms in the American oyster, *Crassostrea virginica*, *Mol. Biol. Evol.* **13**: 1114–1118.

McLachlan, G. J. (1987) On bootstrapping the likelihood ratio test statistic for the number of components in a normal mixture, *J. Roy. Stat. Soc. C-App.* **36**: 318–324.

McLachlan, G. J. and Krishnan, T. (2008) *The EM Algorithm and Extensions*, 2nd edn, John Wiley & Sons, Inc., New Jersey.

Meinhold, R. J. and Singpurwalla, N. D. (1983) Understanding the Kalman filter, *Am. Stat.* **37**: 123–127.

Millar, R. B. (2004) Simulated maximum likelihood applied to non-Gaussian and nonlinear mixed effects and state-space models, *Aust. NZ. J. Stat.* **46**: 543–554.

Millar, R. B. and Willis, T. J. (1999) Estimating the relative density of snapper in and around a marine reserve using a log-linear mixed effects model, *Aust. NZ. J. Stat.* **41**: 383–394.

Navidi, W. (1997) A graphical illustration of the EM algorithm, *Am. Stat.* **51**: 29–31.

Neuhaus, J. M., Kalbfleisch, J. D. and Hauck, W. W. (1991) A comparison of cluster-specific and population-averaged approaches for analyzing correlated binary data, *Int. Stat. Rev.* **59**: 25–35.

Neyman, J. and Pearson, E. S. (1933) On the problem of the most efficient tests of statistical hypotheses, *Phil. Trans. Roy. Soc. Lond. A* **231**: 289–337.

Neyman, J. and Scott, E. L. (1948) Estimates based on partially consistent observations, *Econometrica* **16**: 1–32.

Nocedal, J. and Wright, S. J. (2006) *Numerical Optimization*, 2nd edn, Springer, New York.

Noh, M. and Lee, Y. (2007) REML estimation for binary data in GLMMs, *J. Multivariate Anal.* **98**: 896–915.

Pace, L. and Salvan, A. (1997) *Principles of Statistical Inference from a Neo-Fisherian Perspective*, World Scientific, River Edge, NJ.

Patronek, G. J., Waters, D. J. and Glickman, L. T. (1997) Comparative longevity of pet dogs and humans: implications for gerontology research, *J. Gerontology* **52A**: B171–B178.

Patterson, H. D. and Thompson, R. (1971) Recovery of inter-block information when block sizes are unequal, *Biometrika* **58**: 545–554.

Pawitan, Y. (2001) *In All Likelihood: Statistical Modelling and Inference Using Likelihood*, Oxford University Press, Oxford.

Petersen, C. G. J. (1896) The yearly immigration of young plaice into the Limfjord from the German Sea, *Report of the Danish Biological Station* **6**: 1–48.

Pinheiro, J. C. and Bates, D. M. (2000) *Mixed-effects Models in S and S-PLUS*, Springer, New York.

Pledger, S. (2000) Unified maximum likelihood estimates for closed capture-recapture models using mixtures, *Biometrics* **56**: 434–442.

Pollock, K. H. (1975) A *K*-sample tag-recapture model allowing for unequal survival and catchability, *Biometrika* **62**: 577–583.

Popper, K. (1959) *The Logic of Scientific Discovery*, Routledge, London.

Press, W. H., Teulolsky, S. A., Vetterling, W. T. and Flannery, B. P. (2007) *Numerical recipes: The Art of Scientific Computing*, 3rd edn, Cambridge University Press, New York.

Proschan, F. (1963) Theoretical explanation of observed decreasing failure rate, *Technometrics* **5**: 375–383.

R Development Core Team (2010) *R: A Language and Environment for Statistical Computing*, R Foundation for Statistical Computing, Vienna, Austria. http://www.R-project.org. ISBN 3-900051-07-0.

Rao, C. R. (1948) Large sample tests of statistical hypotheses concerning several parameters with application to problems of estimation, *Proc. Cambridge Phil. Soc.* **44**: 50–57.

Raspe, R. E. (1948) *Singular Travels, Campaigns and Adventures of Baron Münchausen: with an Introduction by John Carswell*, Cresset Press, London.

Ronchetti, E. (1990) Small sample asymptotics: a review with applications to robust statistics, *Comput. Stat. Data An.* **10**: 207–223.

Royall, R. M. (1997) *Statistical Evidence: A Likelihood Paradigm*, Chapman & Hall, London.

Royston, P. and Parmar, M. K. B. (2002) Flexible parametric proportional-hazards and proportional-odds models for censored survival data, with application to prognostic modelling and estimation of treatment effects, *Stat. Med.* **21**: 2175–2197.

Rue, H., Martino, S. and Chopin, N. (2009) Approximate Bayesian inference for latent Gaussian models by using integrated nested Laplace approximations (with discussion), *J. Roy. Stat. Soc. B* **71**: 319–392.

SAS Institute (1999) *SAS/STAT User's Guide, Version 8*, SAS Institute Inc., Cary, NC.

SAS Institute (2008) *SAS/STAT User's Guide, Version 9.2*, SAS Institute Inc., Cary, NC.

Schafer, J. L. (1997) *Analysis of Incomplete Multivariate Data*, Chapman & Hall, London.

Schall, R. (1991) Estimation in generalized linear models with random effects, *Biometrika* **78**: 719–727.

Scheipl, F., Greven, S. and Küchenhof, H. (2008) Size and power of tests for a zero random effect variance or polynomial regression in additive and linear mixed models, *Comp. Stat. Data. Anal.* **52**: 3283–3299.

Schwarz, G. (1978) Estimating the dimension of a model, *Ann. Stat.* **6**: 461–464.

Seber, G. A. F. (1982) *Estimation of Animal Abundance and Related Parameters*, 2nd edn, Blackburn Press, New Jersey.

Seber, G. A. F. and Lee, A. J. (2003) *Linear Regression Analysis*, 2nd edn, John Wiley & Sons, Inc., New Jersey.

Self, S. G. and Liang, K. Y. (1987) Asymptotic properties of maximum likelihood estimators and likelihood ratio tests under nonstandard conditions, *J. Am. Stat. Assoc.* **82**: 605–610.

Severini, T. A. (1994) Approximately Bayesian inference, *J. Am. Stat. Assoc.* **89**: 242–249.

Severini, T. A. (2007) Integrated likelihood functions for non-Bayesian inference, *Biometrika* **94**: 529–542.

Simpson, J. A. and Weiner, E. S. C. (eds) (1989) *The Oxford English Dictionary*, 2nd edn, Oxford University Press, Oxford, U.K.

Skaug, H. J. and Fournier, D. A. (2006) Automatic approximation of the marginal likelihood in non-Gaussian hierarchical models, *Comput. Stat. Data An.* **51**: 699–709.

Smyth, G. K. and Verbyla, A. P. (1996) A conditional likelihood approach to residual maximum likelihood estimation in generalized linear models, *J. Roy. Stat. Soc. B* **58**: 565–572.

Sprott, D. A. (1975) Marginal and conditional sufficiency, *Biometrika* **62**: 599–605.

Stefanski, L. A. and Boos, D. D. (2002) The calculus of m-estimation, *Am. Stat.* **56**: 29–38.

Stewart, J. (1999) *Calculus*, 4th edn, Brooks Cole, Pacific Grove, CA.

Stiratelli, R., Laird, N. M. and Ware, J. H. (1984) Random-effects models for serial observations with binary response, *Biometrics,* **40**: 961–971.

Stram, D. O. and Lee, J. W. (1994) Variance component testing in the longitudinal mixed effects model, *Biometrics,* **50**: 1171–1177.

Trivedi, P. K. and Zimmer, D. M. (2007) Copula modeling: an introduction for practitioners, *Found. Trends Econometrics* **1**: 1–111.

Tuyl, F., Gerlach, R. and Mengersen, K. (2009) Posterior predictive arguments in favor of the Bayes-Laplace prior as the consensus prior for binomial and multinomial parameters, *Bayesian Anal.* **4**: 151–158.

Van Deusen, P. C. (2002) An EM algorithm for capture-recapture estimation, *Environ. Ecol. Stat.* **9**: 151–167.

Venzon, D. J. and Moolgavkar, S. H. (1988) A method of computing profile-likelihood based confidence intervals, *J. Roy. Stat. Soc. C-App.* **37**: 87–94.

von Bertalanffy, L. (1938) A quantitative theory of organic growth, *Human Biol.* **10**: 181–213.

Vonesh, E. F. (1996) A note on the use of Laplace's approximation for nonlinear mixed-effects models, *Biometrika* **83**: 447–452.

Wald, A. (1943) Tests of statistical hypotheses concerning several parameters when the number of observations is large, *Tran. Am. Math. Soc.* **54**: 426–482.

Wand, M. P. (2003) Smoothing and mixed models, *Computation. Stat.* **18**: 223–249.

Warton, D. I. (2005) Many zeros does not mean zero inflation: comparing the goodness-of-fit of parametric models to multivariate abundance data, *Environmetrics* **16**: 275–289.

Wedderburn, R. W. M. (1974) Quasi-likelihood functions, generalized linear models and the Gauss-Newton method, *Biometrika* **61**: 439–447.

Wedderburn, R. W. M. (1976) On the existence and uniqueness of the maximum likelihood estimates for certain generalized linear models, *Biometrika* **63**: 27–32.

Wei, W. W. S. (2006) *Time Series Analysis: Univariate and Multivariate Methods*, 2nd edn, Addison Wesley, New York.

Wild, C. J. and Seber, G. A. F. (2000) *Chance Encounters: A First Course in Data Analysis and Inference*, John Wiley & Sons, Inc., New York.

Wilks, S. S. (1938) The large-sample distribution of the likelihood ratio for testing composite hypotheses, *Ann. Math. Stat.* **9**: 60–62.

Williams, D. A. (1986) Interval estimation of the median lethal dose, *Biometrics* **42**: 641–645.

Willis, T. J. and Millar, R. B. (2005) Using marine reserves to estimate fishing mortality, *Ecol. Lett.* **8**: 47–52.

Wilson, E. B. (1927) Probable inference, the law of succession, and statistical inference, *J. Am. Stat. Assoc.* **22**: 209–212.

Wolfinger, R. (1993) Laplace's approximation for nonlinear mixed models, *Biometrika* **80**: 791–795.

Wolfinger, R. and O'Connell, M. (1993) Generalized linear mixed models: a pseudo-likelihood approach, *J. Stat. Comput. Sim.* **48**: 233–243.

Wu, C. F. J. (1983) On the convergence properties of the EM algorithm, *Ann. Stat.* **11**: 95–103.

Yan, J. (2007) Enjoy the joy of copulas: with a package `copula`., *J. Stat. Softw.* **21**: 1–21.

Zheng, X. and Loh, W.-Y. (1995) Consistent variable selection in linear models, *J. Am. Stat. Assoc.* **90**: 151–156.

Index

Maximum Likelihood Estimation and Inference: With Examples in R, SAS and ADMB, First Edition. Russell B. Millar.

STATISTICS IN PRACTICE

Human and Biological Sciences

Berger – Selection Bias and Covariate Imbalances in Randomized Clinical Trials
Berger and Wong - An Introduction to Optimal Designs for Social and Biomedical Research
Brown and Prescott - Applied Mixed Models in Medicine, Second Edition
Carstensen – Comparing Clinical Measurement Methods
Chevret (Ed) – Statistical Methods for Dose-Finding Experiments
Ellenberg, Fleming and DeMets – Data Monitoring Committees in Clinical Trials: A Practical Perspective
Hauschke, Steinijans & Pigeot – Bioequivalence Studies in Drug Development: Methods and Applications
Källén – Understanding Biostatistics
Lawson, Browne and Vidal Rodeiro – Disease Mapping with WinBUGS and MLwiN
Lesaffre, Feine, Leroux & Declerck - Statistical and Methodological Aspects of Oral Health Research
Lui – Statistical Estimation of Epidemiological Risk
Marubini and Valsecchi - Analysing Survival Data from Clinical Trials and Observation Studies
Millar – Maximum Likelihood Estimation and Inference: With Examples in R, SAS and ADMB
Molenberghs and Kenward – Missing Data in Clinical Studies
O'Hagan, Buck, Daneshkhah, Eiser, Garthwaite, Jenkinson, Oakley & Rakow – Uncertain Judgements: Eliciting Expert's Probabilities
Parmigiani – Modeling in Medical Decision Making: A Bayesian Approach
Pintilie – Competing Risks: A Practical Perspective
Senn - Cross-over Trials in Clinical Research, Second Edition
Senn - Statistical Issues in Drug Development, Second Edition
Spiegelhalter, Abrams and Myles – Bayesian Approaches to Clinical Trials and Health-Care Evaluation
Walters - Quality of Life Outcomes in Clinical Trials and Health-Care Evaluation
Whitehead - Design and Analysis of Sequential Clinical Trials, Revised Second Edition
Whitehead – Meta-Analysis of Controlled Clinical Trials
Willan and Briggs – Statistical Analysis of Cost Effectiveness Data
Winkel and Zhang - Statistical Development of Quality in Medicine

Earth and Environmental Sciences

Buck, Cavanagh and Litton – Bayesian Approach to Interpreting Archaeological Data
Chandler and Scott – Statistical Methods for Trend Detection and Analysis in the Environmental Statistics

Glasbey and Horgan – Image Analysis in the Biological Sciences
Haas – Improving Natural Resource Management: Ecological and Political Models
Helsel – Nondetects and Data Analysis: Statistics for Censored Environmental Data
Illian, Penttinen, Stoyan, H and Stoyan D–Statistical Analysis and Modelling of Spatial Point Patterns
McBride – Using Statistical Methods for Water Quality Management
Webster and Oliver – Geostatistics for Environmental Scientists, Second Edition
Wymer (Ed) – Statistical Framework for Recreational Water Quality Criteria and Monitoring

Industry, Commerce and Finance

Aitken - Statistics and the Evaluation of Evidence for Forensic Scientists, Second Edition
Balding - Weight-of-evidence for Forensic DNA Profiles
Brandimarte – Numerical Methods in Finance and Economics: A MATLAB-Based Introduction, Second Edition
Brandimarte and Zotteri – Introduction to Distribution Logistics
Chan - Simulation Techniques in Financial Risk Management
Coleman, Greenfield, Stewardson and Montgomery (Eds) – Statistical Practice in Business and Industry
Frisen (Ed) – Financial Surveillance
Fung and Hu – Statistical DNA Forensics
Gusti Ngurah Agung - Time Series Data Analysis Using EViews
Kenett (Eds) - Operational Risk Management: A Practical Approach to Intelligent Data Analysis
Jank and Shmueli (Ed.) – Statistical Methods in e-Commerce Research
Lehtonen and Pahkinen - Practical Methods for Design and Analysis of Complex Surveys, Second Edition
Ohser and Mücklich - Statistical Analysis of Microstructures in Materials Science
Pourret, Naim & Marcot (Eds) – Bayesian Networks: A Practical Guide to Applications
Taroni, Aitken, Garbolino and Biedermann - Bayesian Networks and Probabilistic Inference in Forensic Science
Taroni, Bozza, Biedermann, Garbolino and Aitken – Data Analysis in Forensic Science

Printed and bound by CPI Group (UK) Ltd, Croydon, CR0 4YY

16/04/2025

14658546-0003